VISIT US AT

www.syngress.com

Syngress is committed to publishing high-quality books for IT Professionals and delivering those books in media and formats that fit the demands of our customers. We are also committed to extending the utility of the book you purchase via additional materials available from our Web site.

SOLUTIONS WEB SITE

To register your book, visit www.syngress.com/solutions. Once registered, you can access our solutions@syngress.com Web pages. There you may find an assortment of value-added features such as free e-books related to the topic of this book, URLs of related Web sites, FAQs from the book, corrections, and any updates from the author(s).

ULTIMATE CDs

Our Ultimate CD product line offers our readers budget-conscious compilations of some of our best-selling backlist titles in Adobe PDF form. These CDs are the perfect way to extend your reference library on key topics pertaining to your area of expertise, including Cisco Engineering, Microsoft Windows System Administration, CyberCrime Investigation, Open Source Security, and Firewall Configuration, to name a few.

DOWNLOADABLE E-BOOKS

For readers who can't wait for hard copy, we offer most of our titles in downloadable Adobe PDF form. These e-books are often available weeks before hard copies, and are priced affordably.

SYNGRESS OUTLET

Our outlet store at syngress.com features overstocked, out-of-print, or slightly hurt books at significant savings.

SITE LICENSING

Syngress has a well-established program for site licensing our e-books onto servers in corporations, educational institutions, and large organizations. Contact us at sales@syngress.com for more information.

CUSTOM PUBLISHING

Many organizations welcome the ability to combine parts of multiple Syngress books, as well as their own content, into a single volume for their own internal use. Contact us at sales@syngress.com for more information.

SYNGRESS®

SYNGRESS®

Asterisk Hacking

Toolkit and LiveCD

Benjamin Jackson

Champ Clark III

Larry Chaffin and Johnny Long Technical Editors

KEY	SERIAL NUMBER
001	HJIRTCV764
002	PO9873D5FG
003	829KM8NJH2
004	BAL923457U
005	CVPLQ6WQ23
006	VBP965T5T5
007	HJJJ863WD3E
008	2987GVTWMK
009	629MP5SDJT
010	IMWQ295T6T

PUBLISHED BY
Syngress Publishing, Inc.
Elsevier, Inc.
30 Corporate Drive
Burlington, MA 01803

Asterisk Hacking

Printed in the United States of America
1 2 3 4 5 6 7 8 9 0
ISBN: 978-1-59749-151-8

Publisher: Amorette Pedersen
Acquisitions Editor: Andrew Williams
Technical Editors: Johnny Long and Larry Chaffin
Cover Designer: Michael Kavish

Project Manager: Anne B. McGee
Page Layout and Art: Patricia Lupien
Copy Editor: Michael McGee
Indexer: Richard Carlson

For information on rights, translations, and bulk sales, contact Matt Pedersen, Commercial Sales Director and Rights, at Syngress Publishing; email m.pedersen@elsevier.com.

Co-Authors

Benjamin Jackson (Black Ratchet) is a jack of all trades computer guy from New Bedford, MA. Ben holds a BS in Computer Engineering Technology from Northeastern University and spends his days developing applications and doing database administration for the Massachusetts Cancer Registry. By night, he toys with Asterisk, develops security tools, and generally breaks things.

Ben is a co-founder of Mayhemic Labs, an independent security research team, and has lectured at various hacker and professional conferences regarding VoIP and Open Source Software. He has also contributed code to the Asterisk source tree and other open source projects. One of the last true phone phreaks, he also enjoys playing on the Public Switched Telephone Network and spends far too much time making long distance phone calls to far flung places in the world.

Champ Clark III (Da Beave) has been involved in the technology industry for 15 years. Champ is currently employed with Vistech Communications, Inc. providing network support and applications development. Champ is also employed with Softwink, Inc. which specialized in security monitoring for the financial industry. Champ is one of the founding members of "Telephreak", an Asterisk hobbyist group, and the Deathrow OpenVMS cluster. When he's not ripping out code or writing papers, he enjoys playing music and traveling.

Technical Editors

Larry Chaffin is the CEO/Chairman of Pluto Networks, a worldwide network consulting company specializing in VoIP, WLAN, and security. An accomplished author, he contributed to Syngress's *Managing Cisco Secure Networks* (ISBN: 1931836566); *Skype Me!* (ISBN: 1597490326); *Practical VoIP Security* (ISBN: 1597490601); *Configuring Check Point NGX VPN-1/FireWall-1* (ISBN: 1597490318); *Configuring Juniper Networks NetScreen and SSG Firewalls* (ISBN: 1597491187); and *Essential Computer Security: Everyone's Guide to Email, Internet, and Wireless Security* (ISBN: 1597491144). He is the author of *Building a VoIP Network with Nortel's MS5100* (ISBN: 1597490784), and he has coauthored or ghostwritten 11 other technology books on VoIP, WLAN, security, and optical technologies.

Larry has over 29 vendor certifications from companies such as Nortel, Cisco Avaya, Juniper, PMI, isc2, Microsoft, IBM, VMware, and HP. Larry has been a principal architect designing VoIP, security, WLAN, and optical networks in 22 countries for many Fortune 100 companies. He is viewed by his peers as one of the most well respected experts in the field of VoIP and security in the world. Larry has spent countless hours teaching and conducting seminars/workshops around the world in the field of voice/VoIP, security, and wireless networks. Larry is currently working on a follow-up to *Building a VoIP Network with Nortel's MCS 5100* as well as new books on Cisco VoIP networks, practical VoIP case studies, and WAN acceleration with Riverbed.

Johnny Long Who's Johnny Long? Johnny is a Christian by grace, a family guy by choice, a professional hacker by trade, a pirate by blood, a ninja in training, a security researcher and author. His home on the web is http://johnny.ihackstuff.com.

Contents

Chapter 4 Writing Applications with Asterisk 115

## Chapter 1

What Is Asterisk and Why Do You Need It?

Solutions in this chapter:

- **What Is Asterisk?**
- **What Can Asterisk Do for Me?**
- **Who's Using Asterisk?**

☑ **Summary**

☑ **Solutions Fast Track**

☑ **Frequently Asked Questions**

Introduction

For years, telephone networks were run by large companies spending billions of dollars to set up systems that connected to one another over wires, radios, and microwaves. Large machines, filling entire buildings, allowed people to talk to each other over great distances. As the computer revolution progressed, the machines got smaller and more efficient, but still they were almost exclusively the domain of a small sect of companies.

Enter Asterisk… Asterisk has taken the power of the open-source software movement and brought it to the land of telephony. Much like how open source has proven that users don't need to rely on commercial companies for software, Asterisk has proven that users don't need to rely on commercial telephone companies for telephone systems. Open-source software allows you to be free of vendor lock-in, save money on support, use open standards, and change the software to suit your unique problems if the need arises. Looking at the "traditional" Private Branch Exchange (PBX) market, vendor lock-in is all too common, vendors charge exorbitant fees for support, and all too often the PBX you buy is a cookie-cutter solution with little to no customization options. It is common for people to think that their PBX is a black box that handles telephone calls. In reality, it is a bunch of computing equipment running a highly specialized software package. Open-source software can replace that customized software just as easily as it can replace any other software.

Asterisk is a veritable Swiss Army knife of telephony and Voice over Internet Protocol (VoIP). Designed to be a PBX replacement, Asterisk has grown to be all that and more. It boasts the ability to store voice mail, host conference calls, handle music on hold, and talk to an array of telephone equipment. It is also scalable, able to handle everything from a small five-telephone office to a large enterprise with multiple locations.

Thanks to Asterisk and VoIP, it is possible to run a telephone company out of a basement, handling telephone calls for people within a neighborhood, a city, or a country. Doing this only a few years ago would have required buying a large building, setting up large racks of equipment, and taking out a second mortgage. But today, everyone is jumping on the Asterisk bandwagon: hobbyists, telephone companies, universities, and small businesses, just to name a few. But what exactly *is* Asterisk? And what can it do? Let's find out.

What Is Asterisk?

Asterisk is an open-source PBX that has VoIP capabilities. However, this hardly explains what Asterisk is or what it does. So let's delve a little more deeply into PBXes, VoIP, and Asterisk.

What Is a PBX?

Asterisk, first and foremost, is a Private Branch Exchange. A PBX is a piece of equipment that handles telephone switching owned by a private business, rather than a telephone company. Initially in the United States, PBXes were for medium-to-large businesses that would create a lot of telephone traffic starting from, and terminating within, the same location. Rather than having that traffic tie up the switch that handles telephones for the rest of the area, PBXes were designed to be small switches to handle this traffic. Thus, the PBX would keep the internal traffic internal, and also handle telephone calls to and from the rest of the telephone network.

In the United States, thanks in part to the Bell System breakup of 1984, and to the computer revolution shrinking PBXes from the size of a couch to the size of a briefcase, PBXes flooded the market. Hundreds of companies started making PBXes and thousands wanted them. New features started coming into their own: voice mail, interactive menus, call waiting, caller ID, three-way calling, music on hold, and so on. The telecommunications industry grew by leaps and bounds, and the PBX industry kept up. However, with every silver lining comes a cloud. With the proliferation of digital telephone systems, each vendor had a specific set of phones you could use with their PBX. *Company X's* phones would often not work with *Company Y's* PBX. Plus, as with almost every technology, all too often a vendor would come in, set up the telephones, and never be heard from again, leaving the customer to deal with the system when it didn't work.

PBXes are one of the key pieces of hardware in businesses today, ranging from small devices the size of shoeboxes that handle a few lines to the telephone network and five phones in a small office, to a large system that interconnects ten offices across a campus of buildings. However, today's PBXes, when boiled down, all do the same things as their predecessors: route and handle telephone calls, and keep unnecessary traffic off the public switched telephone network.

Asterisk is a complete PBX. It implements all the major features of most commercially available PBXes. It also implements, for free, features that often cost a lot in a commercial installation: Conference calling, Direct Inward System Access, Call Parking, and Call Queues, just to name a few.

Out of the box, Asterisk can be configured to replicate your current PBX install. There have been numerous installs where a company's existing PBX is taken down on a Friday, an Asterisk server is installed and configured on Saturday, wired and tested on Sunday, and is handling calls on Monday. The users only notice a different voice when they grab their voice mail.

What Is VoIP?

Voice over Internet Protocol is one of the new buzzwords of the media today. While VoIP has been around in one incarnation or another since the 1970s, the market and technology has exploded over the past three years. Companies have sprouted up selling VoIP services and VoIP software, and instant messaging services are starting to include VoIP features.

But what exactly *is* VoIP? VoIP is a method to carry a two-way conversation over an Internet Protocol–based network. The person using Vonage to talk to her neighbor down the street? That's VoIP. The person in the United States using Windows Messenger to talk to his extended family in Portugal? That's VoIP. The 13-year-old playing Splinter Cell on his Xbox and talking to his teammates about how they slaughtered the other team? That's VoIP, too.

VoIP has exploded for a number of reasons—a major one being its ability to use an existing data network's excess capacity for voice calls, which allows these calls to be completed at little to no cost. A normal call that uses the standard telephone network compression coder–decoder algorithm (codec), μ–Law, will take up 64 kilobits per second of bandwidth. However, with efficient compression schemes, that can be dropped dramatically. In Table 1.1, we list certain commonly supported codecs, and how many simultaneous calls a T1 can handle when using that codec.

Table 1.1 VoIP Codec Comparison Chart

Codec	Speed	Simultaneous Calls over a T1 Link (1.5 Mbps)	Notes
μ-Law	64 Kbps	24	
G.723.1	5.3/6.3 Kbps	289/243	
G.726	16/24/32/40 Kbps	96/64/48/38	
G.729	8 Kbps	192	Requires license

Continued

Table 1.1 continued VoIP Codec Comparison Chart

Codec	Speed	Simultaneous Calls over a T1 Link (1.5 Mbps)	Notes
GSM	13 Kbps	118	
iLBC	15 Kbps	102	
LPC-10	2.5 Kbps	614	
Speex	2.15 to 44.2 Kbps	714 to 34	"Open" codec

The savings of bandwidth comes at a cost though; the more compression placed on a conversation, the more the voice quality degrades. When using LPC10 (one of the most efficient compression codecs), the conversation, while intelligible, often sounds like two whales making mating calls. If you have no other alternative, it will be sufficient, but it's not a good choice for a business environment.

The other major benefit of VoIP is the mobility. Phone calls can be sent and received wherever a data connection is available, whether it is a residential broadband connection, the office network, or a WiFi connection at a local drinking establishment. This mobility has a many benefits: a company's sales force can be scattered across the country yet have a phone in their home office that is an extension of the company's PBX. They can enjoy a voice mail box, an extension off the company's main number, and all the other features as if they all were in the same building.

It is important to make the distinction that VoIP is not exclusive to Asterisk. There is a growing market of software-based PBXes that tout VoIP as a major feature. Some traditional PBXes are starting to include VoIP features in them, and local phone companies are offering VoIP packages for customers. As a result, the advantages of VoIP have begun to catch the attention of the entire telecom industry.

The History of Asterisk

Mark Spencer, the creator of Asterisk, has created numerous popular open-source tools including GAIM, the open-source AOL Instant Messaging client that is arguably the most popular IM client for Linux, l2tpd, the L2TP tunneling protocol daemon, and the Cheops Network User Interface, a network service manager. In 1999, Mark had a problem though. He wanted to buy a PBX for his company so they could have voice mail, call other offices without paying for the telephone call, and do all the other things one expects from a PBX system. However, upon researching his options, he realized all the commercial systems cost an arm and a leg.

Undaunted, he did what every good hacker would: he set to writing a PBX suitable to his needs.

On December 5, 1999, Asterisk 0.1.0 was released. As the versions progressed, more and more features were added by developers, gathering a following of users, conventions, and everything short of groupies along the way. Asterisk's first major milestone was reached on September 23, 2004, when Mark Spencer released Asterisk 1.0 at the first Astricon, the official Asterisk user and developer's conference. Asterisk 1.0 was the first stable, open-source, VoIP-capable PBX on the market. Boasting an impressive set of features at the time, it included a complete voice conferencing system, voice mail, an impressive ability to interface into analog equipment, and the ability to talk to three different VoIP protocols reliably.

Development didn't stop there though. Asterisk continued to grow. On November 17, 2005, Asterisk 1.2 was released, which addressed over 3000 code revisions, included major improvements to the core, more VoIP protocols, and better scalability. Also, this release introduced Digium's DUNDi (Distributed Universal Number Discovery) protocol, a peer-to-peer number discovery system designed to simplify interconnecting Asterisk servers across, and in between, enterprises.

The latest release of Asterisk, Asterisk 1.4, was released December 27, 2006. This release featured major changes in the configuration process, optimized applications, simplified the global configuration, and updated the Call Detail Records for billing purposes. Also new in this version was better hardware support, an improved ability to interface with legacy equipment, and better interfacing with Cisco's SCCP VoIP protocol. Also, as with any software project, this update addressed the bugs and issues found since the 1.2 release.

Asterisk Today

Today, Asterisk is one of the most popular software-based VoIP PBXes running on multiple operating systems. Asterisk handles most common PBX features and incorporates a lot more to boot. It works with numerous VoIP protocols and supports many pieces of hardware that interface with the telephone network. Asterisk is currently at the forefront of the much talked-about "VoIP revolution" due to its low cost, open-source nature, and its vast capabilities.

The company Mark Spencer wrote his PBX for is now known as Digium, which has become the driving force behind Asterisk development. Digium sells hardware for interfacing computers into analog telephone lines and Primary Rate Interface (PRI) lines. Digium also offers Asterisk Business Edition, an Enterprise-ready version of Asterisk, which includes commercial text-to-speech and speech recognition product capabilities, and has gone through stress testing, simulating hundreds of thousands of simultaneous phone calls. Finally, Digium offers consulting for Asterisk installations and maintenance, and trains people for its Digium Certified Asterisk Professional certification.

Notes from the Underground...

Digi-wha?

Many companies spend millions of dollars with marketing firms to create a new name for their company. When Bell Atlantic and General Telephone and Electric (GTE) merged in 2000, they thought long and hard about their new name, and when they revealed it, millions scratched their head and said "What is a Verizon?" Thankfully, not all companies have this problem.

Digium (Di-jee-um) is the company that maintains most of the Asterisk source tree, and tries to show how Asterisk can provide solutions to the general public. According to legend, Digium got its curious sounding name when one of its employees pronounced paradigm as "par-a-did-jem." This became a meme, and "par-a-did-jem" evolved into "did-jem," which then further evolved into "Digium." Just think how much money Fortune 500 companies pay advertising executives to come up with a new name when companies merge.

What Can Asterisk Do for Me?

Asterisk is so multifaceted it's hard to come up with a general catchall answer for everyone asking what Asterisk can do for them. When a friend and I tried to think up an answer that would fit this requirement, the closest thing we could come up with was "Asterisk will do everything except your dishes, and there is a module for that currently in development."

Asterisk as a Private Branch Exchange

Asterisk is, first and foremost, a PBX. Some people seem to constantly tout Asterisk's VoIP capabilities, and while that is a major feature, they seem to forget that Asterisk doesn't need VoIP at all to be a PBX. But even without VoIP, Asterisk has many advantages over traditional hardware-based PBXes.

Advantages over Traditional PBXes

Asterisk has numerous advantages over "traditional" PBXes. These advantages can benefit both larger and smaller businesses. Let's talk about two different scenarios, with two different problems, but one common solution.

Notes from the Underground...

Is Asterisk Right for Me?

Whether they're an individual interested in VoIP or a group of business heads wondering if they should drop their expensive PBX, people frequently ask "Is Asterisk right for me?" The answer, almost always, is a resounding "YES!" Asterisk is many things to many people, and it is malleable enough to be a perfect fit for your setup, too.

Asterisk in a Large Business Environment

Suppose you are the newly hired IT Director for a medium-sized office. While getting a tour of the server room, you happen across the PBX. What you see disturbs you: a system, which handles approximately 200 people, is about the size of two mini fridges, requiring its own electrical circuit separate from the servers, and producing enough heat it has to be tucked in a corner of the server room so as not to overload the air conditioning system. It also seems to be stuck in the early 1990s: The system has abysmal voice-mail restrictions, no call waiting, and no caller ID. Being the go-getter you are, you attempt to "buy" these features from the vendor, but the quote you receive almost gives your purchase officer a heart attack. As if this wasn't enough, you also have a dedicated "PBX Administrator" who handles adding phones to the system, setting up voice-mail boxes, making backups of the PBX, and nothing else.

Asterisk is made for this kind of situation. It can easily fit within a server environment, and will cut costs instantly since you no longer have to cool and power a giant box that produces massive amounts of heat. Also, dedicated PBX administrators, while possibly still necessary for a large environment, can be easily replaced by other administrators, provided they know how to administrate a Linux box. A competent Linux user can be taught how to administer an Asterisk PBX easily. Finally, as stated repeatedly, Asterisk is open source, which really cuts the software upgrade market off at the knees. Plus, if Asterisk lacks a feature a company needs, there are more than a few options available to the firm: they can code it themselves, hire someone to code it for them, or use Asterisk's fairly active bounty system (available at http://www.voip-info.org).

Asterisk in a Small Business Environment

Asterisk provides advantages for small businesses as well. Suppose you are a consultant to a small company that has you come in a few hours every week to fix computer problems. This company has a small, ten-phone PBX that was installed by another vendor before you came into a picture. After a while, one of the phones—the owner's, of course—will no longer work with the voice-mail system. When you dial his extension, it rings his phone, and then drops you to the main voice-mail prompt instead of going directly to his voice-mail box. When he dials his voice mail from his phone, it prompts him for a mailbox rather than taking him directly to his. The vendor no longer returns phone calls, and the owner begs you to take a look at it. You bang your head against the wall for several hours trying to figure the system out. Besides the basic "How to use your phone" info, no documentation is available, there are no Web sites discussing the system, and diagnostic tools are non-existent. Even if you do figure out the problem, you have no idea how to correct it since you don't know how to reprogram it. In other words, you're licked.

Asterisk will fix most of the issues in this situation as well. Documentation, while admittedly spotty for some of the more obscure features, is widely available on the Internet. Asterisk debugging is very complete; it can be set up to show even the most minute of details. Also, in a typical Asterisk installation, vendor tie-in wouldn't be an issue. If the owner's phone was broken, a replacement phone could have been easily swapped in and set up to use the PBX—no vendor needed (see Figure 1.1).

Figure 1.1 Asterisk Can Be as Verbose, or as Quiet, as You Want

```
Asterisk Console on 'miina' (pid 8137)                          _  □  X

 ==  SIP Listening on 0.0.0.0:5060
 ==  Using SIP TOS: lowdelay
 ==  Parsing '/etc/asterisk/sip_notify.conf': Found
 ==  Registered channel type 'SIP' (Session Initiation Protocol (SIP))
 ==  Registered application 'SIPDtmfMode'
 ==  Registered application 'SIPAddHeader'
 ==  Registered custom function SIP_HEADER
 ==  Registered custom function SIPPEER
 ==  Registered custom function SIPCHANINFO
 ==  Registered custom function CHECKSIPDOMAIN
 ==  Manager registered action SIPpeers
 ==  Manager registered action SIPshowpeer
chan_sip.so => (Session Initiation Protocol (SIP))
Asterisk Ready.
*CLI> [Dec 29 19:03:56] NOTICE[8145]: chan_mgcp.c:3382 mgcpsock_read: Got respon
se back on [dlinkgw] for transaction 2 we aren't sending?
[Dec 29 19:03:57] NOTICE[8163]: chan_sip.c:11811 handle_response_peerpoke: Peer
'fromratchet' is now Reachable. (345ms / 2000ms)
[Dec 29 19:03:57] NOTICE[8163]: chan_sip.c:11811 handle_response_peerpoke: Peer
'fromzap' is now Reachable. (342ms / 2000ms)
[Dec 29 19:03:57] NOTICE[8163]: chan_sip.c:11811 handle_response_peerpoke: Peer
'6200' is now Reachable. (344ms / 2000ms)
 --  Saved useragent "Cisco-CP7960G/7.5" for peer 6200
 --  Saved useragent "Cisco-CP7960G/7.5" for peer fromzap
```

Features and Uses

As previously stated, Asterisk has numerous features, some common to almost all PBXes, and some only found in very high-end models. Let's highlight a few. This is by no means a complete list, but just a sampling of the many features Asterisk has to offer.

Conference Calls

Asterisk's conference calling system, called "MeetMe," is a full-featured conferencing system. All the features you would expect in a conferencing system are included, such as protecting conferences with PINs so only approved users can attend, moderating conferences to allow only certain people to speak to the group, recording conferences so you can have a record of it, and playing music before a conference begins so users don't have to wait in silence.

MeetMe is a huge feature for Asterisk, as the price of commercial conferencing services isn't cheap. Let's look at a simple example: We want to conduct an hour-long conference call with ten members of the press concerning our new Asterisk book. A certain reputable conferencing service costs 18 cents per minute per participant. So, doing the math, 13 users talking for 60 minutes at a cost of 18 cents/minute would cost us $140.40. Let's compare that with Asterisk. Using Asterisk, MeetMe, and an average VoIP toll-free provider whose rates are 2.9 cents per minute per call, the same conference would cost us $22.62. That's a savings of $117.78!

Voice Mail

Voice mail has become critical to business in today's market. Many people have developed a reflexive tendency to check the "Message Waiting" indicator on their phone when first entering their workspace. Technically, voice mail is quite simple. It is simply audio files stored on some kind of storage medium, such as a hard drive or flash storage, on your PBX. Some vendors think a two-hour voice-mail storage card, otherwise known as a 128MB Smart Media card, should cost over $200. Asterisk, considering it's run on a PC, affords you an amazing amount of storage space for your company's voice mail. Since it's not locked into a specific storage media, you can add an extra hard drive, flash card, or network share if you have the need to expand.

Asterisk's voice mail also incorporates almost every feature one would expect from a voice-mail system: a complete voice-mail directory, forwarding, and the ability to play different outgoing messages depending on whether the user doesn't pick up their phone, is already on the phone, or is out for a long period of time. Some of the more advanced features include the ability to send the voice mail as an attachment to an e-mail address. This is useful if you are on the road and do not have a phone available to you, but do have access to e-mail. It's also very handy when you have a voice-mail account you do not monitor regularly.

Call Queues

While everyone might not know what a call queue is, almost everyone has experienced one. When dealing with some kind of customer service department, it's not uncommon to wait on hold while a disembodied voice tells you that all the representatives are currently helping other people. That is a call queue. It is used for handling large volumes of calls with a set amount of people answering the phones. When the amount of calls ("callers") exceeds the amount of people answering the phones ("answerers"), a queue forms, lining up the callers till an answerer can attend to each. When one of the answerers becomes available, the first caller in line gets routed to that answerer's phone. Call queues are essential in any kind of call center environment. Asterisk supports both queues in the traditional sense of a call center full of people, and also a virtual call center in which the call agents call in from home and sit on the phone in their house. It supports ringing all agents at once, a round-robin system, or a completely random ring pattern. Asterisk also can assign priorities to callers when they enter a queue. For example, this is commonly done in cell phone companies. Have you ever wondered how when you visit a cell phone store and they call up customer service, they get answered in about 30 seconds? They call a separate number and are thus assigned a higher priority than if you called from your home. Another use of this is if you run a helpdesk and want to assign problems with mission-critical applications a higher priority than others. Users calling the telephone number for the mission-critical applications would thus receive a higher priority than users that call the general helpdesk number.

Asterisk as a VoIP Gateway

Asterisk's biggest and most talked about feature is its VoIP capabilities. Thanks to the expansion of Broadband into almost every company and an ever-increasing number of residences, VoIP has taken off in the past few years. Asterisk has turned out to be a tool no one really knew they needed, but realized what they were missing once they started using it.

Notes from the Underground…

PSTN Termination and PSTN Bypassing

Don't worry, PSTN termination has nothing to do with the PSTN becoming self aware and sending robots after us. PSTN termination providers are companies that allow third parties to transition their VoIP call between the Internet and the PSTN, or vice versa. These companies don't force users to invest in equipment to connect Asterisk to a phone line and are often much cheaper than what a telephone company would charge.

Of course, the cheapest phone call is the one that's free. The Internet Telephony Users Association, a non-profit organization, runs e164.org, which allows users to publish telephone numbers that can be reached directly via VoIP. This allows other VoIP users to dial a regular number and have Asterisk route it over the Internet rather than the PSTN letting the user save money without making an effort.

People have started using Asterisk to augment, and sometimes even replace, their existing telephone setup. Thanks to Asterisk, an abundance of cheap Internet-to-

PSTN-termination providers, and organizations such as e164.org, Asterisk has allowed people to choose the cheapest path to their destination when placing a phone call. Companies with multiple offices can save money on phone calls that are long distance from the originating office but local to one of the other offices by using Asterisk to route them over the Internet to the remote office and having the Asterisk server dial the remote phone line, thus saving them an expensive long-distance bill.

The Possibilities of VoIP

Looking at various trade magazines and Web sites, it is easy to get the feeling that pundits always rant and rave about VoIP, but companies and end users either have no interest in it or do have an interest but no idea what to do with it. Asterisk and VoIP provide many possibilities for both the end user sitting at home and the company looking to cut costs.

Virtual Call Centers and Offices

Before VoIP, when running a call center, the company either needed to pay for a large building to house all the employees, or pay the cost of forwarding the incoming phone calls to the employee's houses. With the advent of VoIP, a third option has emerged: using the employee's broadband connections to handle telephone calls over VoIP.

Thanks to Asterisk, it is possible to run a call center out of a back pocket. The only physical presences the call center needs are servers to handle the routing of the calls, and some way to terminate the incoming phone calls, such as a VoIP provider or PRI(s). The people answering the calls can either use their computer with a softphone and a headset, or some kind of Analog Telephone Adapter to hook up a VoIP connection to a physical phone (more on these later). Agents can then sign into the call queue without tying up their phone line or costing them money. They can also work anywhere a broadband connection is available.

This benefit isn't limited to call centers either. Would you like to save some money on your road warrior's cell phone bills? Or, would you like to have an option for your employees to work from home for a few days a week, but still have the ability to be contacted by phone like they were in their office? The same concept applies. Once a phone signs into Asterisk, it doesn't matter if it's in the office, down the street, or half a continent away, it becomes an extension on your PBX, with all the features and benefits.

Bypassing the Telephone Companies

Another way people have been using Asterisk is to set up their own "VoIP only" telephone network over the Internet. Suppose you have a group of friends you never talk to. With Asterisk, you can essentially set up your own virtual telephone company. After setting up Asterisk and then arranging the connections between your servers, you can establish a telephone network without even touching the PSTN. Plus, thanks to MeetMe, you can conduct conference calls with ease.

Also, while the media and most of the public associate "VoIP" with "phone calls over the Internet" this is only partly the truth. The "IP" in VoIP means "Internet Protocol," and Internet Protocol is Internet Protocol no matter where it is. If your company has data links between buildings, campuses, or regions, but not voice links, Asterisk can be used to send voice conversations over your data links as opposed to the phone lines, saving money and allowing your phone lines to remain free for other purposes.

One of the best hobbyist roll-your-own examples we've seen to highlight Asterisk's ability to act as an inexpensive gateway for telephones over large geographic areas is the Collector's Net at http://www.ckts.info. Founded in 2004, the Collector's Net is a group of telephony buffs who have, over time, collected old telephone switching equipment. For years, this equipment sat in basements and garages collecting dust until one owner had the bright idea of using Asterisk and VoIP to interconnect the gear over the Internet. And so Collector's Net was born. It is growing monthly and now boasts an Asterisk backbone connecting more than a dozen switches over two continents. While it may seem trivial or downright odd to some, this highlights the ability of Asterisk to provide a connection between a group of people who would have hardly spoken to each other had they not set up this network.

Being Your Own Telephone Company

Asterisk can save money, but it can make money as well. It's also simpler than you think. NuFone, one of the first PSTN termination providers that supported Asterisk's Inter-Asterisk eXchange (IAX) VoIP protocol, started as a computer and a Primary Rate Interface (PRI), sitting in the owner's apartment. It's now one of the more popular PSTN termination providers on the Internet.

However, don't start wearing your monocle and lighting cigars with $20 bills just yet. In years past, termination providers were largely flying under the radar of the various regulatory agencies. However, this golden age is rapidly coming to a close, and VoIP providers are slowly becoming more and more regulated. Today, VoIP providers must provide 911 services, are required to contribute to the Federal government's "Universal Service Fund," must handle taps by law enforcement agencies, and are subject to all kinds of regulations.

Asterisk as a New Dimension for Your Applications

The Internet has grown by leaps and bounds over the past ten years. Most companies have mission-critical applications, applications to monitor the applications, and applications to monitor the applications that monitor the applications, ad nauseam. There are also information systems designed to provide important information to the general public. These systems all have something in common: they require the use of a computer.

Computers, while common, aren't used by everyone. People constantly talk about the "digital divide," referring to people who are unable to afford computers. Plus, sizable portions of the populations, for one reason or another, still treat the computer with apprehension.

Phones, however, are very much ubiquitous. Almost every home has a land-based telephone in it, and with pre-paid mobile phones finally showing up in the United States, mobile phones are further penetrating the market. Despite this large market, developing voice-aware applications has always been costly and time-consuming, making them less common and less functional than their Web-based counterparts.

Asterisk can be a bridge between the world of text and the world of speech. Thanks to programs like Sphinx (a program that translates speech to text), Festival (a program that translates text to speech), and Asterisk's own application interface, programs can be written by any competent programmer. Asterisk's interface is simple to learn yet extremely powerful, allowing programs for it to be written in almost any language. Asterisk can be the conduit for taking your applications out of the text that is the Internet and letting them cross over into the voice arena that is the Public Switched Telephone Network (PSTN)

A great example of how telephone-aware systems can benefit the general public is Carnegie Mellon University's "Lets Go!" bus dialog system. It has been developed to provide an interactive telephone program that allows people in Pittsburgh to check the schedule of buses that run in the city. The system has become such a success that the bus company has had its main phone number forward calls to the application during off-hours, allowing callers to access transportation schedules despite the office being closed. Asterisk can also be used to build similar systems with the same tools used by CMU.

Who's Using Asterisk?

Asterisk really started to make a splash on the Internet in late 2003 when it became fairly stable and early adopters started to pick up on VoIP. Since most early adopters were hard-core technophiles who were looking for a program that was free or cheap, and could be easily configured to do everything from the simple and the mundane to the downright odd, Asterisk was in the right place at the right time. To say it caught on like wildfire is a bit of an understatement.

Today, Asterisk is still very active within the hobbyist's realm. Small groups are setting up Asterisk servers for both public and private use, one of them being the Collector's Net previously mentioned. There are also groups of phone phreaks—people who hack on the telephone network—who are taking the leap into the digital realm, setting up projects such as Bell's Mind (http://www.bellsmind.net) and Telephreak (http://www.telephreak.org). For phone phreaks, the ability to run a telephone system in the privacy of one's own home is just as exciting as when the first personal computers became available to computer hackers.

Not only is Asterisk actively thriving in the hobbyist scene, it is also making beachheads into the Enterprise realm. A university in Texas recently replaced their 1600-phone strong mix of Nortel PBXes and Cisco Call Manager installations with Asterisk. The reasons for this were both the cost of licensing each phone to Cisco, and security concerns due to the fact they ran on Windows 2000. A town in Connecticut recently deployed a 1500-phone Asterisk system, where each department customized it for its own needs, such as the school department's automated cancellation notification system.

Not only is Asterisk making it easy for companies to replace their existing telephone systems, it is making it easy for telephone companies to have the ability to handle VoIP. Numerous Competitive Local Exchange Carriers (CLECs) are jumping onto the VoIP bandwagon and setting it up to handle VoIP from the consumer side (or handle it internally) for either a value-added service or a cost-saving measure.

Summary

PBXes and VoIP have been around for decades: PBXes since the early part of the century, and VoIP since the 1970s. However, despite the vast market and the fact that they are used by almost every business, PBXes not only still cost thousands of dollars, but one vendor's equipment is often incompatible with another vendor's.

Asterisk, created in 1999 because Mark Spencer found commercial PBXes hideously expensive, has put the power of telephony in the hands of the masses. It can be many things to many people, and can be configured to fit into many roles in an Enterprise. From saving money on telephone calls, to making voice-enabled applications, Asterisk can be configured to fit in where it's needed.

Asterisk can augment, or entirely replace an existing telephone system, whether the user is a hobbyist with a single telephone line, or an executive running a large call center with multiple PRIs. An existing PBX installation can be swapped out with ease, and most, if not all functionality can be retained. Asterisk also has numerous advantages over traditional PBXes in the areas of cost, reliability, usability, and hardware support.

Asterisk is not only a traditional PBX, but can also handle Voice over IP telephone calls. This allows users to take advantage of the numerous advantages VoIP provides: low-cost telephone calls, the ability to communicate with remote offices using the Internet rather then the PSTN, or using existing data links instead of connecting buildings with telephone lines.

Asterisk also allows you to integrate existing applications into the world of telephony. Users can interact with existing applications over telephones, rather than their current interface—such as a Web page or a data terminal. This has advantages in both usability and flexibility.

In the current market, Asterisk is being utilized by both large and small companies. It lets small companies find a PBX that won't tie them down to a vendor and incur a hefty initial investment, while large companies see a way of leveraging their existing infrastructure that saves them money by not having to rely on the telephone company.

Solutions Fast Track

What Is Asterisk?

☑ Asterisk is an open-source Private Branch Exchange that replicates, for free, many expensive features found in expensive high-end PBXes.

☑ Created in 1999 by Mark Spencer, it was initially made because commercial PBXes were far too expensive for his company. Today, his company is the driving force behind Asterisk.

☑ Asterisk's current version, 1.4, boasts a load of new features over its predecessors.

What Can Asterisk Do for Me?

☑ Asterisk can be fit into both the large and small business environment, saving time and money in the workplace. It can also be useful to the hobbyist.

☑ Asterisk can replace your traditional hardware PBX and replicate most of its features. It can also bring many new features to the table to replace other telephony services you currently use.

☑ Thanks to the advantages provided by VoIP, Asterisk allows you to run virtual call centers and bypass the telephone company for phone calls. It also lets you be your own telephone company.

☑ With the ubiquity of voice communication channels, Asterisk lets you bring a whole new dimension to your current suite of applications.

Who's Using Asterisk?

☑ Asterisk took the market by storm by being in the right place at the right time, and by also being free.

☑ Hobbyists are using Asterisk to set up their own private telephone playlands, complete with voice conferences, voice-mail systems, and voice bulletin boards.

☑ Companies both large and small are using Asterisk to replace their current PBX systems and are saving themselves both time and money in the process.

Links to Sites

- **Asterisk (http://www.asterisk.org)** Here, you can download the source, keep up-to-date on Asterisk-related news, read developer weblogs, and generally get your daily dose of Asterisk scuttlebutt.

- **Digium (http://www.digium.com)** These folks are the driving force behind Asterisk. Get trained, buy hardware, and find out about developer programs.

- **Collector's Net (http://www.ckts.info)** This is an inventive group of old Bell System workers and telephone system collectors who have hooked together their antique equipment using Asterisk. Not as much Asterisk stuff here, but a cool enough group of people that warrant a mention, and it shows that Asterisk can be used to do almost anything.

- **Bell's Mind (http://www.bellsmind.net)** A project that provides information regarding various telephone systems, and a PBX for public use.

- **Telephreak (http://www.telephreak.org)** Telephreak is a free voice-mail and conferencing service run *for* phone phreaks and computer hackers *by* phone phreaks and computer hackers.

Frequently Asked Questions

The following Frequently Asked Questions, answered by the authors of this book, are designed to both measure your understanding of the concepts presented in this chapter and to assist you with real-life implementation of these concepts. To have your questions about this chapter answered by the author, browse to **www. syngress.com/solutions** and click on the **"Ask the Author"** form.

Q: What is Asterisk?

A: Asterisk is an open-source PBX. Built by Digium Incorporated and developers across the globe, it is at the forefront of VoIP usage.

Q: How much does Asterisk cost?

A: While Asterisk itself is completely free, the cost of a complete install depends greatly upon your existing installation, what you want to use Asterisk for, and what kind of hardware you are willing to invest in. As always, your mileage may vary.

Q: I currently have a PBX, what advantage is there for me to move to Asterisk?

A: Asterisk has a lot of features that your current PBX likely does not have. It also has numerous advantages over a "traditional" PBX, such as the support of open standards, not being tied down to a specific vendor, and the common advantages of being open source.

Q: Do I need to move to VoIP to use Asterisk?

A: No. Asterisk supports numerous hardware devices, allowing you to use both analog phones and analog telephone lines with the system.

Q: What companies can most benefit from Asterisk?

A: There is no right kind of company for an Asterisk setup. Safe to say, if you have a PBX already, you can, and probably should, run Asterisk.

## Chapter 2

Setting Up Asterisk

Solutions in this chapter:

- **Choosing Your Hardware**
- **Installing Asterisk**
- **Starting and Using Asterisk**

Related Chapters: 3, 7

☑ **Summary**

☑ **Solutions Fast Track**

☑ **Frequently Asked Questions**

Introduction

Setting up and installing any kind of PBX server isn't easy. Adding Asterisk to the mix does simplify some areas, but further complicates others. Asterisk is flexible, but this flexibility creates many options that can overwhelm a novice. Everything from picking out a server, picking a phone setup, to picking an install method can leave you in awe of the options available. Let's not sugarcoat it: Asterisk is hard.

Choosing hardware is a key decision and not one that can be taken lightly, because if something goes wrong with the server or the phones, productivity is lost. Making the proper decision on a server, choosing phones for the users, and selecting the network configuration can mean the difference between a happy user base and a group of angry users outside your office with pitchforks and torches.

Even choosing a method to install Asterisk is filled with options, such as Live CDs, Asterisk Linux distributions, binaries for your operating system, or compiling from scratch. And there is no "correct" option either. Each method has benefits and drawbacks, and each one suits certain situations differently than others. Making sure you choose the right method of installing can save you a lot of heartburn later.

If you're scared right now, don't be. While Asterisk isn't easy, it is nowhere near impossible. While Asterisk may have a high learning curve, once you become familiar with its intricacies, everything suddenly starts to make sense.

Choosing Your Hardware

One of the first things to do when setting up Asterisk is to figure out your hardware needs. Hardware is a bit of a catch-all term and refers to the server, the phones, and the connections between them. There is no standard ratio for Asterisk that dictates "To support A calls over a B period of time, you need a server with X megabytes of RAM, a processor faster than Y, and a hard drive bigger than Z" or that "If you are in a call-center environment, X brand phones is the best choice." To figure out what is the correct fit for your situation, research is required.

Picking the Right Server

Picking the right server is a key decision when running Asterisk. The last thing a company wants to hear is that their phone system is down. Asterisk *can* run on obsolete hardware, but you will get what you pay for. Reliable, capable equipment is the foundation for any reliable, capable PBX system.

Processor Speed

Processor speed is the most important feature when looking at a server to run Asterisk. The more processing power, the more responsive the system will be when it is placed under heavy call loads. Asterisk runs well on any modern processor, handling moderate call loads without any issue. However, this does depend on how the system is configured to handle calls.

Transcoding and Protocol Translation

Transcoding is when the server is handling a conversation that is coming in with one codec and coverts it on-the-fly to another. This happens a lot more than thought, as most VoIP telephones transmit in μ-Law, which is the standard codec for telephone conversations. If the server is using the GSM codec for outbound calls, it needs to "transcode" the conversation and convert it from μ-Law to GSM. This, by itself, is pretty simple; however, when the server starts having to transcode multiple conversations simultaneously, more processing time is required. If a performance bottleneck develops, the conversations will start to exhibit delays in the conversation, more commonly referred to as "lag."

Protocol translation is the same problem as transcoding, except instead of converting the audio codec, it needs to translate the protocol used. This is also common with VoIP providers who only offer access to their networks via specific protocols.

RAM

RAM usage on Asterisk is pretty low. Asterisk can easily fit within a 64MB footprint even on a fairly large install. Since Asterisk is modular, trimming RAM consumption is as easy as removing modules from the startup sequence. A bare bones Asterisk startup can fit within a memory footprint of fewer than 30MB.

Storage Space

Storage space is probably one of the least important choices when choosing a server for Asterisk. Hard drives keep getting larger and cheaper with each passing month, allowing even a low-end computer to have massive amounts of space. Asterisk, by itself, hardly takes up any room; however, when voice prompts for Interactive Voice Response (IVR) menus and voice mail start being added to the system, Asterisk's footprint starts growing. Hard drive size needs to be determined by the amount of users on the system and the amount of voice mail expected.

For example, a sound file encoded with μ-Law takes up about a megabyte a minute. While this may not seem like a lot at first, consider that a person can average about five voice mails a day in a busy office. If each of those messages is about a minute each, and there are 100 people in the office, that's 500MB of storage per day! When you calculate the math per year, we're talking almost 13GB! Plus, other factors exist as well. Let's say a team leader sends a five-minute group message to his or her team of ten people. That 5MB message just copied across the system into ten separate mailboxes consumes 50MB. Also, don't forget to factor in saved messages, people on extended absences, and group mailboxes that may be accessed by the public.

Asterisk, like any high-demand server application, benefits from Redundant Arrays of Independent Disks (RAIDs). RAIDs are very important in any kind of high-availability environment. They are a system in which multiple disks are grouped together in a redundant fashion, allowing the computer to write data across all the disks at once. The upside of this is that it allows for one disk to fail within the group but let the computer still function. Using a RAID allows Asterisk to continue to handle phone calls and voice mails despite one of the server's hard drives no longer functioning.

Picking the Right Phones

Phones are arguably the most important part of a PBX setup. This is how most users interface with the PBX system. Picking the proper phone is key to a successful PBX deployment. There have been instances where users were ready to give up on Asterisk solely because they hated their phones. Thankfully, changing phones is easy and these users quickly changed their opinions once new phones were installed.

Soft Phones

The easiest phone to set up with Asterisk is a soft phone. A soft phone is a computer program that emulates a phone on your PC. Soft phones are easy to set up and can be configured in a matter of minutes. They're usually very easy to use, often displaying a telephone-like interface on the screen. Soft phones utilize the computer's sound card for transmitting and receiving audio, or optionally a "USB phone," which is a phone-like device that plugs into the computer's USB port. Soft phones are inexpensive (often free) and USB phones generally cost less than $50. Figure 2.1 shows iaxComm.

Figure 2.1 iaxComm, an IAX2-Compatible Soft Phone

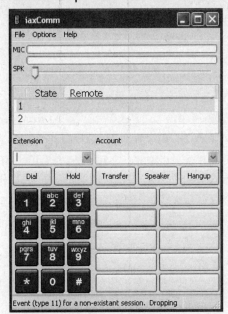

Soft phones have the advantage in price and ease of setup and configuration, but that's about it. It's common to see people preferring some kind of physical device rather than a program that runs on a PC. USB phones sometimes can help, but they usually aren't geared for a business environment. In addition, these users are tethered to a PC. If the PC crashes, no phone calls.

Soft phones are handy though if a user wants to make VoIP calls while on the road without wanting to lug another device with them. Simply install and configure the soft phone on the user's laptop with a headset and they're ready to go—all they need is an Internet connection. However, soft phones are not fit for most tasks common to a business environment.

Hard Phones

The alternatives to soft phones are hard phones—the phones we've used the past 125+ years: a physical device that sends and receives telephone calls. Hard phones are on the opposite side of the spectrum from soft phones: they're expensive and often harder to set up than their software counterparts. However, most users prefer a hard phone; it's what they're accustomed to.

The most common hard phones include IP phones: analog phones connected to an Analog Terminal Adapter (ATA) and analog phones connected via interface cards.

Each of these has their advantages and disadvantages, which we'll discuss in the following sections.

IP Phones

IP phones are one of the most common solutions you'll see for VoIP in a business environment. They plug in to an Ethernet connection and emulate a regular analog phone. They're made by numerous companies, including Cisco Systems, Polycom, Aastra, and Siemens, just to name a few. The price and quality of these phones run the gamut, but the general rule of "you get what you pay for" applies here. In today's market, a good IP phone will cost you at least $150 per unit, like the Cisco 7960 IP Phone shown in Figure 2.2.

Figure 2.2 A Cisco 7960 IP Phone

Analog Telephone Adapters

ATAs are the bridge between the world of analog telephones and the world of VoIP. They are small devices, usually in the form of a small plastic cube, with a power port, one or more telephone jacks, and an Ethernet port. An analog phone connected through an ATA can participate in phone calls on a VoIP network.

ATAs are cheaper then IP phones, mainly because they are slightly simpler. ATAs are often offered by the same companies that make IP phones and range in price from $50 to $100 depending on the protocols they support, the number of ports, and, of course, the number of features. Some ATAs have both a port for a phone and a port for an outside phone line, allowing a quick and easy way to interface Asterisk with both your phone and the public switched telephone network.

ATAs work with most phones, the exceptions being proprietary phones from digital PBXs and older rotary dial phones. Digital phones are nearly impossible to support due to their complexity and the differences between one manufacturer and another. Rotary phones aren't supported by most ATAs because most developers consider, somewhat correctly, that pulse dialing is an obsolete protocol. Figure 2.3 shows a D-Link analog telephone adapter controlling two older analog phones.

Figure 2.3 A D-Link Analog Telephone Adapter Controlling Two Older Analog Phones

Interface Cards

Analog phones do not always need an ATA. Asterisk supports multiple interface cards that allow analog phones to connect directly to an internal port on the server. Digium sells numerous cards supported by its Zaptel drivers. These cards support anywhere from 1 to 96 phones depending on how they are configured. There are also other cards that support anywhere from a single phone line to an entire PRI.

PRIs can be attached to a device called a "channel bank," which will split the PRI's 24 channels into 24 separate interfaces, allowing a single interface card to support up to 24 phones. Cards also aren't limited to a single PRI interface, either. And some cards out there can support four simultaneous PRIs.

Digium also sells cards that sustain up to four modular sockets that can either support telephone lines or telephones depending on the modules purchased. While these are rather pricey, they are cheaper than PRI cards and will allow you to avoid purchasing a channel bank on top of a card.

Sadly, interface cards do not support digital phones either. Another issue when considering these is that there needs to be wiring run between the phones and the cards, which can be difficult in an existing server setup. The good news is that most of these cards support pulse dialing, allowing older equipment to interface into the system.

Configuring Your Network

A network is like a car. You can use it every day and not notice it until the day it breaks down. This is even truer when the network is also the phone system's backbone. For most folks, phone service is much more important than Internet access.

When looking at it from a network management standpoint, VoIP conversations using the μ-Law codec are 8KB/s data transfers that run for the duration of the calls. While this amount of traffic is negligible if designing a network for an office of ten people, it starts to add up quickly when designing the network for a voicemail server serving 10,000 people. For example, if there are 2500 simultaneous phone calls connecting to and from the server, that would be a constant stream of 20 megaBYTES per second being transferred across the network.

When designing networks for VoIP, virtual local area networks (VLANs) are a big help. VLANs are a software feature in networking switches that allow managers to set up virtual partitions inside the network. For example, you can set up a switch to have even-numbered ports on VLAN A and odd-numbered ports on VLAN B. When plugging networking equipment into the switch, equipment on VLAN A won't be able to connect to equipment on VLAN B, and vice versa, allowing the two VLANs to be independent of one another. VLANs help immensely for a VoIP network since they keep voice and data traffic separate from each other. The last thing you want is a giant multicast session DoS-ing your phones. By keeping the computers on separate VLANs, computer traffic will not interfere with voice traffic, allowing a user to make a large file transfer and not see any degradation of the voice quality on their phone.

Notes from the Underground...

Who's Listening to Your Phone Calls?

VLANs not only help immensely with traffic management, but also with security. Much like how attackers can sniff your existing traffic via ARP poisoning and other attacks, they can do the same with your VoIP traffic. Automated tools such as VoMIT and Cain and Abel allow attackers to sniff and record all voice traffic they intercept.

The most secure solution to this is to set up a second Ethernet network or VLAN on your network and limit the connections to the phones only. While this is not a completely foolproof solution, since attackers on the network can spoof MAC addresses, thus bypassing the restrictions, this will keep random script kiddies from recording the boss's phone calls to his mistress.

WAN links are another part of the chain. WAN links can vary from a simple DSL connection to a massive Optical Carrier connection, but they each have something in common, they are a link to the outside world. When thinking about setting up a WAN connection or making changes to your current one, you need to figure out what the current bandwidth consumption is, and how much more bandwidth will be consumed by adding VoIP to the equation. If the link's free bandwidth during lunch is under 100Kb/s, it will be able to support one μ-Law encoded VoIP call during that timeframe without running into issues. If there are usually five simultaneous telephone calls during that timeframe, that's a major issue.

While with WANs it's impossible to have a VLAN, it is possible to shape the bandwidth. Bandwidth shaping is when a device, called the bandwidth shaper, gives certain traffic priority over others. Numerous ways exist to do this, the most common being to dedicate a portion of bandwidth exclusively to VoIP, or giving priority to VoIP traffic. Each has their pluses and minuses: dedicating a portion of the bandwidth to VoIP allows you to guarantee there will always be a set amount of bandwidth for telephone calls. While this may seem desirable, this is inefficient; if there is no voice traffic but the data portion is at 100-percent utilization, the voice portion will sit idle while the data portion suffers. The alternative, giving priority to VoIP traffic, allows the WAN link to fluctuate how much bandwidth is being used for data and how much is being used for voice. This allows data to use 100 percent of the bandwidth if there is no voice traffic, but still permits voice traffic to get through if the need arises. This is accomplished by letting the bandwidth shaper dynamically allocate bandwidth for the voice traffic when a conversation starts: if a voice packet and a data packet reach the bandwidth shaper at the same time, the bandwidth shaper gives the priority to the voice packet over the data packet. This does have a downside though: in certain shaping schemes, if voice packets keep reaching the bandwidth shaper faster than it can send data packets, the data packets will take longer and longer to get through. This will result in the data connections timing out and failing.

Installing Asterisk

So, you've purchased your server, installed an operating system, and you're ready to plunge head first into Asterisk. Determining the "right way" to install Asterisk depends on your situation. If you just want to try Asterisk out and are worried about messing up an existing system, the Live CD would likely be your best route. If you are not too familiar with Linux installation, but are looking to set up a dedicated Asterisk system, you may want to look at a CD distribution of Asterisk. If you are an experienced Linux administrator and want to configure Asterisk to fit into a custom environment, you'll likely just want to compile it from scratch. Finally, if you are either a Mac OS X or Microsoft Windows user, and you just want to use your existing operating system for an Asterisk install, you'll likely just want to use the packages for your operating system.

Asterisk's ability to be customized isn't just limited to the final setup; it starts at the installation phase of the system. You can easily make it fit almost any environment.

Using an Asterisk Live CD

Live CDs are bootable CDs that contain a complete operating system. After booting, your machine will run the operating system from the CD without installing it to the hard drive. If something goes wrong, you can turn off the computer, eject the CD, reboot, and boot back to the operating system installed on your hard drive. Although this installation method is not recommended for most production environments, it is a perfect way for a novice user to try out the features of an operating system without altering the boot machine in any way. In this section, we'll take a look at one of the more popular live Asterisk CDs: SLAST.

SLAST

SLAST (SLax ASTerisk) is an Asterisk-ready version of the Slackware-based SLAX Live CD. Maintained by the Infonomicon Computer Club, SLAST was designed to help educate people about the advantages of Asterisk and allow them to set up a simple Asterisk server in the easiest way possible.

Getting SLAST

SLAST is available at http://slast.org. The ISO image is available from their download page. The download size comes in at just a bit over 100MB, so any broadband connection should make quick work of the download. Once the ISO is downloaded, the disk image can be burned to a CD using the "image burn" feature of most popular CD recording programs.

Booting SLAST

Booting SLAST is as simple as inserting the CD into an Intel-based machine, and rebooting. Depending on how your machine is configured, you may need to press a key during startup to instruct the machine to boot from a CD. Once the CD is booted, the SLAST screen is displayed, as shown in Figure 2.4.

Once SLAST loads the system into memory, the login screen is displayed. The login screen has a quick "cheat sheet" of sorts showing file locations of Asterisk configuration files, Asterisk sounds, Asterisk modules and the SLAST documentation. The root password is also displayed. Log in with the username *root* and the password *toor*, and you will be presented with a root shell, as shown in Figure 2.5.

Figure 2.4 The SLAST Splash Screen Booting SLAST

```
      Welcome to the Slast Live CD.
   Hit Enter to continue booting, press F1/F2 for help or F3 to contribute
   To load Slast into RAM type "slast copy2ram"

boot:
Loading boot/vmlinuz............
```

Figure 2.5 The SLAST Login Screen

```
                 Slast, now with more Asterisk 1.4

Username/Password.
root / toor <--change this to something good

Useful Commands:
mcedit ............ This is your text editor, for editing config files
asterisk -cvvv& .... Starts Asterisk 1.4
asterisk -r ........ The Asterisk CLI, only works after you start asterisk
setup_sshd ........ Setup SSH Remote Access
configure_network .. Configure Network Script

Useful File Locations:
/etc/asterisk .............. Asterisk configuration files
/var/lib/asterisk/sounds ..... Sounds for Asterisk to use
/user/lib/asterisk/modules ... Asterisk modules

Saving Your Changes:



slast login: _
```

Configuring the Network

While a network connection isn't specifically required for Asterisk, unless the target system has hardware to connect it directly to a phone, some kind of network connection will likely be necessary if you want to connect to something besides the local computer. SLAST, as with most live distributions, does a pretty good job at detecting any and all hardware on the target system. If everything is plugged in and turned on, SLAST should have no issues setting up the hardware. However, SLAST, like other Live CDs, may have trouble detecting networks settings. If you're running a DHCP server, Asterisk should automatically configure your settings. However, if manual intervention is required to configure these settings, you may need to rely on *ifconfig*, the InterFace Configurator.

Running *ifconfig* without any arguments will display any configured network interfaces on the system. Ethernet interfaces will be shown labeled by their abbreviations *ethX*, where X is a number starting at 0 for the first interface. Next to the name will be fields for the IP address labeled as "inet addr," the broadcast address labeled as "Bcast," the network mask labeled as "Mask," along with various statistics regarding the interface. See Figure 2.6.

Figure 2.6 Running the ifconfig Utility to See Your Configured Network Interfaces

```
slast login: root
Password: ****

root@slast:~# ifconfig
eth0      Link encap:Ethernet  HWaddr 00:0C:29:95:4D:4E
          inet addr:192.168.248.132  Bcast:192.168.248.255  Mask:255.255.255.0
          UP BROADCAST NOTRAILERS RUNNING MULTICAST  MTU:1500  Metric:1
          RX packets:4 errors:0 dropped:0 overruns:0 frame:0
          TX packets:3 errors:0 dropped:0 overruns:0 carrier:0
          collisions:0 txqueuelen:1000
          RX bytes:808 (808.0 b)  TX bytes:1240 (1.2 KiB)
          Interrupt:11 Base address:0x1400

lo        Link encap:Local Loopback
          inet addr:127.0.0.1  Mask:255.0.0.0
          UP LOOPBACK RUNNING  MTU:16436  Metric:1
          RX packets:0 errors:0 dropped:0 overruns:0 frame:0
          TX packets:0 errors:0 dropped:0 overruns:0 carrier:0
          collisions:0 txqueuelen:0
          RX bytes:0 (0.0 b)  TX bytes:0 (0.0 b)

root@slast:~# _
```

If the Ethernet connection is not displayed when running *ifconfig* without arguments, it is either not configured, or it has not been detected on your system. To determine this, run the command *ifconfig eth0*. This will show the first Ethernet interface on the system, configured or not. If no text is displayed, SLAST has not found your Ethernet card and it will need to be manually set up. However, if text is displayed similar to the preceding figure, but missing the text regarding the IP address, the Ethernet interface is set up, just not configured with an address. SLAST provides a script to perform this configuration.

The *configure_network* script allows the system's network interface to be configured with minimal user interaction. The user can run the script by entering *configure_network* at the prompt and pressing **Enter**. The script will execute, prompting you for information regarding your desired network configuration, as shown in Figure 2.7.

Figure 2.7 Running the *configure_network* Script to Configure Your Network

```
root@slast:~# configure_network

Configure Networking on Slast
-----------------------------

What network device do you want to configure?

eth0

What is the ip address you want to use?

192.168.248.42

What is your subnet mask?

255.255.255.0

What is your gateway IP?

192.168.248.1

What is your primary DNS Server?

4.2.2.1_
```

The *configure_network* script first prompts for the name of the interface you are looking to configure. This will most likely be the first Ethernet interface, eth0. In case the system has multiple Ethernet interfaces, this could be eth1 or eth2, depending on which card was detected first and how many Ethernet interfaces are installed. After entering the desired interface, *configure_network* will prompt you for the desired IP address, followed by the network's subnet mask. These are very important to configure correctly since entering incorrect values would at best cause the system to be unable to access the network, and at worst cause the entire network to be taken offline! The next piece of information *configure_network* needs is the network's gateway IP address. If the system is on a standalone network—that is, a network without a connection to the Internet—leave this blank. Finally, *configure_network* will prompt for the network's DNS primary and secondary DNS servers.

After entering the entire network configuration, the script will prompt you to confirm all the settings entered. If the network settings are correct, the script will apply the changes. Otherwise, it will return you to the root prompt. This script can be run later, allowing you to change any of the information, and it can be aborted at any time by pressing **Ctrl + C**.

Saving Your Changes

One major advantage of a Live CD is that they do not make any permanent changes to your system, allowing you to undo any changes simply by rebooting your computer. This, while handy if you mess something up, can become a problem in certain situations: if the computer restarts for any reason, all the configuration changes are lost. SLAST, because it is based on SLAX, has two utilities that address this problem: *configsave* and *configrestore*. These utilities allow a user to back up and restore any changes they made. One of the more interesting ways to save the changes is to do so to a USB memory stick. This way, you can easily carry around the bootable CD and any configuration changes made to it, allowing you to essentially take your Asterisk server with you in your pocket.

To save your configuration changes, use the command *configsave*, followed by the name of a file to save to. For example, to save to a USB memory stick, run the command *configsave /mnt/sda1/asteriskconfigs.mo*. SLAST will then save any changed file from the /var, /etc, /home, and /root directories.

To restore your changes, use *configrestore* with the same syntax. If you saved your configurations to a USB memory stick, as in the preceding example, you can restore them by booting SLAST, inserting the memory stick, and then running *configrestore /mnt/sda1/asteriskconfigs.mo*. This will restore the files saved in the file. Remember, after you restore your files, if you make changes, you will need to run *configsave* again.

Installing Asterisk from a CD

Four Linux distributions focus on Asterisk: PoundKey, a Linux distribution supported by Digium; Evolution PBX, a distribution made for small businesses with commercial support; Elastix, a distribution supported by a commercial company; and trixbox.

trixbox was released in 2005 as "Asterisk@Home," a simple and easy way to install Asterisk on a computer. Self contained within a bootable CD, Asterisk@Home focused on ease of use and ease of install, allowing someone with little to no Linux experience to start playing with Asterisk. In 2006, Asterisk@Home was acquired by Fonality, a California-based VoIP services firm, who renamed the new version of Asterisk@Home to "trixbox." Today, trixbox is one of the leading Asterisk Linux distributions. With over 30,000 downloads a month, it takes its place among the "heavy hitters" of Asterisk distributions.

The trixbox CD contains numerous add-ons to Asterisk: freePBX, a Web-based configuration manager; HUDLite, a cross-platform operator panel; and SugarCRM, a complete Customer Relationship Manager suite. All of these are configured to run out of the box with trixbox, allowing a complete suite of tools for managing and maintaining your Asterisk installation.

Getting trixbox

trixbox is available at www.trixbox.org. The most up-to-date version at the time of this writing is trixbox 2.0 which contains Asterisk 1.2.13. The download size clocks in at a hefty 550MB, so you may want to put on a pot of coffee before you start downloading. Like the live CD's discussed earlier, the downloaded image can be burned with the "image burn" function of any standard CD recording program.

Tools & Traps…

Getting Messed Up by Old Asterisk Versions

It's common to think "Hmmmm… You know, I don't NEED the latest version of Asterisk" if you're looking at installing it from a binary package or an installer CD. However, watch out. Sometimes the differences between the versions are pretty big, and while what this book covers will work in Asterisk 1.4, it may not work in earlier versions.

Booting trixbox

After burning the trixbox CD, use it to boot the machine you will be installing to. Again, as mentioned in the earlier "Booting SLAST" section, the computer may need some kind of setting changed to boot from a CD. Once the CD is booted, the trixbox boot screen is displayed, as shown in Figure 2.8.

Figure 2.8 trixbox Booting

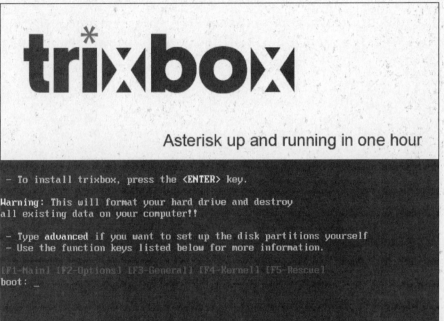

```
- To install trixbox, press the <ENTER> key.

Warning: This will format your hard drive and destroy
all existing data on your computer!!

- Type advanced if you want to set up the disk partitions yourself
- Use the function keys listed below for more information.

[F1-Main] [F2-Options] [F3-General] [F4-Kernel] [F5-Rescue]
boot: _
```

Tools & Traps...

Behold trixbox, Destroyer of Data

The trixbox CD is an Installer CD, not a Live CD. Installing trixbox onto a system will wipe out all existing data. If you are using a current system, it would be wise to make sure it has no data you want to keep, or that you have good backups of that data. The alternative is to use someone else's system, preferably someone you do not like.

After about five seconds, the CentOS installer will start loading up, as shown in Figure 2.9.

Figure 2.9 Anaconda, the CentOS Installer, Loading Drivers for SCSI Hardware

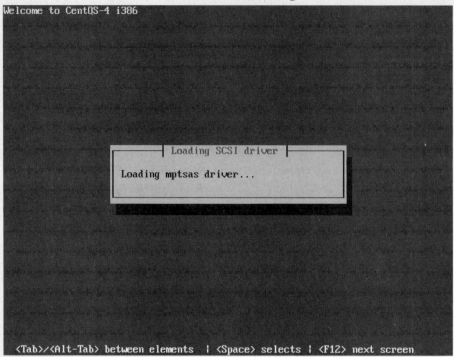

After all the system's hardware is detected, the installer will start prompting you for questions regarding keyboard layouts and time zones. Answer these as appropriate to your system. Once done with that, it will prompt you for a root password. Once enough information is gathered, the installer will start formatting your hard drive and the installation will begin, as shown in Figure 2.10.

Tools & Traps…

Excuse Me… Your Users Are Showing….

trixbox, allows Secure Shell (SSH) by default. This by itself is not much of a security issue, but root access is allowed from remote terminals. This means that if your trixbox system is publicly accessible on the Internet, anyone can

Continued

log in to your system if they guess your root password. This may seem unlikely, but it's common for script kiddies to scan entire networks looking for badly configured servers that allow root access and have common root passwords. So, either have an excellent root password, keep your system behind a firewall that disallows inbound port 22 traffic, or read up on how to disable root logins via SSH.

Figure 2.10 trixbox Installing CentOS Packages to the System

The trixbox installer will copy files, reboot, and begin to install specific packages on the system (see Figure 2.11).

After installation, trixbox will reboot one last time and display a login prompt. Log in with the username *root* and the password you specified in the setup process and you will be presented with a root shell. After logging in, the URL of the Web management interface will be displayed, as shown in Figure 2.12.

Figure 2.11 trixbox Installing the trixbox Packages

```
--------------------------------------------------
¦    Installing trixbox                          ¦
¦                                                ¦
¦    This can take some time...                  ¦
¦    System will reboot when installation in complete  ¦
--------------------------------------------------


Installing trixbox...

Sun Feb 18 16:04:26 EST 2007

**********************************************************
** install addon ****************************************
**********************************************************
Adding group asterisk...
adding user asterisk...
------------------------------------------
installing misc RPM
------------------------------------------
warning: /var/trixbox_load/rpms/lame-3.96.1-4.el4.rf.i386.rpm: V3 DSA signature:
NOKEY, key ID 6b8d79e6
Preparing packages for installation...
lame-3.96.1-4.el4.rf
_
```

Figure 2.12 Logging In to trixbox

```
CentOS release 4.4 (Final)
Kernel 2.6.9-34.0.2.EL on an i686

asterisk1 login: root
Password:
Last login: Sun Feb 18 17:10:38 on tty2

Welcome to trixbox
------------------------------------------------

For access to the trixbox web GUI use this URL
http://192.168.10.129

For help on trixbox commands you can use from this
command shell type help-trixbox.

[root@asterisk1 ~]# _
```

Configuring trixbox

trixbox, like SLAST, should configure its network automatically if there is a DHCP server on the network. If it didn't, or if the DHCP address is not the address you want for the server, you can run the *netconfig* utility to manage network settings, as shown in Figure 2.13.

Figure 2.13 The Main netconfig Screen

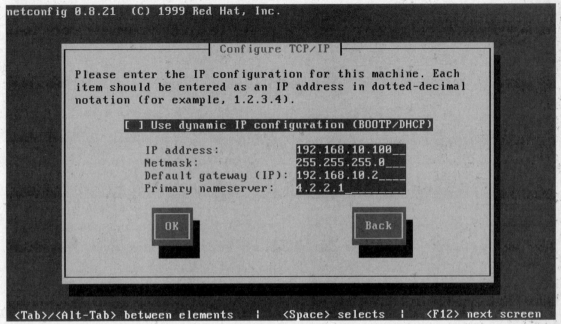

netconfig will prompt you for the IP address, netmask, gateway, and nameserver of your network. Enter these as appropriate for the system. After confirming these settings, the utility will exit. Reboot the system, and the new network settings will take effect.

trixbox's Web Interface

One of trixbox's nicer features is a Web interface that allows you to manage the system through a Web browser. It uses PHPConfig Asterisk config editor, which allows you to edit the files directly, in addition to using freePBX, which is a standardized interface for managing certain Asterisk features.

Tools & Traps…

The Danger with Frameworks

freePBX is an amazing system for simplifying the Asterisk configuration process. However, as with any framework, you are constrained by what the framework supports. Trying to go beyond what the framework supports is often a tedious process. So, while freePBX lowers the bar for learning Asterisk, you can grow out of it quickly.

By entering the system's IP address into your Web browser, you'll be greeted with trixbox's home page. You'll see links for the system's Asterisk Recording Interface which manage the ability to record audio conversations on Asterisk, scripts to manage Asterisk's recordings, voice mail, and call monitoring recordings; the MeetMe management system, a system to manage MeetMe conferences; Flash Operator Panel, a phone operator panel for Asterisk written in Flash; and SugarCRM customer relationship management software. In the upper right, you'll see a link to switch into "Maintenance" mode. Clicking the link will prompt you for a username and password. Log in with the username **maint** and the password **password**.

Tools & Traps…

I See What You Did There…

trixbox doesn't use an SSL-encrypted Web session when maintaining the system. This means anyone sniffing the network can see exactly what you are doing on the Web page, including any usernames and passwords you may enter.

The trixbox management system is very full featured, and a book could be written on these two systems alone, so let's just take a (very) quick tour of the two major configuration editors on the system: The PHPConfig Asterisk config editor and the freePBX system. Figure 2.14 shows the trixbox system default page.

Figure 2.14 The trixbox System Default Page

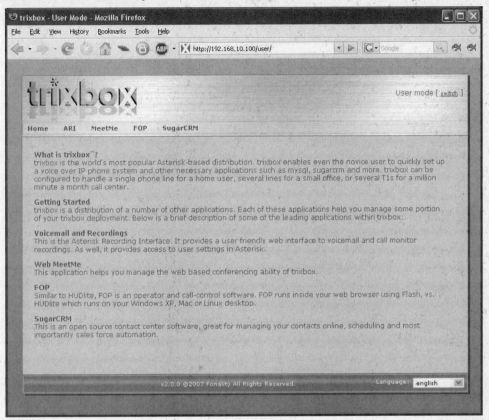

freePBX

freePBX is accessed by clicking the *Asterisk* link of the main menu, and then clicking the *freePBX* link. freePBX will greet you with a welcome screen and a list of menu options on the top. From here you can access the setup options, system tools, call activity reports, Flash Operator Panel, and the Asterisk recording interface. Clicking *Setup* will take you to the setup main page. The main page has a list of options on the left, which will allow you to administer user accounts, extensions, and general Asterisk settings; configure dial plans; and set up and control inbound and outbound trunks. See Figure 2.15.

Figure 2.15 Setting Up an SIP Account in freePBX

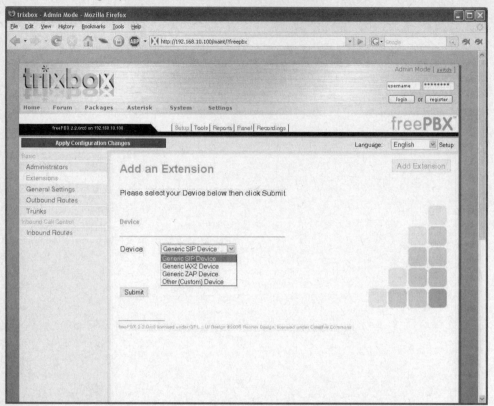

PHPConfig

PHPConfig is a great way to edit configuration files without having to deal with a shell terminal. It allows you to edit files just like they were in a text editor, but without having to learn how to use a Linux shell. It provides the best of both worlds. PHPConfig can be accessed by clicking the Asterisk link on the maintenance home page and then clicking the Config Edit link. Afterward, PHPConfig lists all the files in the Asterisk configuration directory. Clicking the name of one of these files brings the file up in an edit window. To the left of the edit window, PHPConfig lists all the sections it reads from that file, allowing you to quickly jump to and edit the section you wish to work on. When finished editing, click the **Update** button below the edit window. PHPConfig will then write the file to disk. The changes are not immediately reflected in Asterisk though. To reload all the configs, you will need to click the **Re-Read Configs** link at the top of the page. This tells Asterisk to perform a "reload" command that will reload all the configuration files. If there are no errors, PHPConfig will then display "reset succeeded." See Figure 2.16.

Figure 2.16 Editing extensions.conf in PHPConfig

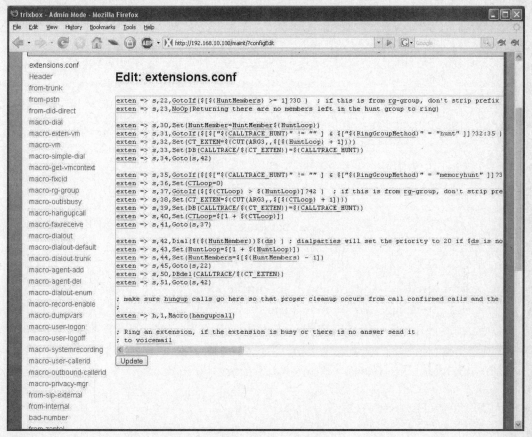

Installing Asterisk from Scratch

Before there were live CDs and distributions, there was source code. Asterisk's availability of source code is one of its biggest features, allowing anyone to "poke under the hood," see the internal workings, and rewrite portions if needed. Compiling Asterisk from its sources gives you the greatest amount of control as to what files are installed, and where they are installed. Unneeded options can be removed entirely, allowing a leaner Asterisk install. However, as always, there is a downside. Compiling anything from source is intimidating if you aren't used to doing it. However, it's terribly once you figure it out.

The Four Horsemen

When compiling Asterisk from source, there are four major pieces to the puzzle: LibPRI, Zaptel, Asterisk-Addons, and Asterisk.

Asterisk is, you guessed it, the PBX itself. This package contains the code for compiling the PBX and all its modules. You aren't going to get far compiling Asterisk without this package.

LibPRI is a library for handling the PRI signaling standard. The PRI standard was created by the Bell System back in the 1970s and is now an ITU standard. LibPRI is a C implementation of the standard. This package may be required depending on the hardware installed on the system.

Asterisk-Addons is a package that contains certain optional "bells and whistles," such as an MP3 player so Asterisk can handle sound files encoded in MP3, and modules for logging calls to a MySQL database. While these modules are completely optional, they are good to have, especially the MP3 player, and the resources they take up are minimal. Installing them is recommended.

Zaptel is the package that contains the driver and libraries for Asterisk to talk to Zapata telephony hardware, which are the telephone interface cards discussed earlier. This is a handy package to install, even if there is no Zaptel hardware on the system, since the conferencing software requires it for timing purposes.

Asterisk Dependencies

Before you start compiling Asterisk, you must make sure you have all the requirements satisfied. First off is the compiler. If you don't have a compiler like GNU C Compiler (gcc) installed, you aren't going to get very far compiling the source code. Next, make sure you have the libraries required to compile, otherwise you will likely have some kind of odd error at compile time. Asterisk has three dependencies: ncurses (www.gnu.org/software/ncurses/), a library for text-based "graphical" displays; OpenSSL (www.openssl.org/), an open-source library of the TLS and SSL protocols; and zlib (www.zlib.net/), a data compression library.

Asterisk requires both the library itself and the associated include files. These are included automatically if you compile from source. However, if you install the libraries from a binary repository, you will need to include the development packages as well. For instance, you would need to get both *zlib* and *zlib-devel*.

Getting the Code

Links to all of the Asterisk code are available at http://www.asterisk.org. Clicking the **Downloads** tab will take you to a page with links to grab all the necessary files. The links to get Asterisk provide options for downloading either Asterisk 1.2 or Asterisk 1.4 directly, or visiting the source archive. Grabbing Asterisk directly only downloads the Asterisk package, so you'll want to download the LibPRI, Zaptel, and Asterisk-Addons separately. The latest versions of each package should end in *-current*. Since there are multiple source archives, it is best to put all of them in a common subdirectory wherever the system's source code directory is located (for example: /usr/local/src/asterisk/). See Figure 2.17.

Figure 2.17 Getting the Source Archives via *wget*

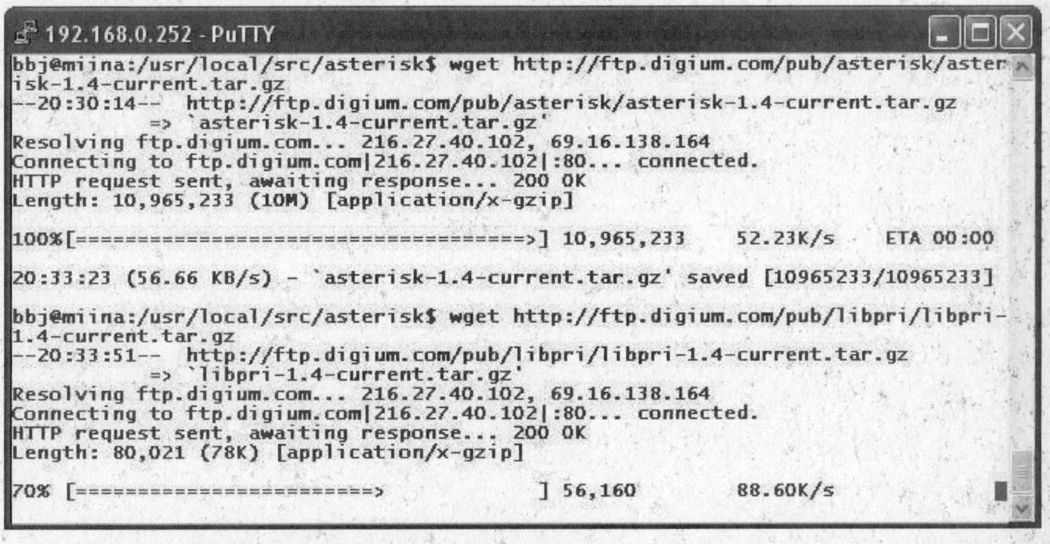

```
192.168.0.252 - PuTTY
bbj@miina:/usr/local/src/asterisk$ wget http://ftp.digium.com/pub/asterisk/aster
isk-1.4-current.tar.gz
--20:30:14--  http://ftp.digium.com/pub/asterisk/asterisk-1.4-current.tar.gz
           => `asterisk-1.4-current.tar.gz'
Resolving ftp.digium.com... 216.27.40.102, 69.16.138.164
Connecting to ftp.digium.com|216.27.40.102|:80... connected.
HTTP request sent, awaiting response... 200 OK
Length: 10,965,233 (10M) [application/x-gzip]

100%[====================================>] 10,965,233    52.23K/s    ETA 00:00

20:33:23 (56.66 KB/s) - `asterisk-1.4-current.tar.gz' saved [10965233/10965233]

bbj@miina:/usr/local/src/asterisk$ wget http://ftp.digium.com/pub/libpri/libpri-
1.4-current.tar.gz
--20:33:51--  http://ftp.digium.com/pub/libpri/libpri-1.4-current.tar.gz
           => `libpri-1.4-current.tar.gz'
Resolving ftp.digium.com... 216.27.40.102, 69.16.138.164
Connecting to ftp.digium.com|216.27.40.102|:80... connected.
HTTP request sent, awaiting response... 200 OK
Length: 80,021 (78K) [application/x-gzip]

70% [========================>         ] 56,160       88.60K/s
```

Gentlemen, Start Your Compilers!

Compiling is simpler than one might think. Often, all that's required is three commands: *./configure*, *make*, and *make install*. Once you have these three commands memorized, you'll do fine.

Compiling LibPRI

The first step is to compile LibPRI. This is required if you have a PRI interface hooked into the system, but optional if you do not. First, expand the archive.

```
tar xvzf libpri-1.4-current.tar.gz
```

This will expand the source archive into a directory. At the time of this writing, it is `libpri-1.4.0/`. After the file is done expanding, change to the LibPRI directory.

```
cd libpri-1.4.0/
```

LibPRI doesn't have a configuration command yet, so the only two steps are to compile it via the *make* command, wait until it finishes, and then run *make install*.

It is important to run the *make install* command as a root user, otherwise the library will not be installed correctly due to permission errors. Once everything is done, you can exit the LibPRI directory.

```
cd ../
```

Compiling Zaptel

Compiling Zaptel more or less follows the same steps that compiling LibPRI did. However, there are a few changes. First though, expand the archive.

```
tar xvzf zaptel-1.4-current.tar.gz
```

Next, enter the Zaptel directory:

```
cd zaptel-1.4.0/
```

This is where things change from LibPRI. Zaptel is a bit more complicated than LibPRI, so it includes a configuration script. (See Figure 2.18.) You can run this by executing

```
./configure
```

Figure 2.18 The Zaptel Configure Script

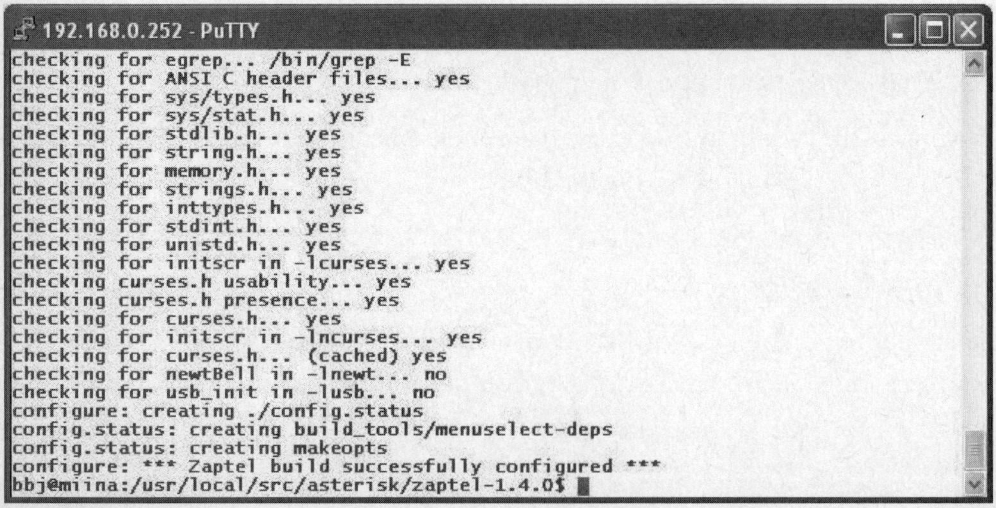

```
192.168.0.252 - PuTTY
checking for egrep... /bin/grep -E
checking for ANSI C header files... yes
checking for sys/types.h... yes
checking for sys/stat.h... yes
checking for stdlib.h... yes
checking for string.h... yes
checking for memory.h... yes
checking for strings.h... yes
checking for inttypes.h... yes
checking for stdint.h... yes
checking for unistd.h... yes
checking for initscr in -lcurses... yes
checking curses.h usability... yes
checking curses.h presence... yes
checking for curses.h... yes
checking for initscr in -lncurses... yes
checking for curses.h... (cached) yes
checking for newtBell in -lnewt... no
checking for usb_init in -lusb... no
configure: creating ./config.status
config.status: creating build_tools/menuselect-deps
config.status: creating makeopts
configure: *** Zaptel build successfully configured ***
bbj@miina:/usr/local/src/asterisk/zaptel-1.4.0$
```

The configure script will make sure all the dependencies are fulfilled and that Zaptel knows where to look for all the libraries. Once the configure script is done, the next step is to run the following command:

```
make menuselect
```

This will compile and execute the menuselect utility. menuselect is a new feature in Asterisk 1.4 that allows you to choose which modules to compile and install, permitting you to "trim the fat" of any software not required in your particular situation. For example, if you do not have a Digium TDM400, you can deselect the *wctdm* module during *menuselect* and that module will not be compiled or installed. See Figure 2.19.

Figure 2.19 The Initial Zaptel menuselect Menu

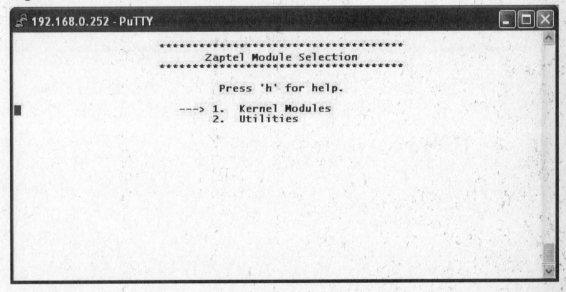

You can navigate through menuselect with the arrow keys—up and down scroll through the menu, left exits to the previous menu. Pressing **Enter** or the **Spacebar** will select/deselect a module or enter a menu. **F8** will select all the modules, and **F7** will deselect all the modules. To save and quit, press **x**, and to quit without saving, press **q**. If you forget any of the keys, press **h** and the help screen will be displayed, as shown in Figure 2.20.

Figure 2.20 The Zaptel Kernel Module Menu

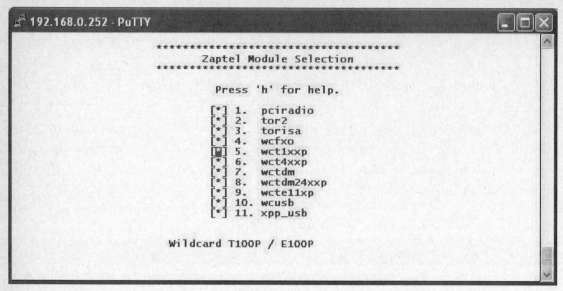

Menuselect lists a description at the bottom of the screen that explains which module supports which card. You can safely deselect any cards your system does not have installed. If a dependency is broken, menuselect will inform you of this and allow you to correct the configuration.

Once you are done trimming modules from the menu, exit and save. This will bring you back to the shell. Next, compile the Zaptel modules. This is done in one of two ways. If the system is running a 2.4.X kernel, simply run:

```
make
```

However, if the system is running a 2.6.X kernel, run:

```
make linux26
```

After the modules are done compiling, regardless of the system kernel version, run the installation command as a root user:

```
make install
```

And so the Zaptel modules will install. Finally, once everything is done compiling, move back up to the asterisk subdirectory:

```
cd ../
```

Compiling Asterisk

Believe it or not, Asterisk is just as easy to compile as LibPRI and Zaptel. Despite the menuselect system being more complex and the compile taking a bit longer, compiling the code more or less follows the same process as Zaptel. First, expand the archive:

```
tar xvzf asterisk-1.4-current.tar.gz
```

Next, enter the Asterisk directory:

```
cd asterisk-1.4.0/
```

Asterisk has a configure script, same as Zaptel. Run it by issuing the same command:

```
./configure
```

Next, compile and execute the menuselect utility:

```
make menuselect
```

The Asterisk menuselect is fairly more involved than the Zaptel one because the amount of options available for Zaptel pale in comparison to those for Asterisk. You can poke around and see if there are things you want to skip, but remember to be careful about choosing what modules to include. As the old saying goes "It is better to have it and not need it, then need it and not have it."

Once you are done with the menuselect process, start compiling Asterisk:

```
make
```

Compile time varies from system to system. Once completed, the next step is to install Asterisk onto the system.

```
make install
```

Sample programs, demos, and configuration references can then be (optionally) installed.

```
make samples
```

Finally, move back up into the source subdirectory.

```
cd ../
```

Compiling Asterisk-Addons

Same steps, different package. First, expand the archive:

```
tar xvzf asterisk-addons-1.4-current.tar.gz
```

Next, enter the Asterisk directory:

```
cd asterisk-addons-1.4.0/
```

Run the configure script:

```
./configure
```

Next, compile and execute the menuselect utility:

```
make menuselect
```

Once you done with the menuselect process, start the compile.

```
make
```

And, finally, install:

```
make install
```

Installing Asterisk with Binaries

Another option available for Linux users is to install Asterisk via an installer package. Installer packages are files that install software packages onto a Linux distribution. Installer packages vary from distribution to distribution: For example, a Debian's DPKG format will not install on a Fedora system, nor will Fedora's RPM format install correctly on a Debian system.

Asterisk installer packages exist in various forms for the various distributions of Linux, Windows, and Mac OS X. While these packages are maintained by third parties, and are sometimes not completely up-to-date, these provide an almost completely painless way to install Asterisk.

Installing Asterisk on Windows

AsteriskWin32 is a version of Asterisk compiled for Windows. Created by Patrick Deurel, it is currently the only real option for running Asterisk on Windows. However, AsteriskWin32 suffers from the same issues as Asterisk on Mac OS X, namely, the inability to keep up with Asterisk development. While the current version of Asterisk at the time of this writing is currently at Version 1.4.0,

AsteriskWin32 is at 1.0.10, being two major revisions behind. However, it has the advantage of being the only game in town, so it can choose its own pace.

Getting AsteriskWin32

The installer package is available for download in the download section of http://www.asteriskwin32.com/. The latest version is 0.56 which is based on Asterisk 1.0.10.

Installing AsteriskWin32

After downloading the installer package, locate the downloaded file and execute it. Click **Next**, as shown in Figure 2.21.

Figure 2.21 Welcome Window to AsteriskWin32 Setup

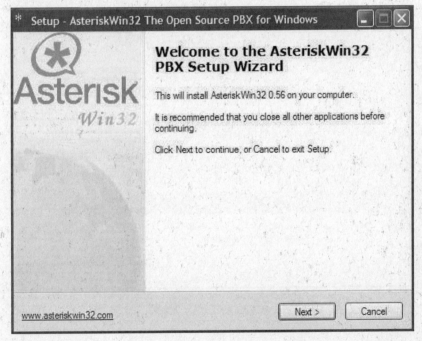

Scroll through the license agreement (Figure 2.22), read it carefully (You always read the license agreements carefully, right?) and click **Next**. After an "Information" screen that further disclaims the author from any issues his program may cause, the installer prompts you to choose a directory for it to install its files to.

Figure 2.22 License Agreement

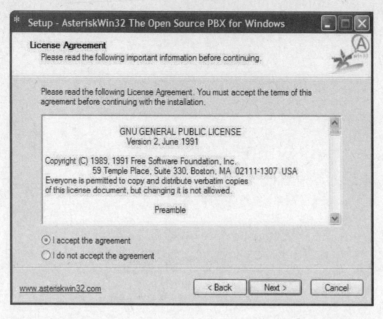

Since this version of Asterisk is compiled with Cygwin (a Windows port of many popular Linux commands), the main install directory is c:\cygroot. Asterisk will be installed as a subdirectory within this directory. See Figure 2.23.

Figure 2.23 Selecting Destination Location

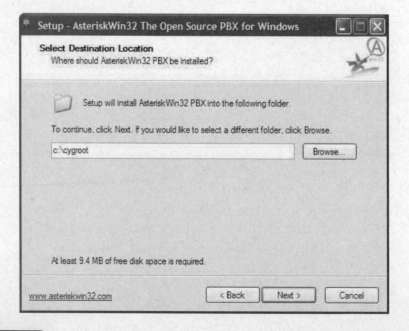

Unless the system has a working Asterisk configuration installed on it already, it is best to keep both options selected, as shown in Figure 2.24. The sample configuration files guarantee that Asterisk will find everything it needs to start itself up correctly.

Figure 2.24 Additional Tasks Selection

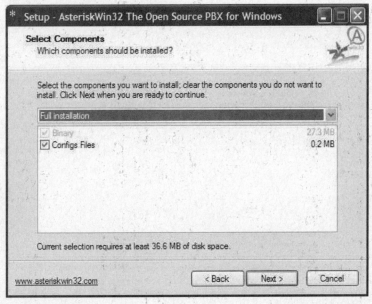

Next, the installer will prompt you as to whether to create a shortcut to the PBX console on your desktop, as shown in Figure 2.25.

Figure 2.25 Components Installation Selection

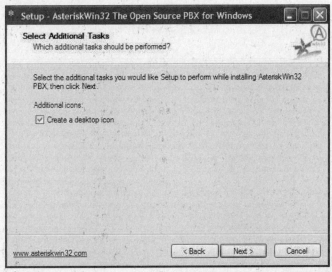

Choosing this option is purely personal preference. The installer will create a group under **Start | Programs** that will have all the necessary shortcuts. Click **Next**. AsteriskWin32 will start to copy files over. Finally, Asterisk will be installed (Figures 2.26 and 2.27). Pat yourself on the back. Wasn't that easy?

Figure 2.26 Installation of AsteriskWin32

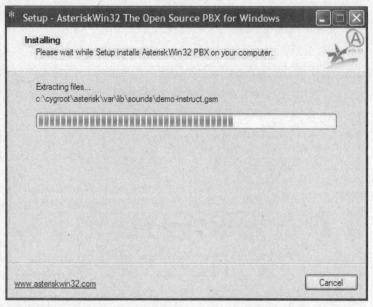

Figure 2.27 AsteriskWin32 Setup completion

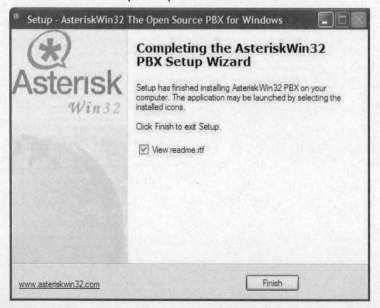

Starting AsteriskWin32

AsteriskWin32 has three different "consoles": The PBX Manager & Console, the AsteriskWin32 Console, and AsteriskWin32 GUI. All of these serve the same purpose: to start, run, and manage the Asterisk server. However, each of these has slightly different abilities and caveats.

The AsteriskWin32 Console can be started by choosing **Start | Programs | AsteriskWin32 | AsteriskWin32 Console**. This is the standard Asterisk console that is part of every Asterisk install. You'll be met with the same exact console if you start up Asterisk in Linux. When executed, the Asterisk process starts up and never goes into the background, leaving the console up on the screen. From here, you can interact with Asterisk just as you would anything else. However, when that console is closed, Asterisk does not continue running. Because it never put itself in the background, it will exit when the console closes.

Another option is AsteriskWin32's GUI. This is a GUI frontend to the Asterisk CLI. While it behaves similarly to the CLI console, it has the advantage of being able to minimize itself to the system tray, keeping itself running while not having to be up on your screen. However, just like the CLI console, if the window is closed, the server will stop running.

Finally, AsteriskWin32 has its own PBX manager, which is designed to automate the starting and stopping of the Asterisk process. This is available under **Start | Programs | AsteriskWin32 | PBX Manager & Console**. When the console starts, it will try to connect to Asterisk. If Asterisk is running on the system, it will connect and display "Connected to Asterisk" and start displaying system messages in the main window. However, if Asterisk isn't running, it will display "Unable to connect to remote Asterisk" in the main window. To start Asterisk, select **PBX Tools | Start** and the console will start the Asterisk GUI minimized to your system tray. After the server is started, it will connect to it.

Key differences exist between the PBX managers and the consoles. The biggest difference is that when the manager is closed, the server process continues to run separately, be it in the form of the GUI or the CLI console. There are also some rudimentary options for controlling voice-mail boxes, loaded modules, call parking, and the call manager system. While these do simplify the process, and let you avoid editing the configuration files directly, they only hit on the basic options and do not let you configure the advanced capabilities.

Starting and Using Asterisk

Congratulations, you are the proud owner of a full-fledged Asterisk installation. Feel free to pass out cigars in the office. If you're under 18, make sure they're candy cigars. After Asterisk is installed, the next step is to start it. Thankfully, if you installed the sample configuration files, Asterisk should run out of the box without any additional changes.

Starting Asterisk

Starting asterisk is easy, just run *asterisk -vvvc*, which will execute the server. These options tell the server not to run in the background and to run at a verbosity level of three, which means all the important messages will be displayed and enough less important ones so as to not overwhelm and that the user will see all diagnostic messages. While many messages will quickly scroll by on the screen, most of these are simple initialization messages that can be ignored. If any fatal errors occur, Asterisk will stop and exit so the message remains on the screen. Asterisk will display "Asterisk Ready" when it has successfully run, as shown in Figure 2.28.

Figure 2.28 Congratulations! You're Running Asterisk!

```
    -- Added extension '600' priority 1 to demo
    -- Added extension '600' priority 2 to demo
    -- Added extension '600' priority 3 to demo
    -- Added extension '600' priority 4 to demo
    -- Added extension '8500' priority 1 to demo
    -- Added extension '8500' priority 2 to demo
    -- Registered extension context 'default'
    -- Including context 'demo' in context 'default'
pbx_config.so => (Text Extension Configuration)
    == Parsing '/etc/asterisk/dundi.conf': Found
    == Using TOS bits 0
    == DUNDi Ready and Listening on 0.0.0.0 port 4520
    == Registered custom function DUNDILOOKUP
pbx_dundi.so => (Distributed Universal Number Discovery (DUNDi))
pbx_loopback.so => (Loopback Switch)
pbx_realtime.so => (Realtime Switch)
pbx_spool.so => (Outgoing Spool Support)
    == Registered application 'DeadAGI'
    == Registered application 'EAGI'
    == Registered application 'AGI'
res_agi.so => (Asterisk Gateway Interface (AGI))
res_clioriginate.so => (Call origination from the CLI)
res_convert.so => (File format conversion CLI command)
Asterisk Ready.              █
*CLI> _
```

The other way to run Asterisk is to start the daemon by running the *asterisk* command without any arguments at the command prompt. This will start the server in the background. Starting in the background as opposed to the foreground has advantages and disadvantages. While the server won't tie up a terminal or exit when the terminal is closed, it will not display any diagnostic messages to the terminal during startup either. Running Asterisk in the background is the most common way to run Asterisk since normally an Asterisk process would be running at all times. One would want to run Asterisk in the foreground if diagnosis information is needed.

To connect to an already running Asterisk process, run the command *asterisk -vvvr*. This will duplicate the verbosity settings to the *above asterisk -vvvc* command, except it will not start the server process, only attempt to connect to an existing one.

Restarting and Stopping Asterisk

Every beginning has an ending. Asterisk can be stopped and restarted many ways, from the immediate and abrupt stop, to the slow and graceful shutdown. While stopping and restarting is usually not required in the normal course of operation, occasionally it is required.

The ways to stop and restart Asterisk are syntactically similar. You can issue the *stop* or *restart* command to Asterisk in three ways. When issuing the *restart* or *stop* command, you can tell Asterisk to do it *now*, *gracefully*, or *when convenient*. These control how the server will go about shutting down.

now is the proverbial "neck snap" when shutting down or restarting. The server process is shut down immediately, without any concerns for activity. Any active calls are terminated and all active threads are killed. This is not normally the way to shut down the server in a production environment. However, if the server needs to be quickly downed, this is the command to issue.

gracefully is a much cleaner way to shut down or restart. After the command is issued, Asterisk stops answering all new calls. However, unlike *now*, Asterisk does not terminate calls currently in process. While this is much better in a production environment, this can also be undesirable since it leaves calls unanswered.

Stopping or restarting *when convenient* solves this problem. After issuing this command, Asterisk continues functioning normally, the server restarts or stops when there are no active calls within the system. While this is the best when talking in terms of lost productivity, if the system constantly has active calls on it, the system will never stop or restart.

Updating Configuration Changes

Configuration changes are one of those day-to-day changes Asterisk faces. Users are added, voicemail boxes are deleted, extensions change. Every time you edit one of the configuration files, the changes aren't immediately reflected by the system. Restarting Asterisk allows these changes to be loaded, but on a high-traffic system, this will either stop phone calls, or possibly wait a long time. *reload* fixes that. Rather than shut down the Asterisk process and restart it, *reload* reloads all the configuration files on-the-fly without interrupting system activity.

Checklist

- Make sure voice and data networks are separated either physically or by VLANs. VLANs allow you to control both reliability and security. If the voice and data networks are not separated, it is possible for an attacker to monitor all telephone calls on the network.

- Make sure that trixbox is isolated from the public Internet or that root logons are disabled from remote hosts via SSH.

- Ensure that precautions are taken when entering passwords for trixbox's Web management software since these passwords will go over the wire in plain text.

Summary

Setting up Asterisk is a tedious process. Servers need to be designed to handle the expected call load. Figuring this out requires figuring out if the calls must be transcoded or have protocol translation, along with storage space for the voice prompts and voice mail. In addition to the server, networks also must be redesigned in order to provide reliability and security for the phone conversations.

Installing Asterisk can be done one of many ways. Live CDs are the easiest way to try Asterisk, just boot a CD and the system is running Asterisk. Installation CDs allow you to install Asterisk onto a clean system and set up a working system. Compiling Asterisk permits you to have maximum flexibility as to how Asterisk is set up and installed. Binaries can allow you to set up a system quickly and easily, but that system may be a few versions behind. How you set up Asterisk depends on your situation.

Starting and using Asterisk is mostly done through the command-line interface. The CLI allows you to start and stop Asterisk, along with reloading the configurations. Different options on the *shutdown* and *restart* commands let you control exactly when and how the system will shut down or restart.

Asterisk isn't an easy system to learn, but once you get the hang of it, it's a breeze to work with.

Solutions Fast Track

Choosing Your Hardware

- ☑ Choosing a reliable server for a PBX is important, because if the server goes down, the telephones go down.

- ☑ Choosing the proper RAM and processing speed will allow a server to handle multiple calls without overtaxing the processor, including situations where transcoding and protocol translation are required.

- ☑ Two types of phones are in use today: soft phones, which are software-based telephones; and hard phones, which are physical hardware devices or interfaces that emulate an analog phone system.

- ☑ VLANs are important for both security and network management reasons

☑ Different types of bandwidth management have both their pros and cons.

Installing Asterisk

☑ There are numerous ways to install Asterisk, Live CDs, Asterisk distributions, binaries, and compiling from scratch.

☑ Live CDs, such as SLAST, are great if you want a system where you can try out Asterisk without fear of screwing something up.

☑ Asterisk Linux distributions, such as trixbox, provide a simple and easy way to install Asterisk on a new system. trixbox also comes with numerous bells and whistles such as CRM software and a Web-based configuration editor.

☑ Compiling from scratch permits you to take the most control over the installation of Asterisk, allowing you to determine what modules are compiled and installed.

☑ Binaries allow you to set up Asterisk easily and quickly, but you are at the mercy of the package maintainer.

Starting and Using Asterisk

☑ Asterisk has both a debug and a remote console, allowing you to run it in the foreground when needed and keep it in the background when it is running.

☑ You can start and stop an Asterisk server in three ways: *now*, *gracefully*, and *when convenient*. Each method controls when Asterisk will restart.

☑ Reloading Asterisk lets you reread configuration files without restarting the system.

Links to Sites

■ http://slast.org – The SLAST home page.

■ www.infonomicon.com – The Infonomicon Computer Club, maintainers of SLAST.

■ www.trixbox.org – The trixbox home page.

- www.centos.org – CentOS, the Linux distribution trixbox is based upon.

- www.gnu.org/software/ncurses/ – The NCurses home page, a dependency of Asterisk.

- www.openssl.org/ – The OpenSSL project, a dependency of Asterisk.

- www.zlib.net/ – The ZLib compression library, a dependency of Asterisk.

- www.asteriskwin32.com/ – The AsteriskWin32 home page.

- www.imgburn.com/ – A free ISO burner for Microsoft Windows.

Frequently Asked Questions

The following Frequently Asked Questions, answered by the authors of this book, are designed to both measure your understanding of the concepts presented in this chapter and to assist you with real-life implementation of these concepts. To have your questions about this chapter answered by the author, browse to **www.syngress.com/solutions** and click on the **"Ask the Author"** form.

Q: What is the best way for me to install Asterisk?

A: There is no "best way" to install Asterisk. It depends heavily on your situation. Different methods are better for different situations. If you want to just test the waters, however, perhaps use a Live CD on your personal workstation. If you want to task an existing server to store voice mail for your company, you might want to consider compiling Asterisk from scratch.

Q: How much should I spend on phones?

A: Phones follow the "you get what you pay for" rule. If you're cheap when it comes to phones, you will get cheap phones. A good VoIP phone should cost about $150.

Q: I have Windows. How can I burn an ISO?

A: The Windows XP CD burning system does not support burning ISOs to disk. However, there is a freeware utility that will burn ISOs called ImgBurn, which is available at http://www.imgburn.com/.

Q: How can I make my computer boot from a CD?

A: This depends greatly on your computer. Certain BIOSes, in order to get the computer to boot from a CD, may need a special key pressed during startup, or a setting may need to be configured within the BIOS itself.

Q: Is there a disadvantage to running Asterisk 1.0 versus Asterisk 1.4?

A: Yes! Asterisk 1.4 has many major bug fixes and feature additions. Plus, since this book is based on Asterisk 1.4, certain descriptions in the book may not work on Asterisk 1.0.

## Chapter 3

Configuring Asterisk

Solutions in this chapter:

- **Figuring Out the Files**
- **Configuring Your Dial Plan**
- **Configuring Your Connections**
- **Configuring Voice Mail**
- **Provisioning Users**
- **Configuring Music on Hold, Queues, and Conferences**

Related Chapters: 1, 2

- ☑ **Summary**
- ☑ **Solutions Fast Track**
- ☑ **Frequently Asked Questions**

Introduction

Installing Asterisk is only half the battle. The other half is configuring it. Asterisk configuration can be just as difficult as installing the program, so don't think you're in for an easy ride. Configuring Asterisk depends heavily on how exactly you want your PBX to function and what features you want available to users.

Configuring Asterisk can be somewhat of an adventure. Asterisk, like many Unix utilities, has many small configuration files all interconnected to one another. This has its pros and cons: While it adds a level of complexity to the system by requiring you to remember what feature is in which specific file, it allows you to make a mistake in one file and not have the entire proverbial house of cards come crashing down.

The plus about configuring Asterisk is that once you get the hang of it, you can easily start flying through configuration files and tackle larger and more complex problems. Asterisk's configuration files have a certain way of doing things and once you figure it out, picking up the advanced stuff is easy.

Figuring Out the Files

If you enter into your Asterisk configuration directory, /etc/asterisk, you'll see 62 files by default. If you're taking over a previous installation administered by someone else, you may see more than that. Looking at the file names, you'll see they have cryptic labels like rtp.conf, or file names that seem to be the same thing, like *asterisk.adsi* and *adsi.conf*. When trying to configure your system, finding the right file to edit can be like unearthing the proverbial needle in a haystack. (See Table3.1 for information on what each file controls.)

Table 3.1 Asterisk Configuration Files

Filename	Role
adsi.conf	Controls Asterisk Analog Display Services Interface settings
adtranvofr.conf	Contains settings related to Voice over Frame Relay and AdTran equipment
agents.conf	Contains settings for call agents that work call queues
alarmreceiver.conf	Contains settings for the Alarm Receiver application
alsa.conf	Contains settings for the CLI sound system if using ALSA sound drivers

Continued

Table 3.1 continued Asterisk Configuration Files

Filename	Role
amd.conf	Contains settings for answering machine detection on outbound calls
asterisk.adsi	Asterisk Analog Display Services Interface script
cdr.conf	Contains settings for Call Detail Records (CDRs)
cdr_custom.conf	Contains settings for custom Call Detail Record mappings
cdr_manager.conf	Contains settings for sending CDRs to the Asterisk Management Interface
cdr_odbc.conf	Contains settings for storing your CDRs into a database connected via ODBC
cdr_pgsql.conf	Contains settings for storing your CDRs into a PostgreSQL SQL database
cdr_tds.conf	Contains settings for storing your CDRs into a FreeTDS database
CODECs.conf	Contains CODEC settings
dnsmgr.conf	Contains settings about Domain Name System (DNS) lookups done by Asterisk
dundi.conf	Controls Distributed Universal Number Discovery connections and settings
enum.conf	Controls Telephone Number Mapping/E164 connections and settings
extconfig.conf	Contains mappings for external database connections for configuration settings
extensions.ael	Contains the dial plan settings, written in Asterisk Extension Language
extensions.conf	Contains the dial plan settings
Features.conf	Contains settings for call parking
festival.conf	Contains settings for the connection between Asterisk and the Festival TTS Engine
followme.conf	Contains settings for the FollowMe application
func_odbc.conf	Contains settings for template-based SQL functions accessed via ODBC
gtalk.conf	Controls Google Talk connections and settings
h323.conf	Controls H323 Protocol connections and settings

Continued

Table 3.1 continued Asterisk Configuration Files

Filename	Role
http.conf	Contains settings for Asterisk's integrated HTTP server
iax.conf	Controls Inter Asterisk eXchange Protocol Connections and Settings
iaxprov.conf	Contains settings for IAXy provisioning
indications.conf	Contains settings for the system's Ring, Busy, Reorder, and Special Information tones
jabber.conf	Controls Jabber Protocol connections and settings
logger.conf	Contains settings about where and what to log
manager.conf	Contains settings for the Asterisk Management Interface
meetme.conf	Contains settings for the MeetMe conferencing system
mgcp.conf	Controls Media Gateway Control Protocol connections and settings
misdn.conf	Controls Integrated Serial Digital Networks (ISDNs) connections and settings
modem.conf	Controls ISDN modem settings
modules.conf	Controls which applications and modules are loaded when the server is started
musiconhold.conf	Contains Music on Hold settings
muted.conf	Contains settings for the Mute Daemon
osp.conf	Controls settings and connections for the Open Settlement Protocol
oss.conf	Contains settings for the CLI sound system if using OSS sound drivers
phone.conf	Contains settings for Linux Telephony devices
privacy.conf	Contains settings for the PrivacyManager application
queues.conf	Contains settings for call queues
res_odbc.conf	Contains settings for external database connections for configuration settings
res_snmp.conf	Contains Simple Network Management Protocol settings for the SNMP application
rpt.conf	Controls settings of the app_rpt application, which enables radio systems to be linked via VoIP
rtp.conf	Contains Real-time Transport Protocol settings

Continued

Table 3.1 continued Asterisk Configuration Files

Filename	Role
say.conf	Contains string settings for the various say_* applications
sip.conf	Controls Session Initiation Protocol (SIP) connections and settings
sip_notify.conf	Contains settings for SIP's NOTIFY command
skinny.conf	Controls Skinny Client Control Protocol connections and settings
sla.conf	Controls Shared Line Appearance connections and settings
smdi.conf	Contains settings for the Simplified Message Desk Interface
telcordia-1.adsi	Default Telcordia Analog Display Services Interface script
udptl.conf	Contains settings for UDPTL, one of the transports for Faxing over IP Networks
users.conf	A file that controls a combination of settings, allowing for easier user maintenance
voicemail.conf	Contains voice mail settings and mailbox details
vpb.conf	Contains settings for VoiceTronix hardware
zapata.conf	Controls settings for Zapata hardware

To say Asterisk has a lot of settings would be an understatement of mammoth proportions. While this is a plus when you want to tweak Asterisk to fit your needs exactly, it is a bit overwhelming. However, don't fret. Certain configuration files can be ignored if you don't have certain hardware, and other files can be ignored if you do not need to enable certain features of Asterisk.

Configuring Your Dial Plan

The dial plan is the logic behind how phone calls are routed through your Asterisk installation. Asterisk runs every incoming call, every outgoing call, and every call in between extensions through the dial plan logic in order to determine where it should go and whether or not it should be completed. The dial plan is contained in *extensions.conf*, and therefore it can be said that *extensions.conf* is easily the most important configuration file in Asterisk. Removing *extensions.conf* is similar to removing a traffic light at a complicated intersection; cars will want to enter and cars will want to leave, but there will be no way to direct them.

extensions.conf is a bit more complicated than a typical configuration file. On top of the usual sections and settings, there is a logical flow similar to a program. Essentially, *extensions.conf* is one giant script. The sooner you keep this in mind, the easier it will be for you to write a good dial plan.

Contexts, Extensions, and Variables! Oh My!

extensions.conf can be broken down into three major parts: contexts, extensions, and variables. Each has their own unique and important function and needs to work together for a good dial plan to function.

Contexts

To put it simply, contexts are the fences that keep your extensions from getting tangled up in a big mess. A context is a simple way for grouping extension commands based on what the user has dialed. To begin a context, put the name of a context by itself in square brackets. Each context then contains a list of commands. In *extensions.conf* there are two special contexts called *[general]* and *[globals]* in which certain settings can be set.

general has a few special settings that define how *extensions.conf* behaves. First off is the *static* setting. This, can be set to either *yes* or *no*, but for some reason, only yes has been implemented. This will eventually control Asterisk from rewriting the *extensions.conf* every time an extension is added or deleted. The next setting is *writeprotect*. This can also be set to either *yes* or *no*, and this controls the ability of someone at the CLI to rewrite your dial plan via the *save dialplan* command. This may seem handy, but doing so will delete all comments in the file.

Each extension follows a similar syntax. *exten => EXTENSION,PRIORITY,COMMAND(ARGS). exten =>* precedes every extension. This is a directive that tells Asterisk to define an extension, as opposed to a context. The next three parts of an extension are *EXTENSION, PRIORITY, and COMMAND()*. Let's cover these three portions.

Extensions

Extensions can be broken down into three types: a constant extension, a wildcard extension, and a special extension. A constant extension is an extension that when coded to a literal constant is the dial plan. A wildcard extension is a context that uses wildcards to match multiple possibilities for the extension. Wildcards can be either internal Asterisk wildcards or RegEx-like patterns (see Table 3.2).

Table 3.2 Extension Wildcards Used in Asterisk

Wildcard	Patterns Matched
[0126-9]	Any digit within the pattern. (In this case: 0,1,2,6,7,8, and 9).
X	Any number 0 through 9. The equivalent of [0-9].
Z	Numbers between 1 through 9. The equivalent of [1-9].
N	Numbers between 2 through 9. The equivalent of [2-9]. This scheme is used most commonly in Area Code and Prefix assignments.
.	Any number, one or more times.

So with Wildcard extensions, it is simple to reroute numerous extensions with one line of code. Let's say a department in your building, the ever-important widget department, have moved to another division and wanted to leave a message at their old extensions informing callers that they had moved. They previously occupied Extensions 300 through 329 on your PBX. Rather than rewrite 30 lines; you can add a single extension of

```
exten => 3[0-2]X,1,Playback(WidgetDeptHasMoved)
```

This will have any caller dialing the department's former extensions greeted by a message informing them of the move. Playback is a command that plays back a sound file stored on the system; we'll cover it and its counterparts later.

In addition to wildcard and literal extensions, there are also special extensions that correspond to special events in the dial plan (see Table 3.3).

Table 3.3 Special Extensions Used in Asterisk

Extension	Name	Description
S	Start	Used when a caller is put in a context before dialing a number.
I	Invalid	Used when a caller dials an extension not defined in the current context.
H	Hangup	Used when a caller hangs up.
T	Time Out	Used when a caller does not respond within the response timeout period

Continued

Table 3.3 continued Special Extensions Used in Asterisk

Extension	Name	Description
T	Absolute Time Out	Used when a caller does not respond within the about timeout period
O	Operator	

Extensions do not necessarily need to be numbers either. They can be made with any type of text. While extensions like "fuzzybunnydept" cannot be dialed by a caller if included in your context, it can be used internally by your dial plan. We'll see how this can come in handy later in the chapter.

Priorities

PRIORITY controls the flow in which commands are executed. For each extension, this is either controlled by an increasing number or a special *n* syntax. The *n* syntax tells Asterisk to execute the extension one line after the other:

```
[incomingcall]
exten => s,1,Answer()
exten => s,n,Playback(mainmenu)
exten => s,n,Hangup()
```

In this example, any call being routed to the "incomingcall" extension in Asterisk would have its call answered, a menu would then play, and then the call would be terminated. After Asterisk finishes executing one line, the next line would be executed. Numbering the steps provides greater flexibility with the dial plan since it is possible to control the flow logically rather than line by line. For example, the extension shown earlier could be rewritten with a numbered sequence

```
[incomingcall]
exten => s,2,Playback(mainmenu)
exten => s,1,Answer()
exten => s,3,Hangup()
```

Asterisk still answers, plays the menu, and hangs up because it executes by line number rather than by the order in which the lines appear. It executes step 1, followed by steps 2, and then 3. These steps could be scattered throughout the context and intertwined with hundreds of extensions. As long as they are numbered correctly, Asterisk will execute them in order for that context.

Dial Plan Commands

The commands are the heart of any dial plan. They are what actually cause Asterisk to answer the call, ring the phone, transfer the call, play the menu, and do numerous other things. See Table 3.4 for a look at some of the more common ones.

Table 3.4 Common Commands in Asterisk

Command	Description
Dial(CHANNEL)	Dials a channel
Answer()	Answers a ringing channel
Playback(FILE)	Plays a sound file in the foreground
Background(FILE)	Plays a sound file in the background, while waiting for the user to input an extension
Hangup()	Hangs up the call
SayDigits(NUMBER)	Says a number, digit by digit

Notes from the Underground…

Channels vs. Extensions

It's easy to get confused when people start tossing around terms like "extensions" and "channels" when the two words seem interchangeable. Sometimes, people do use them as if they are identical, but don't be one of these people. Channels and extensions are two separate and completely different things. Extensions are the physical numbers assigned to a device, while channels, on the other hand, are the connections to the devices themselves. For example, you can have a phone at your desk set up to ring on three separate extensions; however, each of these extensions will ring the same channel—namely, your phone.

Variables

Variables in extensions.conf are nothing special. They act like variables in any other language. Variables are set via the *Set()* command and are read via the variable name encased in *${}*:

```
[example]
exten => s,1,Set(TEST=1)
exten => s,2,NoOp(${TEST})
```

Variables are common in simple dial-plan applications and Asterisk uses certain variables for internal functions, but their use is somewhat uncommon in regular dial plan usage.

Tying It All Together

All of these pieces of dial plans make little to no sense when thinking about them in the abstract, so you may be scratching your head right now. Let's take a look at how all of these would be used in an everyday environment, by looking at a simple *extensions.conf*:

Example 3.1 A Very Simple *extensions.conf*

```
[default]
exten => s,1,Answer()
exten => s,2,Background(thank-you-for-calling-conglomocorp)
exten => s,3,Background(conglomocorp-mainmenu)
exten => s,4,Hangup()

exten => 100,1,Dial(SIP/10)
exten => 200,1,Dial(SIP/20)
```

When a call enters the *[default]* context, it is answered by Asterisk. Asterisk then starts playing the mainmenu sound file while waiting for the caller to enter digits. At this point, the caller can either enter 100 and be connected to the channel SIP/10 or 20 and be connected to the channel SIP/20. If the menu finishes playing and the user has not entered any digits, the call will be hung up on.

Using Special Extensions

Now, hanging up on your caller if they wait to listen to the whole menu seems kind of rude, doesn't it? So let's take the file we had before and use some special extensions to have the menu replay if the user hasn't entered an extension and inform them if the extension they entered is invalid.

Example 3.2 A Very Simple *extensions.conf* with Special Extensions

```
[default]
exten => s,1,Answer()
exten => s,2,Background(thank-you-for-calling-conglomocorp)
exten => s,3,Background(conglomocorp-mainmenu)

exten => t,1,Goto(s,2)

exten => i,1,Playback(sorry-thats-not-valid)
exten => i,2,Goto(s,2)

exten => 100,1,Dial(SIP/10)
exten => 200,1,Dial(SIP/20)
```

That's much nicer. Now the behavior of the dial plan is the same, up until the main menu ends. At that point, the menu repeats. Also, now if the caller dials an incorrect extension, the dial plan will play a menu that informs them the extension they entered is not valid.

Creating Submenus

Normally, most small to medium-sized companies only require a single menu, but let's say your boss wants to have a support menu that allows customers to direct their questions to the appropriate support group. We can accomplish this by creating a second context that contains the appropriate menu and extensions. Let's build on the previous example again and add a second menu that allows callers to be connected to the Blivet, Widget, or Frob support lines.

Example 3.3 Creating Submenus in *extensions.conf*

```
[default]
exten => s,1,Answer()
exten => s,2,Background(thank-you-for-calling-conglomocorp)
exten => s,3,Background(conglomocorp-mainmenu)

exten => t,1,Goto(s,2)

exten => i,1,Playback(sorry-thats-not-valid)
exten => i,2,Goto(s,2)
```

```
exten => 3,1,Goto(s,1,supportment)
exten => 100,1,Dial(SIP/10)
exten => 200,1,Dial(SIP/20)

[supportmenu]
exten => s,1,Background(conglomocorp-supportmenu)

exten => 1,1,Dial(SIP/blivetsupportline)
exten => 2,1,Dial(SIP/widgetsupportline)
exten => 3,1,Dial(SIP/frobsupportline)
exten => #,1, Goto(s,2,default)

exten => t,1,Goto(s,1)

exten => i,1,Playback(sorry-thats-not-valid)
exten => i,2,Goto(s,1)
```

In this example, we've added a third option to the main menu. If a caller dials 3, they are connected to the *[supportmenu]* context with a *Goto()* statement. *Goto()* can be called many different ways. You can jump between priorities in the same extension by just specifying *Goto(priority)* or you can jump between extensions in the same context by specifying *Goto(priority,extension)*. Lastly, you can switch contexts by specifying *Goto(context, extension, priority)*.

Tools & Traps...

Watch Your Spaces!

Goto() is a bit finicky with its syntax and whitespace. For example: *Goto(supportmenu,s,1)* will behave differently than *Goto(supportmenu, s, 1)*. In the first example, *Goto* will behave as expected and jump to the "s" extension, priority 1. However, in the second example, *Goto* will jump to the "s" extension, priority 1. Note how there is a space that precedes the "s". This can be a source of frustration if you don't know to look for it.

Including Other Contexts within the Current One

It's important to note that when creating another context, the settings and extensions from one context do not propagate to another. Setting up these extensions over and over again can be tedious and will lead to a duplication of code and effort. Thankfully, Asterisk permits other contexts to be joined together via the *include =>* directive. This allows other contexts to be *include*-ed into the current context and act as one giant context.

Let's go back to our example. The *t* and *i* context are duplicated in both the *[default]* and *[supportmenu]* contexts. With a couple of small changes, we can make a separate context with just the *t* and *i* extensions and *include =>* them into both contexts.

Example 3.4 Using *includes* in *extensions.conf*

```
[default]
include => specialextensions
exten => s,1,Answer()
exten => s,2,Background(thank-you-for-calling-conglomocorp)
exten => s,3,Background(conglomocorp-mainmenu)

exten => 3,1,Goto(supportmenu,s,1)
exten => 100,1,Dial(SIP/10)
exten => 200,1,Dial(SIP/20)


[supportmenu]
include => specialextensions
exten => s,1,Background(conglomocorp-supportmenu)

exten => 1,1,Dial(SIP/blivetsupportline)
exten => 2,1,Dial(SIP/widgetsupportline)
exten => 3,1,Dial(SIP/frobsupportline)
exten => #,1,Goto(s,2)


[specialextensions]
exten => t,1,Goto(s,1)

exten => i,1,Playback(sorry-thats-not-valid)
exten => i,2,Goto(s,1)
```

Okay, pop quiz time. Did you notice the difference between this example and the previous one? Don't worry if you didn't, it's pretty subtle. Because we are including the same *t* and *i* context between two files, the same code will be executed between both. Namely, they will be going to step 1 of the *s* extension in both contexts. Previously in the *[default]* context, the *t* and *i* extension went to step 2 of the *s* extension, bypassing the *Answer()* command. What does this change? Not a single thing. Technically, you're adding an extra step every time a caller times out or enters an invalid extension, which may affect performance if this happens repeatedly in a very high-traffic environment, but, in the grand scheme of things this extra step will not be perceptible. *Answer()* only answers the call if the call is in an unanswered state. It ignores being called if the call is already in answered.

Writing Macros

include-ing (other contexts within the current one is a handy way to save lines of code and duplication of code. Another easy way to increase efficiency and decrease code duplication is through Asterisk's macro abilities. Macros can be described as special contexts that accept arguments. They allow for more flexibility than contexts, and allow common tasks to be automated and not repeated.

In our previous examples, if someone dialed an extension, it rang a channel. It would continue ringing the channel until someone picked up, or the call terminated. What happens if we want to have that extension drop to voice mail playing the user's "I'm not here" message after 20 seconds of ringing, or playing the user's "I'm currently on the phone" message if the phone line is busy?

Example 3.5 Creating Voice Mail Support for Existing Extensions without the Use of Macros

```
[default]
include => specialextensions
exten => s,1,Answer()
exten => s,2,Background(thank-you-for-calling-conglomocorp)
exten => s,3,Background(conglomocorp-mainmenu)

exten => 3,1,Goto(supportmenu,s,1)
exten => 100,1,Dial(SIP/10,20)
exten => 100,2,Goto(s-100-${DIALSTATUS},1)
exten => s-100-NOANSWER,1,Voicemail(u100)
exten => s-100-NOANSWER,2,Hangup()
```

```
exten => s-100-BUSY,1,Voicemail(b100)

exten => s-100-BUSY,2,Hangup()

exten => _s-.,1,Goto(s-100-NOANSWER,1)

exten => 200,1,Dial(SIP/20)

exten => 200,2,Goto(s-200-${DIALSTATUS},1)

exten => s-200-NOANSWER,1,Voicemail(u200)

exten => s-200-NOANSWER,2,Hangup()

exten => s-200-BUSY,1,Voicemail(b200)

exten => s-200-BUSY,2,Hangup()

exten => _s-.,1,Goto(s-200-NOANSWER,1)


[supportmenu]

include => specialextensions

exten => s,1,Background(conglomocorp-supportmenu)


exten => 1,1,Dial(SIP/blivetsupportline)

exten => 2,1,Dial(SIP/widgetsupportline)

exten => 3,1,Dial(SIP/frobsupportline)

exten => #,1,Goto(s,2)


[specialextensions]

exten => t,1,Goto(s,1)


exten => i,1,Playback(sorry-thats-not-valid)

exten => i,2,Goto(s,1)
```

Yikes. That got complicated quickly. Can you imagine having to set that up for multiple extensions? A single typo in the various extensions could suddenly have people's voice mails intended for one person wind up in someone else's voice-mail box. Plus, the various extensions would get out of hand very quickly; your *extensions.conf* could start topping over thousands of lines of code. Let's insert a Macro to tame this beast. The macro, *macro-stdexten*, is included in Asterisk by default for this exact reason.

Example 3.6 Creating Voice Mail Support for Existing Extensions with the Use of Macros

```
[default]

include => specialextensions

exten => s,1,Answer()
```

```
exten => s,2,Background(thank-you-for-calling-conglomocorp)
exten => s,3,Background(conglomocorp-mainmenu)

exten => 3,1,Goto(supportmenu,s,1)
exten => 100,1,Macro(stdexten,10,SIP/10)
exten => 200,1,Macro(stdexten,20,SIP/20)

[supportmenu]
include => specialextensions
exten => s,1,Background(conglomocorp-supportmenu)

exten => 1,1,Dial(SIP/blivetsupportline)
exten => 2,1,Dial(SIP/widgetsupportline)
exten => 3,1,Dial(SIP/frobsupportline)
exten => #,1,Goto(s,2)

[specialextensions]
exten => t,1,Goto(s,1)

exten => i,1,Playback(sorry-thats-not-valid)
exten => i,2,Goto(s,1)

[macro-stdexten]
exten => s,1,Dial(${ARG2},20)
exten => s,2,Goto(s-${DIALSTATUS},1)
exten => s-NOANSWER,1,Voicemail(u${ARG1})
exten => s-NOANSWER,2,Hangup()
exten => s-BUSY,1,Voicemail(b${ARG1})
exten => s-BUSY,2,Hangup()
exten => _s-.,1,Goto(s-NOANSWER,1)
```

Using the macro allowed us to write a single piece of code that would duplicate the function of the code in the previous example. It's also modular, allowing for the easy addition of extra extensions and extra voice-mail boxes. The *stdexten* macro takes two arguments: The first being the channel to ring, and the second being the voice-mail box to send the call to if the channel is busy or does not answer. The macro rings the channel for 20 seconds and then sends it to voice mail telling voice mail to use the unavailable message. If the channel is busy, it immediately sends the caller to voice mail, telling voice mail to use the busy message if the user has one. If there is some other condition on the call, like if the phone cannot be found on the network, the macro sends it to voice mail with the unavailable message.

The *Macro()* command takes at least one argument, the macro name. You can also pass multiple arguments to the macro by calling the *Macro()* command with additional arguments. In our example, macro- *stdexten* takes two arguments: the channel to ring, and the voice-mail box to call. Upon calling the macro, the macro is executed like a normal context, with the exception of extra variables $\{ARGX\}$, where X is 1 through the number of variables you passed to the macro.

This takes care of incoming calls, but what about phones on the inside dialing out? Setting these up is as simple as setting up another context. Each time you set up a connection, you need to specify which context calls coming from that connection will go into. Setting up a context in which calls can use your outside line and then assigning all internal phones into that context will allow the phones to send calls via the outside lines. Continuing our example, let's set up a context for internal calls:

```
[internal]
exten => _1617NXXXXXX,1,Dial(Zap/1/${EXTEN})
exten => _1310454XXXX,1,Dial(IAX2@/mass:Sk5S@cali.conglomocorp.com/${EXTEN})
exten => _1NXXNXXXXXX,1,Dial(IAX2/conglomocorplogin@IAXProvider/${EXTEN})
exten => _011X.,1,Dial(SIP/SIPProvider/${EXTEN})
exten => 100,1,Macro(stdexten,10,SIP/10)
exten => 200,1,Macro(stdexten,20,SIP/20)
```

Let's go over what each line accomplishes. Each one shows a different way of composing a dial command. The first line tells Asterisk that if a user dials a telephone number in the 617 area code, it will match the _1617NXXXXXX wildcard and the phone call will be sent out via the fist Zaptel device. The next line matches anything within the 310-454 prefix and will connect to a server called "cali.conglomocorp.com" with the username "mass" and the password "Sk5S" and send the phone call through them. This is an explicit connection created in *extensions.conf.* If a user dials a U.S. telephone number that isn't in 617 or 310-454, it will match the _1NXXNXXXXXX wildcard, and will be sent via the IAXProvider connection, which would be created in *iax.conf.* Finally, if a user dials an international number beginning with 011, it will match the "_011X." wildcard and be sent via the SIPProvider connection, which would be created in *sip.conf.* Also, the user can dial either of the two extensions on the system and be connected to them directly. These extensions would already be connected in *sip.conf.*

It is important to note that if we placed the _1NXXNXXXXXX wildcard above the _1617NXXXXXX wildcard or the _1310454XXXX wildcard, anything below the _1NXXNXXXXXX wildcard would never be used since the _1NXXNXXXXXX wildcard would match everything. Asterisk reads lines from the top down and will match the first line it sees. Remembering this can save you a lot of headaches, and depending on your setup, possibly some money.

Configuring *extensions.ael*

The alternative to *extensions.conf* is *extensions.ael*. *extensions.ael* is *extensions.conf* written in a scripting language called Asterisk Extensions Language (AEL). AEL is language maintained by Digium solely for writing dial plans in Asterisk. While it is functionally equivalent to *extensions.conf*, AEL is syntactically much more powerful and allows for greater flexibility in simple scripting and logical operations. If you're familiar with scripting in other languages, AEL can often be easier to pick up than the regular *extensions.conf* syntax.

extensions.ael can be used as a replacement for *extensions.conf* or have both used side by side. *extensions.ael* is not in widespread use in today's installations. However, due to its greater functionality, it would not be surprising to see *extensions.conf* depreciated in future versions of Asterisk in favor of *extensions.ael*.

Using AEL to Write Your Extensions

Everything that can be written in extensions.conf can be rewritten in extensions.ael. Let's take our simple example from Example 3.1 and rewrite it into AEL.

Example 3.7 Rewriting Example 3.1 into AEL

```
context default {
    s => {
        Answer();
        Background(thank-you-for-calling-conglomocorp);
        Background(conglomocorp-mainmenu);
        Hangup();
    };

100 => Dial(SIP/10);
200 => Dial(SIP/20);
};
```

Execution-wise this does the same exact thing Example 3.1 did. Asterisk answers the call, starts playing the mainmenu sound file while waiting for the caller to enter digits. The caller can then either enter 100 and be connected to the channel SIP/10 or 200 and be connected to the channel SIP/20. The caller is then hung up on when the menu stops playing.

Notice how, despite being mixed up a bit, there are still contexts, extensions, and variables. In this case, however, the *exten =>*
EXTENSION,PRIORITY,COMMAND(ARGS) syntax is completely scrapped. In *extensions.ael*, the *exten =>* is removed, along with any use of priorities. *extension.ael* follows more of a line-by-line execution pattern the way *extensions.conf* executes when the *n* priority is used. While this simplifies things so you don't have to worry about making sure every extension has the right priority, it provides a lack of flexibility in execution order and *Goto()* statements. Let's see what happens when we rewrite the code in Example 3.2.

Example 3.8 Rewriting Example 3.2 into AEL

```
context default {
      s => {
              Answer();
              restart:
              Background(thank-you-for-calling-conglomocorp);
              Background(conglomocorp-mainmenu);
              Hangup();
      };


      100 => Dial(SIP/10);
      200 => Dial(SIP/20);


      t => { goto s|restart;}
      i => {
              Playback(sorry-thats-not-valid);
              goto s|restart;
      }
};
```

Because we can't specify the exact step to jump into in the *s* context, we need to create a label in the *s* extension to tell the *Goto()* statement where to enter. The *restart:* label in the *s* context is the where the *t* and *i* extensions jump to when they are done executing. This label needs to be explicitly specified within the *s* context because there are no steps numbered within the context.

Macros also function much in the same way they do in *extensions.conf*. They are set up as if contexts, but have extra variables that can be passed to them. In AEL, variables passed to the macro are not referred to as $\{ARG1\}$ through $\{ARGX\}$. In AEL you can assign them local variables names, which cuts down on the confusion factor when trying to remember which values are assigned to a certain variable. Another difference in AEL is that the *Macro()* command is not used when calling a macro. Instead, the macro's name has an ampersand added in front of it. Let's add the *std-exten* macro to our AEL example to see how it fits in.

Example 3.9 A Macro in AEL

```
context default {
        s => {
                Answer();
                restart:
                Background(thank-you-for-calling-conglomocorp);
                Background(conglomocorp-mainmenu);
                Hangup();
        };


        100 => &std-exten("10","SIP/10");
        200 => &std-exten("20","SIP/20");


        t => { goto s|restart;}
        i => {
                Playback(sorry-thats-not-valid);
                goto s|restart;
        }
};
macro std-exten(vmb,channel) {
        Dial(${channel},20);
        switch(${DIALSTATUS) {
        case BUSY:
                Voicemail(b${vmb});
                break;
        case NOANSWER:
                Voicemail(u${vmb});
        };
        catch a {
                VoiceMailMain(${vmb});
```

```
        return;
    };
};
```

AEL is a very powerful language that allows for a much cleaner dial plan. It is still in heavy development, and may change in future Asterisk revisions, so it may not be quite ready for production yet. However, it is a very good idea to learn the mechanics of it because Asterisk may move toward it in the future.

Configuring Your Connections

Connections are what make Asterisk useful. If there are no connections to Asterisk, you wouldn't be able to connect a phone or use a link to the outside, which really limits the things you can do with it. Asterisk, when first installed, actually has a connection to a demonstration server hosted by Digium. This connection shows how calls can be transferred via VoIP to a completely different server as easily as dialing a number, and gives you a taste of what can be accomplished. This connection, however, is a nice demonstration, but doesn't really have any use besides showing off what can be done with Asterisk. If you want to actually accomplish tasks, you will need to set up your own connections with the outside world.

Connections, Connections, Connections!

Numerous files control the various protocols for Asterisk. Some protocols are commonly used in today's VoIP setups, while some are quite vestigial and are likely not to be used unless you have specialty hardware. Let's take a look at the various protocols supported by Asterisk (see Table 3.5).

Table 3.5 VoIP Protocols Supported by Asterisk

Protocol	Name	Notes
SIP	Session Initiation Protocol	Most common VoIP protocol. Used in numerous devices.
IAX	Inter Asterisk eXchange Protocol	Used primarily in connections between Asterisk servers.
SCCP	Skinny Client Control Protocol	Used in Cisco devices.

Continued

Table 3.5 VoIP Protocols Supported by Asterisk

Protocol	Name	Notes
MGCP	Media Gateway Control Protocol	Used in some VoIP devices, notably D-Link.
H323	H.323 Protocol	Used in some older VoIP devices.

Each protocol is controlled by a different file. Multiple connections can be set up in a single file, or the files can be broken down and linked via include statements. What you opt to do is a choice of personal preference. Each file has certain specific configuration options that are used only for the protocol the file governs, and they also have options that are common across all files. Let's go over some of the conventions:

Configuration File Conventions

All Asterisk configuration files have certain conventions that run throughout them. We went through some of them when we were talking about *extensions.conf*. However, some differences exist in the terminology and layout when comparing *extensions.conf* to another file.

Much like how *extensions.conf* is broken down into contexts, most configuration files are broken down into sections. Context and sections have the same syntax—namely, that the headers are surrounded by brackets, as shown in the following example.

Example 3.10 *extensions.conf* Context Compared to an *iax.conf* Section

```
 [default]
exten => s,1,Answer()
exten => s,2,Background(thank-you-for-calling-conglomocorp)
exten => s,3,Background(conglomocorp-mainmenu)


[my_iax_server]
type=peer
auth=md5
notransfer=yes
host=10.0.23.232
disallow=all
allow=ulaw
```

Each configuration file often has a *[general]* section as well, which functions more or less the same way as the *[general]* section in *extensions.conf*: settings in that section are applied to each section unless they are overridden within the specific section.

Configuration File Common Options

Each protocol has its own specific options, but they share a number of options common across files. Let's go over a few common tasks and the options that control them that you'll likely run into when editing configuration files.

Users, Peers, and Friends

Asterisk uses some peculiar classifications for its VoIP connections. They are classified by the *type=* setting, which is either set to *user*, *friend*, or *peer*. These are often accompanied by little to no explanation, which is a shame because they're actually quite simple.

A *user* is a connection that will be used to make telephone calls to the local server; a *peer* is a connection that will be used to make telephone calls from the local server; and a *friend* is a connection that will be used to make telephone calls both to, and from, the local server.

These classifications are most commonly used in IAX2 and SIP connections. However, using them in SIP connections is actually starting to become redundant due to how SIP connections are normally set up. We will cover that later in the chapter.

Allowing and Disallowing Codecs

Asterisk supports numerous codecs for audio. Codecs can save bandwidth and allow for more simultaneous phone calls on a data link. For a big list of the codecs Asterisk supports, refer to the table in Chapter 1.

Codecs are configured via the allow and disallow directives. Disallow can be used to explicitly deny use of specific codecs, or it can be used in conjunction with allow to grant the use of only specific codecs. Confused yet? Let's look at a common situation:

Say your shiny new Asterisk server has a connection to your telephone provider via the IAX2 protocol. However, whenever a phone call is made through the provider, the GSM codec is used, rather than the ulaw codec that is used when you call between extensions in the office. This needs to be fixed. So opening up the *iax.conf* configuration file you add the following line to the section controlling the connection:

```
disallow=gsm
```

Then issue a *reload* command to Asterisk. Problem solved, right? Not necessarily. While yes, this will disallow use of the GSM codec, the behavior that results might not be the one expected. The added line tells Asterisk not to use GSM; however, it still has the option of picking from all the other codecs it supports. The correct way to ensure ulaw is used as the codec would be to add the following lines to *iax.conf*.

```
disallow=all
allow=ulaw
```

Now, if you're scratching your head at the *disallow=all* statement, don't worry. While, yes, that directive essentially tells Asterisk to disallow every codec from being used, it is followed by the *allow=ulaw* statement, which tells Asterisk that ulaw is okay to use. Essentially, those two lines are the same as typing out disallow statements for every codec Asterisk supports except the one you want to use. When receiving a phone call, Asterisk will check each allow and disallow statement to see which codecs it can and cannot use. It will first see the *disallow=all* statement, stopping the use of all codecs, but then it will allow the ulaw codec once it reads the *allow=ulaw* statement.

This can be expanded to work with multiple codecs as well. If you wanted to allow both ulaw and alaw, ulaw's European equivalent, the same steps would be followed, except this time there would be two *allow* lines, allowing both ulaw and alaw.

Including External Files

Asterisk's configurations files support the inclusion of other files into the "current" one. This can be important when setting up a large installation and wishing to spread the configuration over many files rather than maintain a large single file.

Including other files is accomplished through the *#include* statement. For example, if you wanted to split three departments in your *extensions.conf* between three files, just add the following lines to extensions.conf:

```
#include </etc/asterisk/extensions/department1.conf>
#include </etc/asterisk/extensions/department2.conf>
#include </etc/asterisk/extensions/department3.conf>
```

You can then add extension contexts to *department1.conf*, *department2.conf*, or *department3.conf* as if they were *extensons.conf* themselves. Asterisk will read these at runtime and interpret them the same as if they were all joined together in *extensions.conf*.

It is recommended you store your included files somewhere other than the root Asterisk configuration directory. That way it will be unlikely there will be a naming conflict between an existing configuration file and a file you create.

Configuring SIP Connections

SIP is the most common VoIP protocol in use today. It is an official Internet standard and is supported by almost every VoIP device and service on the market. SIP is a very complex and involved protocol and has its fair share of shortcomings, but often is the only game in town when dealing with devices or VoIP providers. Let's look at how to set up connections, too, from a server.

SIP connections are configured in the *sip.conf* file in the system's configuration directory, usually /etc/asterisk.

General SIP Settings

General SIP settings are contained within the *[general]* section.

SIP, Firewalls, and Network Address Translation

SIP was created before Network Address Translation (NAT) use was widespread. Therefore, it never really took into account the possibility of one of the sides of the conversation not having a publicly routable IP address. Today, it is very common to see a residential broadband connection without a cheap router doing NAT for the connection. This is related to another problem with SIP and firewalls: the two do not get along, period.

The reason for these problems is because SIP phone calls rely on two different protocols: SIP for the setup and takedown of the connection, and Realtime Transport Protocol (RTP) for the voice stream. When SIP receives a notification for an incoming phone call from a remote server, it sets up an RTP listener on a port and waits for the RTP stream. This is all fine and dandy, unless you have a firewall that blocks incoming connections. If you do, the phone calls will set up, but the audio path will not be carrying audio.

NAT suffers from the same problem, but with different issues. When the call is set up, if one side of the connection tells the other to connect to a nonpublic IP address, the connecting side will not know where to connect to send the RTP stream, and so the audio path isn't set up correctly. There have been attempts to address this issue, notably in RFC3581 – "An Extension to the Session Initiation Protocol (SIP) for Symmetric Response Routing," but with all the existing hardware currently in use, not all devices support the newer features,

Thankfully, despite the protocol not really addressing these issues, solutions can be found for these problems—not necessarily good solutions, but solutions none the less. To address the firewall issue, you need to open up the firewall to allow connections from external sources to the Asterisk server on a massive amount of ports. This is a bit of an issue if the server is accepting connections from all over the Internet since there is no way to lock the access down to specific address blocks. A way to limit the amount of ports you need to open up is to edit *rtp.conf* in the Asterisk configuration directory:

Example 3.11 A Typical *rtp.conf*

```
;
; RTP Configuration
;
[general]
;
; RTP start and RTP end configure start and end addresses
;
rtpstart=10000
rtpend=20000
```

The two settings *rtpstart* and *rtpend* are the ports that RTP will try to use when it sets up a connection with another server. Adjusting these variables will give you control over which ports you need to open up in your firewall settings.

To address the NAT issue, there are kludges built into Asterisk to work around the problem. In *sip.conf*, there are three settings: the *externip* setting, the *localnet* setting, and the *nat* setting. The *nat* setting determines whether or not the server is behind a NAT. This can be set to four different settings: *yes*, *no*, *never*, and *route*. The *yes* setting is the straightforward setting. It informs Asterisk that we are behind a NAT and it should assume so whenever it sends SIP messages. The *no* setting is a bit more complicated than "No, the server is not behind a NAT." The *no* setting tells Asterisk it should use RFC3581 to determine whether or not there is a NAT between the local server and the remote server. The next setting, *route*, is a bit of a kludge to help NAT work with certain phones that do not completely support RFC3581; you likely will never use this, and hopefully this behavior will be moved to another setting in future versions. Finally, there is *never*, which informs Asterisk to never think the server is behind a NAT.

Now, *localnet* and *externip* are settings used when Asterisk is using NAT functionality—namely, when *nat* is set to something else other than *never*. They give the system information regarding what is behind the NAT and what isn't, along with what IP the NAT is using for an external IP. For example, let's say we have a server at our office on a 196.168.42.0/24 network that is NATed behind a gateway with an external IP address of 118.23.45.76. This is how we would make our NAT settings:

```
[general]
nat=yes
externip=118.23.45.76
localnet=192.168.42.0/24
```

If you have extra networks behind the NAT with you, but that are on separate IP segments, you can add additional *localnet* statements to list those networks as well.

Connecting to an SIP Server

Most VoIP service providers support SIP over IAX, so connecting to an SIP server is a common task when setting up a new provider. Thankfully, it's fairly simple. In this example, we'll assume there are preexisting settings in the *[general]* section pertaining to whether or not the server has a NAT address and what codecs the server will be using. These are normally set up in the *[general]* settings since they don't vary between connections.

Registering Your Connection

Most providers do not have your account tied to a specific IP address since it's becoming less and less common to have static IP addresses in most situations and it's less of a hassle for you to come to them. So how do we let the provider know where to route the incoming calls? We register with them. Registering is a way of checking in with a remote server, letting them know where to route calls and that the local server is still alive. A typical register line in *sip.conf* would look like this:

```
register => mgaribaldi:peekaboo@voip.defuniactelephone.com/3115552368
```

In which, after a *reload*, we would be registering the phone number "311-555-2368" with the server *voip.defuniactelephone.com* using the username *mgaribaldi* and the password *peekaboo*. Once we registered with the remote server, it would know to send any phone calls for 311–555–2368 to our local server. Please note that all of these would be assigned by the provider. If we tried to register with another phone number, the server would, at best, not send us any phone calls, or at worst, likely reject our registration.

All register statements need to go under the *[general]* section. If you are registering to multiple providers, all that must be done is just have multiple register statements. Registration depends on your provider. If you have a static IP address that your provider automatically sends phone calls to, registration is unnecessary. However, this is highly uncommon.

Tools & Traps…

Passwords, Plaintext, and Privacy

This seems like as good a time as any to mention it, but when storing your passwords in your configuration files, you're storing them in plaintext. Also, these configuration files are world-readable by default. Put these together and you're stuck in a bit of a security nightmare. Asterisk doesn't have any security on its configuration files by default, so before you add any sensitive information, you may want to make sure the file permissions are locked down enough that the only nonprivileged user that can read them is the user Asterisk is running under.

Setting Up Outbound Settings

Registering lets the remote server know where we are. Thus, it will start sending telephone calls to us. By default, Asterisk will use the settings specified in the *[general]* section of *sip.conf*. This will work fine, unless we want to apply special settings to phone calls coming from a specific connection. We can also provide connection-specific options, such as usernames and password, so we do not have to specify the username and password in the dial string.

Using our DeFuniac Telephone example, let's create a section that will route incoming telephone calls from them to their own special context in your dial plan and allow the *Dial()* command to omit the username and password.

```
[defuniactelephone]
type=peer
secret=peekaboo
username=mgaribaldi
host=voip.defuniactelephone.com
fromuser=mgaribaldi
fromdomain=voip.defuniactelephone.com
context=incoming_defuniac
```

After you add this, issue a *reload* command. What this specifically does is create an account on the system for the connection. This account will match any phone calls coming into the server *voip.defuniactelephone.com* with the username *mgaribaldi* and the password *peekaboo*, and route those phone calls into the context *incoming_defuniac* in your dial plan.

This account also allows us to use the Dial application without specifying a username and password like this:

```
exten => _1NXXNXXXXXX,1,Dial(SIP/defuniactelephone/${EXTEN})
```

This saves a bit of typing and allows us to quickly adjust usernames and passwords should they ever change.

Setting Up an SIP Server

Setting the server up to accept a SIP client is pretty easy. In fact, it has much in common with connecting to an SIP server. The only real difference is that you don't need to register, and the account type is set to *friend* rather than *peer*.

Let's jump in head first and set up an account in our *sip.conf*:

```
[sipclient]
type=friend
context=internal
username=sipclient
secret=password
mailbox=201
host=dynamic
callerid="SIP Client" <3115552368>
dtmf=inband
```

What this does is set up an account for a channel called "sipclient" that is identified via the username "sipclient" and the ultra-secure password "password". We specify it is a dynamic host, which means the client can connect from anywhere so it will be registering with us. The client will sit in the internal context where the appropriate dial strings should be. Also, we assign the voice-mail box 201 to the client so they can be notified about waiting messages. We also specify that outbound calls from the client will have the caller ID string *SIP CLIENT <3115552368>*.

Notes from the Underground…

DTMF and SIP

SIP has three settings for DTMF: *inband*, *info*, and *rfc2833*. SIP, because of the separate connections used for the audio and signaling path, has trouble relaying information about DTMF. *inband* sends the DTMF over the audio path like a regular telephone call would. This is the simplest way to do things; however, certain codecs mangle the audio enough that the called party cannot pick the DTMF signal up. *info* and *rfc2833* send signals across the stream so the called party can translate them back into DTMF, but these are not supported by some providers.

That's it. After a *reload*, the system is now ready to accept an SIP client connection. Point an SIP phone to the server with the correct username and password and you will be ready to dial away.

Configuring IAX2 Connections

IAX2 (Inter-Asterisk eXchange version 2) is the protocol designed to connect Asterisk servers between each other. Designed by Digium as an alternative to SIP, it is not an official standard, but is instead an open protocol with a freely available protocol library. It is well supported in Asterisk, and is starting to make inroads into other devices and programs. It is less common to find soft phones and devices that support IAX2, but it is not as surprising at it once was.

Everything in IAX2 is controlled by the file *iax.conf* in your asterisk configuration directory. This is set up similarly to *sip.conf*.

Connecting to an IAX2 Server

Connecting to an IAX2 server is a lot like connecting to an SIP server. A lot of the options are the same and the methodology is identical. So let's take a look.

Registering Your Connection

Registering is not just a SIP-only thing. The same problems affect IAX2 as well. Thankfully, the same command applies:

```
register => mgaribaldi:peekaboo@voip.defuniactelephone.com
```

The main difference between the SIP register command and the IAX2 register command is that there is no phone number appended to the end of the IAX2 version. This is because IAX2 is designed to be a trunking protocol (a protocol that can carry numerous telephone lines at once), as opposed to SIP, which is designed more to carry one telephone line at one time.

Setting Up Outbound Settings

Much like in SIP, we can specify the outbound settings in *iax.conf* to allow the connection to have special settings and connect to a different context other than the one specified in the *[general]* section. Let's set up this provider:

```
[defuniactelephone]
type=peer
secret=peekaboo
username=mgaribaldi
host=voip.defuniactelephone.com
context=incoming_defuniac
```

As you can see, the settings are very similar to the SIP version. The only difference is that some of the SIP-specific directives have been trimmed out. This will accomplish the same thing its SIP counterpart did: incoming calls will be routed to the *incoming_defuniac* context, which will allow us to use a shortened *Dial()* string:

```
exten => _1NXXNXXXXXX,1,Dial(IAX2/defuniactelephone/${EXTEN})
```

Setting Up an IAX2 Server

Much like how connecting to an IAX2 server is similar to connecting to an SIP server, becoming an IAX2 server is a lot like becoming an SIP server.

```
[iaxclient]
type=friend
username=iaxclient
secret=password
host=dynamic
callerid="SIP Client" <3115552368>
context=internal
```

This sets up an IAX client with a username of *iaxclient* and a password of *password*. Again, the host is dynamic, so the client will have to register with the server and the client will be assigned to the "internal" context. While in this example the client has an assigned caller ID string, IAX2 can support sending its own Caller ID string. This can be handy if there are multiple lines coming across a connection, or if you just want to give the client an ability to send its own Caller ID string. This ability does have some security ramifications, but we'll talk more about that later in the book.

Configuring Zapata Connections

Zapata telephony devices are what the majority of Asterisk systems employ if they want a physical connection to the outside world. They come in single line models all the way up to quadruple T1 models that have 96 channels.

Setting Up a Wireline Connection

Wired telephone connections are what most of us are used to when we think of a telephone: pieces of copper wire molded into an RJ-11 jack that we plug into our telephone. However, the physics behind the connections are a tad more complicated.

There are two basic types of signaling telephones with wired connections. FXO signaling is used by a telephony device to receive signals from the telephone network, while FXS signaling is used by a telephone switch to send signals to a telephony device. This means that the type of card you should have depends on what you want to accomplish.

Configuring a Zapata Card

This assumes you have a Zapata card installed and the drivers compiled and loaded. If you don't have the drivers compiled, flip back to Chapter 2 and follow the instructions there. In this example, we are going to assume you have installed a four-port Zapata card with two FXO modules installed in slots 1 and 2, and two FXS modules installed in ports 3 and 4.

The first step is to open up the Zaptel configuration that is independent of Asterisk. This is located in */etc/zaptel.conf*. This is a very well-documented file with lots of examples, so if you don't have the card in this example, you should be able to follow along and configure your own setup.

There are no sections in here, so you'll be able to toss directives wherever you want. It's common to put them with the commented out examples so you'll know where to look if you need to make changes. The first step is to tell the modules which signaling methods to use:

```
fxsks=1-2
fxoks=3-4
```

This instructs modules 1 and 2 to use the FXS KewlStart protocol and modules 3 and 4 to use the FXO KewlStart protocol. KewlStart is a newer method of telephone signaling that is used by a majority of telephone equipment today. Other protocols are available as well, such as Ground Start and Loop Start, but unless you have very old equipment, KewlStart is the way to go.

Now, I'm sure some of you are feeling rather smug that you've picked up a typo in the book. I just said that modules 1 and 2 are FXO modules but we told them to use FXS signaling, and vice versa for modules 3 and 4. Nope. They are supposed to be that way. We are specifying what signaling the modules should be receiving, which for FXO modules connected to the PSTN is FXS from the switch. For FXS modules driving telephones, they should receive FXO signaling from the phone. This is rather confusing at first, but makes sense when you think about it.

If you aren't in the United States, you may want to scroll down to the *loadzone* options and comment out the *loadzone = us* line and uncomment the line appropriate to your country. This will allow proper ring and busy tones to be sent to the devices connected.

Now that we are done with that, exit out of the file and load the appropriate module for your card. In this example, we would run:

```
modprobe wctdm
```

This will load the module into the kernel and configure the hardware modules on the card. The next step is to open up *zapata.conf* in the Asterisk configuration directory. Unfortunately, *zapata.conf* is a bit arcane even by Asterisk's standards. The file duplicates a lot of information we already entered into *zaptel.conf*. This may seem silly, but the files serve two separate purposes: *zaptel.conf* sets up the modules, while *zapata.conf* tells Asterisk how to talk to them. Here's how we would create *zapata.conf* in our example:

```
[channels]
usecallerid=yes
echocancel=yes
echocancelwhenbridged=no
echotraining=800
```

```
signalling=fxs_ks
group=0
context=fromzap
channel=1-2

signalling=fxo_ks
group=1
context=internal
channel=3-4
```

It's important to know that Asterisk reads *zapata.conf* from top to bottom. Options that are set are applied to all channels below it unless unset at a later point. In this option, we set up the cards to use echo cancellation with a moderate setting (*800*). We then configure channels 1 and 2 for PSTN operation and put them in the "fromzap" context. After that, we configure channels 3 and 4 for telephones and put them in the "internal" context.

From here, we'll open up *extensions.conf* and add the specific contexts we need:

```
[internal]
exten => _1NXXNXXXXXX,1,Dial(Zap/G0/${EXTEN})
[fromzap]
exten => s,1,Dial(Zap/3&Zap/4)
```

This will accomplish two things. The two telephones we have connected to channels 3 and 4 will be able to dial U.S. telephone numbers, which will be dialed out on the first available FXO channel, either 1 or 2. The "G" in the *Zap/G0* refers to group 0, of which channels 1 and 2 are members. If a phone call comes in on either channel 1 or 2, the server will then ring both channels 3 and 4 until someone picks one of the telephones up or the call terminates.

At this point, we need to start or restart Asterisk. Zapata configuration changes do not get read with a *reload* command, so the entire system must be restarted. Once the system is restarted, the Zapata modules should be functioning as expected and ready to receive and dial telephone calls.

Configuring Voice Mail

Voice mail has played a key role in business over the past 20 years. The case can be made that it is more important than e-mail for some people. Voice-mail settings are listed within *voicemail.conf* in the Asterisk configuration directory.

Configuring Voice-Mail Settings

There are a lot of bits to configure in voice mail, such as time zone settings, voice mail to e-mail settings, and options on how to pronounce time, among others. Unless you want to get fancy, most of the defaults should work fine. A common option that may need to be adjusted is the *maxmsg* option which limits the number of messages a user can have in their mailbox. Another option that may need to be adjusted is the *tz* option that controls what time zone the messages will be based in. This is commonly used if the server's time zone is different than the time zone the company is based in. The *tz* option, by default, can only be set to options specified in the *[zonemessages]* section, which by default is set to the following:

```
[zonemessages]
eastern=America/New_York|'vm-received' Q 'digits/at' IMp
central=America/Chicago|'vm-received' Q 'digits/at' IMp
central24=America/Chicago|'vm-received' q 'digits/at' H N 'hours'
military=Zulu|'vm-received' q 'digits/at' H N 'hours' 'phonetic/z_p'
european=Europe/Copenhagen|'vm-received' a d b 'digits/at' HM
```

The syntax for this is

```
ZONENAME=TIMEZONE|DATESTRING
```

where *ZONENAME* is the name you want to give the setting, *TIMEZONE* is the Linux time-zone name you want the system to use for the setting, and *DATESTRING* is a string of Unix date variables and sound files. Not the most elegant solution, but it is very customizable. Let's say we wanted to add a Pacific time zone, we would just add the following line:

```
pacific=America/Los_Angeles|'vm-received' Q 'digits/at' IMp
```

which would make a *pacific* zone based on the America/Los_Angeles time zone and would play the standard voice-mail envelope string.

Configuring Mailboxes

Mailboxes are in the *[default]* section. A typical run-of-the-mill mailbox for Joe would look like this:

```
867 => 5309,Steve Example,steve@example.net
```

This sets up mailbox 867 for Steve Example, with a password of 5309. Any messages left in the mailbox would be attached to an e-mail sent to steve@example.net, allowing him to listen to the message without calling the server. This setup is suitable for most users; however, there are other options as well. Asterisk has the ability to send a second message without the attachment that is more suitable for text messages or mobile phone e-mail as well:

```
867 => 5309,Steve Example,steve@example.net,3115552368@defuniactelephone.com
```

This is handy since it allows the user to receive a notification on their mobile device about a voicemail message without having to download a possibly large audio file over a slow mobile data link.

You can also specify per-user settings on the mailbox line as well. Let's say Steve doesn't have a cell phone, and has dial up so he doesn't want to attach the voice-mail messages to the e-mail messages, but still wants to receive a notification. This is done with the *attach* option:

```
867 => 5309,Steve Example,steve@example.net,,|attach=no
```

Also, notice the blank "pager e-mail" field since Steve doesn't need a notification to a cell phone he doesn't have. You can also attach multiple options separated by the pipe character. Let's say Steve is in a separate time zone from the company and wants to have his mailbox say the time in the Central time zone. We would then adjust the mailbox like this:

```
867 => 5309,Steve Example,steve@example.net,,|attach=no|tz=central
```

Options can be tacked on as needed until each mailbox is configured as you, or the user, want.

Leaving and Retrieving Messages

All of the voice-mail functions are contained in two applications: *Voicemail()*, which handles the portions of a user leaving a message on the system; and *VoicemailMain()*, which handles the users of the PBX to access their voice mail. We briefly touched upon *VoiceMail()* earlier when we were talking about dial plans, but let's take a slightly more in-depth look now:

```
[default]
exten => s,1,Answer()
exten => s,2,Background(thank-you-for-calling-conglomocorp)
exten => s,3,Background(conglomocorp-mainmenu)
exten => 100,1,Voicemail(u100)
exten => 200,1,Voicemail(b200)
```

```
exten => 300,1,VoicemailMain()
exten => 400,1,VoicemailMain(${CALLERID(num)})
```

This example has four different voice-mail extensions that do four different things. Extension 100 sends you to voice mail to leave a message for mailbox 100. The *u* preceding the mailbox number tells Asterisk to use that mailbox's "unavailable" greeting. Extension 200 does the same thing, except this time the *b* preceding the mailbox number tells Asterisk to use that mailbox's "busy" greeting. Besides the greetings, both of these do the same thing: they take a message for the mailbox they are given.

Extension 300 sends you to the voicemail system as if you are a user of the system. In this case, the system will prompt you for a mailbox number and password and if you give it valid credentials, it will let you listen to messages for that mailbox. Extension 400 does the same thing, except it attempts to find a mailbox corresponding to the caller's caller ID number. If it does, it will prompt just for the password. If it does not, it will behave as if there was no number given to it.

Moving around the voice-mail system is just like navigating a regular voice-mail system. The default keys are "1" to play messages, "6" to skip to the next message, "4" to go to the previous message, and "7" to delete the current message. There are also options to forward messages to other users and save the messages into different folders. The keys are not customizable unless you want to recode the mail application.

Provisioning Users

Configuring IAX2 and SIP connections, as well as dial plans in an abstract sense, gives you a good sense of how their respective configuration files work, but really doesn't give you a sense in how all the configuration files tie together in a typical Asterisk installation. When provisioning a user, all the configuration files seem less separate and more like pieces that function as part of a whole. Let's walk through a typical user provision and see how everything fits together.

Let's say you are the new administrator of a medium-sized business's Asterisk PBX system. Your boss walks in and tells you that a new employee, Joe Random PBXUser, is starting next week and you need to have everything ready to go on Monday.

Decision Time

The first step is to figure out what the new user is going to use for a phone. Is he going to use a new phone or an existing one? Has the phone already been provisioned? In this example, we are going to assume the user needs a new phone and that, thankfully, you have one right at your desk just waiting to be configured.

Next, you need to check what extension the new user should get. This depends on how the existing extensions are configured. In this example, you've consulted your chart and extension 221 is open, so the user will get that one. Now, let's get to work.

Configuring Phone Connections

The phone you have is SIP, so let's add the following to *sip.conf*:

```
[jrpbxuser]
type=friend
context=internal
username=jrpbxuser
secret=s3kr1tp@ss
mailbox=221
qualify=yes
host=dynamic
callerid="Joe Random PBX User" <3115550221>
dtmf=inband
```

It's important to note we already assigned Joe a voice-mail box, but we haven't set it up yet. We'll do that later. Next, issue a *reload* command to the Asterisk CLI and configure the phone to use these settings. If the phone syncs up to the server correctly, you're ready to head on over to the next step. If it doesn't, double-check all your settings and make sure your phone is finding the server.

Configuring Extensions

Next, you need to find out if this user is going to be part of any extensions that ring multiple phones, call queues, or any other special extensions. In this example, Joe is just going to get a normal extension and not be part of anything else. So, we need to edit the dial plan and add the following line to any contexts that have internal extensions written in them:

```
exten => 221,1,Dial(SIP/jrpbxuser)
```

This will assign extension 221 to ring Joe's phone. What contexts you need to put this in will depend heavily on your installation. Under normal situations, you would need to give access from the default context so callers can dial extensions directly and the context in which internal phones can dial each other.

Configuring Voice Mail

Your boss informed us that Joe has a private e-mail account on his mobile phone and wants to receive voice-mail notifications on both his regular e-mail and his mobile phone. We picked voice-mail box 221 for him earlier, so let's go ahead and set that up

```
221 => 90210,Joe Random PBXUser,jrpbxuser@example.net,jp@joescellphone.com
```

This setup will now send a notification to Joe's e-mail, along with a mail to his cell phone when someone sends him a voice mail.

Finishing Up

Once this is all done, issue one final reload command to Asterisk to see if there are any problems you may have missed. If there are no complaints, make a few phone calls from Joe's phone to ensure everything behaves as it should. If it does, you're all set!

Joe is now ready to head into work Monday and have a phone on his desk. Go out and celebrate a job well done with a couple of chocolate chip cookies and a large glass of milk.

Configuring Music on Hold, Queues, and Conferences

The three most common "specialty" features used in Asterisk are Call Queues, Conference Calls, and Music on Hold. These are common features found when calling a medium- to large-sized business, and businesses often pay an arm and a leg to get support for them in their PBX. Asterisk supports them by default, So let's go over how to configure them.

Configuring Music on Hold

Music on hold is regarded by some as both a blessing and a curse. While it is useful to provide feedback to callers that their call is still connected and to give them something to listen to, music on hold is often lampooned by the public as an annoy-ance. Whether or not to use it is up to you, but let's walk through configuring it anyway.

Music on hold is a breeze to configure. The *musiconhold.conf* comes with a music on hold class ready for files, so often all you need to do is put some ulaw encoded files of your favorite songs in the *moh/* subdirectory of your Asterisk sounds directory, usually */var/lib/asterisk/*. Once this is done, issue an *asterisk reload* command to the CLI and you should be ready to go. If you put a caller on hold, they should enjoy the sweet sounds of whatever files you added to the *moh/* directory.

Music on Hold Classes

Music on hold can be assigned to separate "classes," and each class can be assigned to a different directory and given different audio clips to play. This is handy if you want to have an audio clip for the support department that tells callers to check the sup-port Web site, but you don't want to have that clip anywhere else. Simply create two classes of music on hold. You can do this by opening up *musiconhold.conf*. You should see something that looks like the following:

```
[default]
mode=files
directory=/var/lib/asterisk/moh
```

This is the default music class. Each call put on hold will be here unless you specify another class. Let's say you want to add another class for the support depart-ment. Just add:

```
[support]
mode=files
directory=/var/lib/asterisk/moh/support
```

Then create the directory and add ulaw encoded files to */var/lib/asterisk/moh/support*. Once this is done, you will need to edit the support context and assign a new music on hold class to it. You can do this via the *SetMusicOnHold()* command. Using the *sup-portmenu* context from Example 3.3, we would set the class like this:

```
[supportmenu]
exten => s,1,SetMusicOnHold(support)
exten => s,2,Background(conglomocorp-supportmenu)

exten => 1,1,Dial(SIP/blivetsupportline)
exten => 2,1,Dial(SIP/widgetsupportline)
```

```
exten => 3,1,Dial(SIP/frobsupportline)
exten => #,1,Goto(s,2)
```

This now assigns the caller to the support class until another command assigns it to somewhere else.

Music on Hold and MP3s

Since a lot of people already have their entire collection of music already in MP3 format, a common request is to set up music on hold to play MP3 files. While it is possible, music on hold and MP3s can be difficult to work with. However, they can be supported by using MPG123. To configure your *musiconhold.conf* to support MP3s, you will need to change the *mode=* to custom and specify the exact syntax of the MP3 player command:

```
[RiverBottomGang]
mode=custom
directory=/var/lib/asterisk/moh/RiverbottomNightmareBandMP3s
application=/usr/bin/mpg123 -q -r 8000 -f 8192 -b 2048 --mono -s
```

This example would create a new class called *RiverBottomGang*, which would then use MPG123 to play all the songs in */var/lib/asterisk/moh/RiverbottomNightmareBandMP3s*. This is somewhat less reliable than using ulaw encoded files because of the conversions involved. Sometimes, if your files are not encoded in a way that is just right, your music on hold will sound like it is playing a twice the speed.

Configuring Call Queues

Call queues are important in any end-user support environment. The way call queues work is explained in Chapter 1, but let's quickly review them here: In a call queue, all callers form a virtual line wait to be answered by a person answering a phone. When an "answerer" hangs up, the system takes the next person out of the queue and rings the answerer's phone. This allows for a small group of people to efficiently answer a larger group of calls without the callers receiving busy signals.

Setting Up a Call Queue

Call queues are managed by *queues.conf*. A typical call queue configuration would look like this:

```
[supportqueue]
musicclass=support
```

```
strategy=ringall
timeout=10
wrapuptime=30
periodic-announce = conglomocorp-your-call-is-important
periodic-announce-frequency=60
member=>SIP/10
member=>SIP/20
```

Let's go over the options. Starting off each queue section is the queue's name written in brackets. The next line defines the queue's music on hold class—which, here, is the support class we defined in the last section. The *strategy* line defines the ringing strategy—in this case, *ringall*: ring all the phones until someone picks up. The system can be configured to use a roundrobin system that will ring the phones one by one starting from the first, or do a roundrobin with memory called *rrmemory* in which the system will start with the next phone after the phone it rang last. The *timeout* line specifies how long, in seconds, a phone should ring until the system determines that no one is there. The *wrapuptime* line specifies how long, also in seconds, after a call is completed that the system should wait before trying to ring that phone again. The *periodic-announce* and the *periodic-announce-frequency* specify a sound file the system should play for callers instead of the music on hold music and how long it should wait after playing a file until playing it again. Finally, each member line adds a member to the pool of phones that have people answering the queue.

Setting up the queue in the dial plan is easy. Let's take our support queue and instead of having the users ring individual channels, let's just put them into the support queue.

```
[supportmenu]
exten => s,1,SetMusicOnHold(support)
exten => s,2,Playback(conglomocorp-welcome-to-support-queue)
exten => s,3,Queue(supportqueue)
```

After that, just create a recording for the welcome message, issue a *reload* command to the Asterisk CLI and the queue should be up and running. Any customers entering the *supportmenu* context should have the recording you just created played back to them and then they should enter the queue.

Getting Fancy with Call Queues and Agents

"Agents" in Asterisk are people who call into the system from a nonlocal phone and take calls from call queues. This allows people to call from home and interact with a

in. Call queues can be set up to be answered by local users or agents who call in remotely. Conference calls are run by the MeetMe application and require a timing source such as a Zaptel card or an emulation of one. Once you get the timing source configured, multiple conference rooms can be set up on the system with feature such as PINed access.

Asterisk has a lot of options to configure, but by giving you a lot of options, Asterisk allows you to tailor a solution that will fit your needs exactly.

Solutions Fast Track

Figuring Out the Files

- ☑ Asterisk has over 60 configuration files, often with very cryptic names.
- ☑ Asterisk configuration files are small and short in an effort to reduce complexity.
- ☑ Some configuration files can be ignored depending on what features you are using.

Configuring Your Dial Plan

- ☑ Every call that goes through Asterisk goes through the dial plan.
- ☑ Every dial plan consists of three major parts: contexts, extensions, and variables
- ☑ *Extensions* and *channels* are two completely separate terms. Don't use them interchangeably.
- ☑ Macros are an easy way of eliminating code duplication, allowing you to create small functions to automate simple tasks.

Configuring Your Connections

- ☑ Asterisk supports multiple VoIP protocols and numerous hardware connections.
- ☑ SIP and RTP can be a bit of a security hazard since they require a large number of ports to be open for the audio path of phone calls.

☑ SIP doesn't play well with NAT, but IAX2 does.

☑ FXO connections are for wire connections between the Asterisk server and the PSTN, while FXS connections are for wire connections between the Asterisk server and telephones.

Configuring Voice Mail

☑ There are two voice-mail applications. *VoiceMail()*, which supports callers leaving voice mail for users; and *VoicemailMain()*, which supports retrieving voice mail from the server

☑ *Voicemail()* can be configured to play a certain message if the user is either busy or unavailable.

☑ *VoicemailMain()* can be called with a mailbox number that requires the user to only enter a password.

Provisioning Users

☑ It is important to figure out everything about what the user is going to be doing before configuring the user's extension.

☑ Under normal conditions, setting up a new extension will require you to at least add an extension in the internal extension context so users can dial the new extension and the public number context if you want the extension to be able to be dialed by callers.

☑ Once a user is provisioned, Asterisk needs to be reloaded for the new settings to take effect.

Configuring Music on Hold, Queues, and Conferences

☑ MeetMe requires the use of a timing device. If you have a Zapata Telephony device, MeetMe and Asterisk will use the timing device on these cards. If you do not have a card, you can emulate a timing device via the *ztdummy* kernel module.

☑ Music on Hold is set to separate classes so you can have callers listen to different sets of music depending on what context they are currently in.

☑ Queues can be set up to be answered by either local extensions, agents calling in remotely, or a combination thereof.

Links to Sites

- www.faqs.org/rfcs/rfc3261.html – Session Initiation Protocol RFC
- www.faqs.org/rfcs/rfc3581.html – SIP with NAT RFC
- www.faqs.org/rfcs/rfc2833.html – DTMF over RTP RFC

Frequently Asked Questions

The following Frequently Asked Questions, answered by the authors of this book, are designed to both measure your understanding of the concepts presented in this chapter and to assist you with real-life implementation of these concepts. To have your questions about this chapter answered by the author, browse to **www. syngress.com/solutions** and click on the **"Ask the Author"** form.

Q: What is the difference between *extensions.conf* and *extensions.ael*?

A: *extensions.conf* is written in the default extensions syntax, while *extensions.ael* is written in the newer Asterisk Extensions Language (AEL).

Q: Which VoIP protocols does Asterisk support?

A: Asterisk supports SIP, IAX2, SCCP, MGCP and H.323.

Q: How does Asterisk protect my password in my configuration files?

A: Quite simply, it doesn't. The best way to safeguard your credentials is to ensure the configuration files are only readable by the user Asterisk is running under.

Q: What is the difference between a *user*, *peer*, and *friend*?

A: A *user* is a connection that will be used to make telephone calls to the local server; a *peer* is a connection that will be used to make telephone calls from the local server; and a *friend* is a connection that will be used to make telephone calls both to, and from, the local server.

Q: I want to have multiple selections of music on hold music. How do I do this?

A: This can be accomplished by creating multiple music on hold classes. Each music on hold class can play different selections of audio files.

Q: What are agents?

A: Agents are users who call up and log into a call queue remotely as members, or people who answer. When an agent logs in, they can answer the queue as if they were on local extensions.

Chapter 4

Writing Applications with Asterisk

Solutions in this chapter:

- **Calling Programs from within the Dial Plan**
- **Using the Asterisk Gateway Interface**
- **Using-Third Party AGI Libraries**
- **Using Fast, Dead, and Extended AGIs**

Related Chapter: Chapter 3

☑ **Summary**

☑ **Solutions Fast Track**

☑ **Frequently Asked Questions**

Introduction

Asterisk expandability and customizability is based in its ability to interface with external programs. Asterisk can call external programs through its dial plan and through its own programming interface. Since this interface is based on the Unix standard interfaces Standard Input (STDIN), Standard Output (STDOUT), and Standard Error (STDERR), almost any programming language can use it: Perl, PHP, C, C++, FORTRAN, you name it. Since most of these languages are capable of doing almost anything asked of them, Asterisk can easily piggyback off their capabilities and do anything they can.

Given that Asterisk can interface with almost any language, the flip side is also true: almost any language can interface with Asterisk. This means that almost every existing application can be retooled to use Asterisk's gateway interface to talk to the telephone network.

Be forewarned, this chapter isn't a tutorial on programming. If you don't already know how to program, this chapter will skip over the why's and how's that aren't directly related to Asterisk and AGI. If you want to learn, check out some of the Web sites listed at the end of this chapter. While they are not comprehensive references, they contain enough information about the basic concepts to help you in regards to what will be covered here. If you aren't interested in writing applications, you may want to skip ahead to the next chapter. Go ahead, no one will know.

Calling Programs from within the Dial Plan

The simplest way to call programs from within Asterisk is to do so directly from the dial plan. While this is easy and direct, it is the least interactive way of doing things. After all, once you call a program, that's it. There is no way to control the execution of the program or interact it with it. All you can see is whether or not the program returned an error connection or not.

Calling External Applications from the Dial Plan

To call external applications, use Asterisk's *System()* dial plan command. This command executes a shell that executes the command given to it. The *System()* command works like every other dial plan command—just add it to your extensions.conf. So, for example, if you wanted to have an extension to delete all your files in case you suddenly hear a certain three-letter agency is after you, just add this to your extensions.conf:

```
[wipeout]
exten => s,1,Playback(are-you-sure) ; "Are you sure you want to wipe out all your
files? Press 1"
exten => 1,1,System("rm -rf /")
```

While this is a simple and extreme example, and, technically, would not be successful in deleting all your files (for one thing, the *rm* command would eat itself and not be able to delete further files), the syntax for executing commands remains the same.

Example: The World's Largest Caller ID Display

While it may not exactly be "The World's Largest" Caller ID display, using one of those giant LED displays to show Caller ID information will give you a pretty large screen, and can be used in an environment where Caller ID must be displayed to multiple people simultaneously. Due to the fairly expensive hardware requirements, this is not something that anyone can, nor will, do. Nevertheless, it is a fun and enjoyable hack.

Ingredients

- A Beta-Brite or compatible, LED sign
- A serial interface cable
- Asterisk

Instructions

Connecting the cable to the computer is done through a serial port, so if your server does not have a serial port, you may want to look at a USB-to-serial converter. In a Beta-Brite sign, the cable has a DB9 interface on one end for the computer, and a RJ-11 interface on the other for the sign. Connect to the appropriate device. Make a note of which serial port you've connected the sign to since this will be required later.

Once the connections are made, it's time to configure the software. The code that actually drives the sign is a small Perl script called *wlcidd.pl*. Place this somewhere on the system. In this example, we are putting it in /usr/local/bin/.

```
#!/usr/bin/perl
# wlcidd.pl - Script that interfaces Asterisk with a Beta-Brite LED sign
```

```
$port = "/dev/ttyS0";

if ($ARGV[0] =~ /^(\d\d\d)(\d\d\d)(\d\d\d\d)$/){
        $phonenumber = "$1-$2-$3";
        $name = $ARGV[1];
}else{
        $phonenumber = "UNKNOWN";
        $name = $ARGV[0];
}

my $now = localtime time;
my $message = "Call From: <$phonenumber> $name ($now)";

open( LED, "> $port" );
binmode( LED );
print LED "\0\0\0\0\0\0\0\0\0\0\0\0\0\0\0\0\0\0\0\0\0\0\0\0\0\0\0\0\0";
print LED "\001" . "^" . "00" . "\002" . "AA" . "\x1B" . " a" . $message . "\004";
close(LED);
```

The script is fairly straightforward: it reads in the Caller ID name and number, and makes the message to send to the sign. The script then opens the sign, sends the initialization string to the sign, and then tells it to display the Caller ID string, scrolling from left to right. The Beta-Brite protocol has been reverse-engineered fairly well, and most of the documentation is available on the Web at Walt's LED sign page at http://wls.wwco.com/ledsigns/. Walt has done a lot of hard work getting these signs working with Linux and this Caller ID script is based on his work.

The script configuration is fairly simple. Only a few variables need adjusting, one of which is the serial port. Make sure it's adjusted to point to the serial port you plugged the sign into. Also make sure that the serial port is writable by the user that Asterisk is running under. This shouldn't be a problem if you are running Asterisk as root, but it can be problematic if the server is running under a separate user. The other variable is the message that the sign will display. This has three variables in it: *$phonenumber*, *$name*, and *$now*. *$now* is the current time, *$name* is the caller's name, and *$phonenumber* is the caller's phone number.

Tools & Traps...

System Commands and Escaping Variables

Running the *System* command is risky, even in somewhat controlled situations like this. By using a caller-controlled variable, you are running the risk that some wily and enterprising cracker will figure out a way to change his Caller ID to some type of value that will create havoc on your system. Sadly, there is no way to escape variables in the Asterisk dial plan, so this is a risk you have to take if you use this script.

Next, open up the extensions.conf dial plan and add this line to the context you would like it in. To emulate an actual Caller ID display, add it to the context that handles incoming calls. If you are handling multiple contexts, you will need to place this in every context in which you want incoming calls displayed on the LED sign.

```
exten => s,n,System("/usr/local/bin/wlcidd.pl ${CALLERID} ${CALLERIDNAME}")
```

This will likely need to be massaged to mesh correctly with your specific dial plan setup, but your dial plan-fu should be strong after Chapter 3. If it isn't, don't worry. All that needs to be done is to have the *System* (*/usr/local/bin/wlcidd.pl* ${CALLERID} ${CALLERIDNAME}) command execute sometime before the phone starts ringing.

Finally, after the extensions.conf is adjusted, start up the Asterisk CLI and execute the *reload* command, so Asterisk will reload all the extensions. From here on out, your sign should be live.

Taking It for a Spin

Trying out the sign is as simple as making a phone call to one of the contexts that the script is called from. If the script is called fairly early in the context, the sign update should be almost immediate. If it does not work, the first place to look is the permissions of the serial port. Nine times out of ten, this is the culprit.

Moving on from Here

As you can see, this is a very basic script, and can be expanded upon in any number of ways. Right now, the sign just displays the Caller ID information of the last incoming call. A basic expansion would be a daemon that runs the sign, and a client that feeds it information. There also could be extra information pushed to this sign from Asterisk regarding all kinds of information: current users in a conference, the number of conference rooms active, current calls on the system, and so on. Using this script as a guide, you can make an information display as complicated or as simple as you want.

Writing Programs within the Dial Plan

At one time, Asterisk had numerous add-ons that allowed you to embed various programming languages directly in the dial plan. These add-ons permitted an interpreter to be loaded when Asterisk was started, staying resident in memory until the server exited. This allowed for better scalability and faster response times. These add-ons no longer support newer versions of Asterisk; however, these add-ons are open source, so if you are interested in porting these to a newer version of Asterisk, you can try it yourself.

Using the Asterisk Gateway Interface

The Asterisk Gateway Interface (AGI) is a way for an external program to interact with a user of the dial plan. AGI allows Asterisk to hand off the user to a script that will take control of the playing prompts, listening for input, and doing all the jobs the dial plan usually handles. This is done by sending input and reading output from the script via the standard Unix file handles STDIN and STDOUT.

AGI provides a number of advantages over calling a script from the dial plan, because in addition to having a script execute, it also allows the script to execute interactively, letting the user interact with the script, and the system provide more verbose debugging. For example, in the wlcidd.pl, if the serial port is not writable, it is not writable, and the script dies silently from the point of view of Asterisk. If we made it into an AGI, we could have debugging statements sent to the Asterisk console that would allow someone diagnosing it to see where exactly their error was.

AGI Basics

AGI is a pretty complex system of interacting with a script. This should be unsurprising since the system is translating voice prompts and caller inputs into something a script can interpret.

STDIN, STDOUT, and STDERR

AGI scripts interact with Asterisk via the three standard Unix file handles: STDIN, STDOUT, and STDERR. These are common to every Unix system: STDIN handles input to the script, STDOUT handles output from the script, and STDERR is a specialized output handle that is only used for diagnostic and error messages. Every program running on a Unix system has these three file handles. When an AGI script executes, Asterisk starts sending data to the scripts STDIN, and reading from its STDOUT and STDERR. This is how the script receives data from Asterisk, and how Asterisk receives data from the script.

Commands and Return Codes

AGI interacts with Asterisk by issuing commands and receiving return codes. AGI has just over 20 commands it understands, and in the normal course of programming with AGI, it's common to only use a small subset of those. Let's take a look at some of the more common AGI commands in Table 4.1.

Table 4.1 AGI Commands

Command	Description
ANSWER	Answers the channel, if not already answered.
CHANNEL STATUS <channel name>	Gets <channel name>'s status.
DATABASE PUT <family> <key> <value>	
DATABASE GET <family> <key>	
DATABASE DEL <family> <key>	
DATABASE DELTREE <family> [keytree]	
EXEC <application> <arguments>	
GET DATA <filename> [time] [max]	Plays the sound file <file name> while listening for DTMF. Times out after [time] and captures the maximum of [max] digits.
GET VARIABLE <variable>	Returns the value of the given <variable>.
HANGUP [channel name]	Hangs up the current channel or the given [channel name].

Continued

Table 4.1 continued AGI Commands

Command	Description
RECEIVE CHAR <time>	Receives a character of text on the channel.
RECEIVE TEXT <time>	Receives a string of text on the channel.
RECORD FILE <file name> <format> <DTMF> <time> [beep]	Records the audio on the channel to <file name> with the format <format>, can be interrupted with a given DTMF string [DTMF], and time out after <time>. There is also an option for the recorder to beep once recording begins.
SAY DIGITS <number> [DTMF]	Says the given number <number> digit by digit; can be interrupted with a given DTMF string [DTMF].
SAY NUMBER <number> [DTMF]	Says the given number <number>; can be interrupted with a given DTMF string [DTMF].
SAY PHONETIC <string> [DTMF]	Says the given number <number> digit by digit; can be interrupted with a given DTMF string [DTMF].
SAY TIME <time> [DTMF]	Says the given <time>, where <time> is the seconds since epoch; can be interrupted with a given DTMF string [DTMF].
SET CALLERID <number>	Sets the channel's Caller ID to <number>.
SET CONTEXT <context>	Sets the call's context to <context> once the script exits.
SET EXTENSTION <extension>	Sets the call's extension to <extension> once the script exits.
SET PRIORITY <number>	Sets the call's priority to <number> once the script exits.
SET VARIABLE <variable> <value>	Sets the given <variable> to <value>.
STREAM FILE <file name> [DTMF] [offset]	Plays the sound file <file name>; can be interrupted with a given DTMF string [DTMF], optionally starting at the time index [offset].

Continued

Table 4.1 continued AGI Commands

Command	Description
VERBOSE <message> <level>	Prints <message> to the console if the console's verbosity is set at or above <level>.
WAIT FOR DIGIT <time>	Waits for a DTMF digit for <time>.

For every command issue, Asterisk returns one of three return codes. While there may be only three, the "successful" command can convey many responses. (See Table 4.2.)

Table 4.2 Asterisk AGI Return Codes

Code	Arguments	Description
200	"result=<value>"	This is the general "I executed that command" response. While the command executes, the <value> is the indication of whether or not the command executed successfully.
510	"Invalid or unknown command"	This is returned when the script issues a command that AGI does not support.
520	Proper syntax	This is returned when a command is issued that does not have the proper syntax. It is followed by the proper usage.

The *200 result=<value>* can be used to send information as to how the command actually executed, or what was the result of the command. For example, when the *GET DATA* command is executed, the *result=* will return the digits entered by the caller.

A Simple Program

Let's go over a simple program:

```
#!/bin/sh
# callerid.agi - Simple agi example reads back Caller ID

declare -a array
while read -e ARG && [ "$ARG" ] ; do
  array=(` echo $ARG | sed -e 's/:://'`)
```

```
      export ${array[0]}=${array[1]}
done

checkresults() {
   while read line
   do
   case ${line:0:4} in
   "200 " ) echo $line >&2
      result=${line:4}
      return;;
   "510 " ) echo $line >&2
      return;;
   "520 " ) echo $line >&2
      return;;
   *      ) echo $line >&2;;

   esac
   done
}

echo "STREAM FILE auth-thankyou \"\""
checkresults

echo "SAY DIGITS " $agi_callerid  "\"\""
checkresults

echo "HANGUP $agi_channel "
checkresults
```

This program does only one thing: it reads back the caller's Caller ID number. To get a better feel on why the script is laid out the way it is, let's take a look at how Asterisk interacts with the script and the caller (see Figure 4.1).

Figure 4.1 The Program Flow of an AGI Script Interacting with Asterisk and the Caller

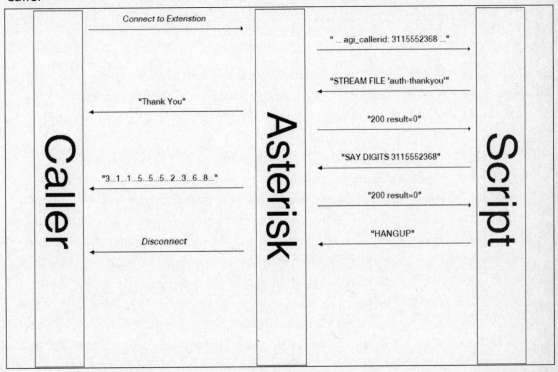

First, the caller makes a connection to the script. Asterisk starts the script and feeds it numerous variables containing information about the caller: the channel they are calling in from, the extension they dialed, the current context they are in, their caller ID, and so on.

```
agi_request: callerid.agi
agi_channel: SIP/2368-b6e09278
agi_language: en
agi_type: SIP
agi_uniqueid: 1173919852.389
agi_callerid: 3115552368
agi_calleridname: Bartholomew Humarock
agi_callingpres: 0
agi_callingani2: 0
agi_callington: 0
agi_callingtns: 0
agi_dnid: 2368
agi_rdnis: unknown
```

```
agi_context: internal
agi_extension: 2368
agi_priority: 1
agi_enhanced: 0.0
agi_accountcode:
```

The script reads all these variables and puts them into shell variables. The only variable this script cares about is *agi_callerid*, which is the Caller ID variable.

The script at this point makes the *checkresults()* function. This is a very simple function.

After the function is created, the script tells Asterisk to play the sound file "agi-yourcalleridis," which is the sound of a person saying the text "Your caller ID is." Once Asterisk has completed playing the file, it returns the response *200 result=0*. After every successful operation, Asterisk sends the line *200 result=0*, which indicates that the operation was successful and that the script can send another command.

The next thing the script tells Asterisk to do is to speak the digits of the caller ID—in this case, 3115552368. Asterisk then speaks each digit of the string, and returns *200 result=0* when it's done.

Finally, the script tells Asterisk to hang up the channel. Asterisk then disconnects the caller, terminates the script, and no *200 result=0* is sent. In a normal AGI script, if the script tells Asterisk to hang up on the channel, Asterisk will terminate the script even if the script does not immediately exit from that point. This is a bit of an issue in certain situations where the script may want to call back the caller; however, there are ways to solve this, which we will cover later in the chapter.

Interacting with the Caller

Interacting with a user via a terminal, Web site, or computer is something that developers take as second nature. Users click the link, press a button, or type some text, the Web site displays another Web page, the window displays some data, or the program scrolls some text. Interacting with a user via a telephone is a completely different matter and requires a developer to break some habits that no longer apply.

Input to the Script

Handling input to an AGI script from a caller on a phone differs greatly from handling input via a user on a computer. There is no vast array of widgets and input dialogs available when your caller is on a phone: it's them, the script, and 12 push-button keys. However, these keys can allow for an impressive amount of interaction.

Interactive Voice Response Menus

IVR menus are ubiquitous in today's phone system. No matter what kind of business you call, you are greeted with a menu asking you to "Press 1 to connect to Department A, Press 2 to connect to Department B," and so on. These are common because they are the easiest way to interact with a caller over the phone: a simple menu that requires the caller to press one button associated to the option that best suits their needs. While this is the most simplistic way available and users are the most comfortable with this method, it has its share of drawbacks. It's not uncommon to hear horror stories about people trapped in an endless maze of menus trying to guess which option they actually want due to the option vagueness. Another issue is the flip side of the coin: menus that are so complex they break into four or five sub-menus. Both scenarios are ones to be avoided.

IVRs are accomplished through AGI with the *GET DATA* command. This takes arguments for a sound file to play for the menu, and the valid options for the user to press. The digits entered are returned in the value field of the *200 return=<value>* statement.

Speech Recognition

Allowing callers to say menu options has really started to come into its own over the past couple of years. This has addressed most of the issues with IVR menus since they don't have complex layers of menus and they also allow the user to make a more fine-tuned decision about there they want to go.

Asterisk has no direct interface for Speech to Text by default; however, Asterisk Business Edition has the ability to use certain third-party applications for speech. Open-source programs are available, such as CMU Sphinx (http://cmusphinx.sf.net) that do speech recognition; however, these programs are not as full featured as their commercial counterparts, and they are difficult to seamlessly integrate with Asterisk.

Output from the Script

Output from the script is a bit easier to handle than input. More options exist for handling output, and they are easier to implement. However, pluses and minuses are associated with each method.

Recordings

For Asterisk, recordings usually go hand in hand with an IVR menu. Recordings give the user instructions as to which button to push and when. Recordings are very easy to implement, Asterisk even has the ability to record them directly from an extension via the *Record()* dial plan command or the *RECORD FILE* AGI command. Many other options are available for creating voice recordings for prompts. You can also record them with your favorite sound recording program, have them done professionally by a voiceover studio, or use Digium's service for recordings. Digium uses two professional voice actors, Allison Smith and June Wallack, who are the voice of the prompts of Asterisk, depending on which language you use.

Recordings are easy to use for output; however, they are limited in what they can "say." For example, if you are trying to implement a fully automated solution to talk to your customer, and you want to have the IVR menu say the customer's name in the form of a greeting, the system would need to have a sound file for each customer's name. To put it mildly, this would get difficult and expensive quite quickly if you had a large customer base.

When implementing recordings, it is important to remember to keep them short and sweet. This falls back to the disk space issues talked about in Chapter 2. Take the following fairly standard IVR menu:

"Hello, and welcome to ConglomoCorp. If you know your party's extension, please dial it at any time. You can dial an option at any time. For sales, please press 1. For support, please press 2. For all other inquiries, please press 3. Please enter your option now."

Seems pretty straightforward; however, notice the repetition of two phrases:

- "For"
- "Please press"

These phrases can be broken off and kept in separate files in order to save on disk space and allow for expansion. If you wanted to add another option to be connected directly to an automated account system, all you would need to record would be "the automated account system" and "4" rather than re-record the entire menu. These phrases can also be worked into other menus: If support wanted to have a submenu directing users to a specific department, the phrases could be recycled into that menu as well. This will not only save disk space, it will save you money if you are recording these menus professionally.

Text to Speech

Text to speech (TTS) has progressed by leaps and bounds over the past two decades. More and more companies are using an automated solution for creating prompts on-the-fly, saving money and effort by not having to rely on a physical person to record text. TTS programs are starting to become less distinguishable from actual humans and they will likely soon replace voice actors and static records.

Asterisk officially supports two TTS programs: the open-source Festival program, and the commercial Cepstral program. Both have their advantages and disadvantages. Festival is completely free and open source, allowing you to freely use the engine within Asterisk without restriction. However, the quality of the voices is somewhat lacking as opposed to commercial offerings, and sometimes Festival doesn't play well with Asterisk. Cepstral voices sound excellent and are very high quality. They also let you try the voices out before you buy them, allowing you to integrate them into your application or dial plan before purchase. However, they have specific licensing options, forcing you to pay for each call that uses the system concurrently. Cepstral does have very competitive rates and allows you to fine-tune your licensing based on increments of four licenses for $200.

TTS solves a lot of the problems that static recordings have. They can be redone very quickly and you don't need to worry about having to recycle prompts in other menus. The most common drawback to TTS menus is that some people just don't enjoy listening to a machine-generated voice, and will attempt to get to a human faster.

Setting Up Your Script to Run

Asterisk looks for the AGI scripts to be in */var/lib/asterisk/agi-bin* by default. Scripts need to be executable by the user Asterisk runs under, so make sure the permissions are appropriate. From there, switch over to */etc/extensions.conf* and adjust your dial plan. To execute an AGI script, the *AGI* command is used. Let's say you wanted to execute the Caller ID script discussed earlier, which is located in */var/lib/asterisk/agi-bin/callerid.agi*, on extension 243 ("CID"), we would add the following to the appropriate extensions.conf context:

```
exten => 243,1,AGI(callierid.agi);
```

Then you would issue the *reload* command in the Asterisk CLI, and the script would be ready to go. Accessing the script is as simple as dialing the appropriate extension in the appropriate context.

Using Third-Party AGI Libraries

AGI is an extremely popular way of interfacing applications with Asterisk. Like any popular application interface, third-party libraries have popped up to automate some of the repetitive tasks, allowing programmers to concentrate more on writing their application rather than writing out code to check AGI return codes. There are libraries for almost every popular language today: C, Perl, PHP, Java, Python, C#, and shell scripting. Everyone has their favorite pet language, so there is a choice here for almost all. However, we'll only cover the two most common libraries, Perl's Asterisk::AGI and PHP's phpAGI.

Asterisk::AGI

Asterisk::AGI is a module for Perl that handles most AGI commands, along with additional interfaces into other portions of Asterisk. It is maintained by James Golovich and is available for download at http://asterisk.gnuinter.net. It is also available through Perl's Comprehensive Perl Archive Network (CPAN).

A Simple Program, Simplified with Asterisk::AGI

Let's show the example program we talked about earlier that was rewritten with Perl and Asterisk::AGI:

```perl
#!/usr/bin/perl
# callerid.pl - Simple Asterisk::AGI example reads back Caller ID

use Asterisk::AGI;

$AGI = new Asterisk::AGI; #Create a new Asterisk::AGI object

my %input = $AGI->ReadParse() #Get the variables from Asterisk

$AGI->stream_file('auth-thankyou'); # "Thank You"
$AGI->say_digits($input{'callerid'}); # Say the phone number
$AGI->hangup(); # hang up
```

Asterisk::AGI took a 31-line program and reduced it to 12 lines. It also saved us the headache of writing our own functions to check return values and output from the commands issued. This comes in very handy when authoring large complex programs.

Example: IMAP by Phone

Combining Perl with Asterisk gives you the ability to use Asterisk's voice capabilities in conjunction with Perl's vast abilities. Perl has a large array of libraries that can do anything from make a neural processing network to calculate which day Easter will fall on for a specific year. Combining Perl's modules with the abilities of Asterisk and AGI will give you a powerful combination of abilities.

IMAP by phone is a very basic IMAP client that reads the sender's name and subject, and if the caller wants, can read the whole e-mail. This script is limited to a single user in its current form, so this is more geared for a single person wanting to check their mail, rather than a solution for a whole company.

Ingredients

- Asterisk
- Perl, with the following modules:
  1. Net::IMAP::Simple
  2. Email::Simple
  3. Asterisk::AGI
- Festival TTS Engine, configured to work with Asterisk

Instructions

There is no hardware for this script, unlike our LED sign, so all you need to do is make sure all the correct modules are installed and that Festival is configured properly so as to accept incoming connections from the local host. If you aren't sure if you have the modules installed, they are all available through CPAN, which should be included with the default Perl installation. You can grab them by running the following command either as root or the user that Asterisk runs under:

```
perl -MCPAN -e 'install <modulename>'
```

Replace *<modulename>* with one of the modules listed earlier. Run this command once for every module. If you have never run CPAN before, the script will prompt you for configuration options. The instructions are fairly straightforward, and the default settings work 99 percent of the time.

If all that is set, place the following script into your AGI directory, which is
/var/lib/asterisk/agi-bin by default.

```perl
#!/usr/bin/perl
# AGI Script that reads back e-mail from an IMAP account.
# Requires the Asterisk::AGI, Net::IMAP::Simple, and Email::Simple modules.

use Net::IMAP::Simple;
use Email::Simple;
use Asterisk::AGI;

my $server = '127.0.0.1'; #INSERT YOUR SERVER HERE
my $username = 'username'; #INSERT YOUR USERNAME HERE
my $password = 'password'; #INSERT YOUR PASSWORD HERE

$AGI = new Asterisk::AGI;

my %input = $AGI->ReadParse();

# Create the object
my $imap = Net::IMAP::Simple->new($server) ||
    die "Unable to connect to IMAP: $Net::IMAP::Simple::errstr\n";

# Log on
if(!$imap->login($username,$password)){
    $AGI->exec('Festival', 'Login failed ' . $imap->errstr);
    $AGI->verbose('Login Failed: ' . $imap->errstr, 1);
    exit(64);
}

# Retrieve all the messages in the INBOX
my $nm = $imap->select('INBOX');
$AGI->stream_file('vm-youhave');
$AGI->say_number($nm);
$AGI->stream_file('vm-messages');

for(my $i = 1; $i <= $nm; $i++){

    my $es = Email::Simple->new(join '', @{ $imap->top($i) } );

    $AGI->stream_file('vm-message');
```

```
$AGI->say_number($i);
$AGI->exec('Festival', $es->header('Subject'));
$AGI->stream_file('vm-from');
AGI->exec('Festival', ('From'));
while($input eq ''){
    $AGI->exec('Festival', "1, Play, 2, Next, Pound, Exit");
    my $input = chr($AGI->wait_for_digit('5000'));
    if ($input eq '1'){
        $AGI->exec('Festival', $es->body);
    }elsif($input eq '2'){
        next;
    }elsif($input eq '#'){
        exit;
    }else{
        $input = ''
    }
}
}

$imap->quit;
```

This script features many methods that have been discussed already in this chapter. It starts off by connecting to the IMAP server and logging in. It finds out how many messages are in the INBOX and then tells the user. From here, the script starts reading the messages, prompting the user to press 1 to read the message, press 2 to go to the next message, or press # to exit the script. It then continues to loop through every message until the user exits, or there are no more messages left.

After placing the script in the directory, make sure the script is executable by the user that the Asterisk process runs under. Then, open up your extensions.conf, which is the context you wish to make this script available to:

```
exten => 4627,n,AGI(imap.pl);
```

You may want to alter the extension, because that line puts it on extension 4627 ("IMAP"). You also might want to place an *Authenticate()* command before it as well since this script doesn't have any kind of password support.

Once you've adjusted your extensions.conf, open up the Asterisk CLI and execute a *reload* command. Now you should be ready to go.

Taking It for a Spin

The script can be accessed by dialing the extension you assigned it, in the context you put it in. If all goes well, you should hear the mechanical voice of Festival start reading your mail to you. If something isn't right, open up the Asterisk CLI and see if any errors are displayed on the console. As mentioned earlier, sometimes Festival doesn't play well with Asterisk and this causes the voice to sound like it is speaking in tongues and your console to start spitting out error messages repeatedly. Usually searching for these error messages on Google will show you how to solve whatever problem it is currently having.

Moving on from Here

This script has very basic functionality, allowing the user to only access their INBOX, and is limited to one user. This could easily be built upon to support a group and allow them to listen to their e-mail from their phone by adding an authentication system and the ability for users to manage their password and other settings. Support for multiple folders could also be added. This script is a fun weekend project just waiting to happen.

phpAGI

phpAGI is an AGI library designed for PHP. PHP started out as a Web-based language, but is slowly starting to creep into shell scripting as more and more people who cut their teeth learning the language start using it for shell work. phpAGI is available at http://phpagi.sourceforge.net/ and is maintained by a group of developers.

A Simple Program, Simplified with phpAGI

Let's look at the example program that is now rewritten with PHP and phpAGI:

```php
#!/usr/bin/php -qn
<?php
    require('phpagi.php'); #Use the phpAGI library

    $agi = new AGI(); #Create a new Asterisk::AGI object

    $agi->stream_file('auth-thankyou'); # "Thank You"
    # Say the phone number
    $agi->say_digits($agi->request['agi_callerid'],'');
```

```
        $agi->hangup($request['agi_channel']); # Hang up
?>
```

   phpAGI gives us another drastic reduction in code, even more than
Asterisk::AGI. phpAGI has a few advantages over Asterisk::AGI, one of them being
its powerful tex2wav function, which replaces executing the internal Festival application with a function that generates the text to a sound file, and then uses Asterisk's
playback system. This is somewhat more reliable and has benefits over scaling since
the sounds are cached in a temporary directory. However, if your script makes
Festival speak many different phrases, disk space could become an issue.

Example: Server Checker

Ingredients

- Asterisk

- PHP

- phpAGI

- Net_Ping PHP Extension and Application Repository (PEAR) module

- Festival TTS Engine, configured to work with Asterisk

Instructions

This is the same as the Asterisk::AGI program. All you need to do is make sure all
the correct modules are installed and that Festival is configured correctly to accept
incoming connections from the local host. If you have PEAR installed, installing the
module is done by running the following command as root:

```
pear install net_ping
```

   If you do not have PEAR installed, the Net_Ping module is available at
http://pear.php.net/package/Net_Ping. Download the package and unzip it in your
AGI directory, which is */var/lib/asterisk/agi-bin* by default.

   Next, download the phpAGI package, unzip it and copy the phpagi.php and
phpagi-asmanager.php files into the AGI directory as well. Also copy phpagi.conf
from the unzipped directory and place that into /etc/asterisk. This contains configuration values for the phpAGI environment.

Once that is all set, place the following script into your AGI directory:

```php
#!/usr/bin/php -qn
//
// AGI Script that ping servers defined in the $server array.
// Requires phpAGI and the Net_Ping PEAR module

<?php
    // Define the servers
    $servers = array (
        1  => array("name" => 'Dev Server',
             "ip" => '192.168.0.1'
             ),
        2  => array("name" => 'Production Server',
             "ip" => '192.168.0.99'
             ),
    );


    require('phpagi.php'); // Use the phpAGI library
    require ("Net/Ping.php"); // Use the Net_Ping PEAR library

    $agi = new AGI(); // Create a new Asterisk::AGI object

    foreach($servers as $server){ // For Every Server...
        $ping = Net_Ping::factory(); // Create a Net_Ping object

        if(!PEAR::isError($ping)){
            // Ping each server, then use Festival to
            // tell the user the status.
            $response = $ping->ping($server['ip']);
            $agi->verbose("moof: " . $response->_received);
            if ($response->_received == $response->_transmitted){
                $text = $server['name']  . " at " .
                $server['ip'] . " is O K";
            }elseif($response->_received == 0){
                $text = $server['name'] . " at "
                . $server['ip'] . " is down";
            }elseif($response->_received < $response->_transmitted){
                $text = $server['name'] . " at " .
                $server['ip'] . " has ping loss";
            }
```

```
    $agi->verbose($text, 1);
    $agi->text2wav($text);
}else{
    // If creating the object failed, tell the console
    $agi->verbose("PEAR Error",1);
}
}


$agi->hangup($request['agi_channel']); // Hang up


?>
```

The next step is to edit the *$servers* array with addresses that fit your network: The *name* variable in the array would be whatever you wanted to call the server, and the *addr* variable would be the server's hostname or IP address. Adding an extra server is easy as well, just adjust the *$servers* variable to this:

```
$servers = array (
    1  => array("name" => 'Dev Server',
         "addr" => '192.168.0.1'
         ),
    2  => array("name" => 'Production Server',
         "addr" => '192.168.0.99'
         ),
    3  => array("name" => 'Another Server',
         "addr" => '192.168.0.42'
         )
);
```

Notice how the comma was added after the second element. You can keep adding servers this way until you have all the servers you want to keep track of listed.

Once the script has been edited, make sure that the script is executable. You may want to run it through your PHP interpreter just to make sure you didn't add any syntax errors when you edited it. If everything checks out, open up your extensions.conf and add the following to the context you wish to make this script available to:

```
exten => 7464,n,AGI(statuscheck.php);
```

If you don't like 7464 ("PING"), feel free to change it.

Once you've adjusted your extensions.conf, open up the Asterisk CLI and execute a *reload* command. You should now be ready to go.

Taking It for a Spin

The script can be accessed by dialing the extension you assigned it, in the context you put it in. The console should be pretty verbose with status messages from phpAGI. Once the script starts executing, Festival should tell you that the servers you listed are OK, down, or suffering packet loss.

Moving on from Here

This script is pretty simple as it stands. You could easily expand it to include other network stats or test for individual services. PEAR, although not as big as CPAN, has a pretty large array of code and has more than a few handy modules, so you don't need to reinvent the wheel.

Using Fast, Dead, and Extended AGIs

Now that we've covered AGIs, let's look at the three "special" variants of AGI used in Asterisk: FastAGIs, DeadAGIs, and EAGIs. Each of these is identical to AGIs and any application written for AGI will work on these "special" AGI types

FastAGI

All AGIs are equally fast, but FastAGI lets you host AGIs on a remote server in order to speed up the execution process. Rather than having one server control both the calls and the AGI execution, FastAGI allows you to offload the AGI scripts onto a separate server and have the other server do script execution.

FastAGI is an open protocol, so any language can implement it. Sadly, FastAGI use is less common than AGI, so the choices of languages for libraries are somewhat limited. FastAGI libraries do exist for Java, Python, Perl, and Erlang.

Setting Up a FastAGI Server with Asterisk::FastAGI

Asterisk has a module called Asterisk::FastAGI that automates much of the setup process of an AGI server. Throughout this example, we will be referring to two servers: the AGI server, which is the server that will be hosting the AGI script; and the Asterisk server, which will be handling the calls.

Starting with the AGI server, we must install Asterisk::FastAGI. The module is also available through CPAN, so you can grab it by running the following command either as root or as the user that Asterisk runs under:

```
perl -MCPAN -e 'install Asterisk::FastAGI'
```

This will run the CPAN module and install Asterisk::FastAGI.

Next, you need to create two files because of the way Asterisk::FastAGI is set up: a perl module that will contain the code to execute when the request AGI request comes in, and the server itself. The first file you should create is the module file. We'll use the "example script" we've used repeatedly in this chapter, but recoded to support FastAGI. Place this anywhere on the AGI server:

```perl
#!/usr/bin/perl
# fastcallerid.pm - Code portion of the simple Asterisk::FastAGI example
# that reads back Caller ID

package AGIExample;

use base 'Asterisk::FastAGI';

sub say_callerid {
   my $self = shift;
   my %input = $self->agi->ReadParse(); #Get the variables from Asterisk

   $self->agi->stream_file('auth-thankyou'); # "Thank You"
   $self->agi->say_digits($input{'callerid'}); # Say the phone number
   $self->agi->hangup(); # hang up

}


return 1;
```

Next, create the server script:

```perl
#!/usr/bin/perl
# fastcallerid.pl - Server portion of the simple Asterisk::FastAGI example
# that reads back Caller ID

use AGIExample;

AGIExample->run();
```

Next, run the server script:

```
perl ReallyFastAGI.pl
```

This should print out text similar to the following:

```
2007/03/18-21:07:45 AGIExample (type Asterisk::FastAGI) starting!
pid(1014)
Port Not Defined.  Defaulting to '20203'
Binding to TCP port 20203 on host *
Group Not Defined.  Defaulting to EGID '0 0'
User Not Defined.  Defaulting to EUID '0'
```

Pay attention to the port number, we'll need that in the next step.

Next, switch back to your Asterisk server and open up your extensions.conf and add the following to the context you wish to make this script available to:

```
exten => 3278,n,AGI(agi://<AGI Server Address>:20203/say_callerid));
```

Make sure you replace <AGI Server Address> with the AGI server's address. As always, you may want to alter the cutesy extension, 3278 ("FAST"), with something you like. Finally, open up the Asterisk CLI and issue a *reload* command.

After Asterisk reloads, dial up the extension and watch your console. Hopefully, you should hear your Caller ID being read back to you. Congratulations! You're running FastAGI!

This was obviously a trivial example, but FastAGI makes sense for applications that use heavy I/O or consume a lot of processor time. Rather than have an AGI script compete with Asterisk for CPU cycles, FastAGI lets you have a separate server handle the heavy processing while Asterisk handles the call load.

DeadAGI

DeadAGIs are AGIs that continue to function after the channel has hung up. As stated previously, Asterisk terminates the AGI when the *HANGUP* command is given or if the caller hangs up on the script, no questions asked. DeadAGIs continue to execute after the channel is in the Hung Up state. This is useful if you want to call the caller back at a number given to confirm it's their number, or if you just want the script to do some additional cleanup before executing.

Using DeadAGI is easy. Let's say we wanted to use the IMAP by Phone script as a DeadAGI rather than an AGI. We would simply replace the existing AGI command:

```
exten => 4627,n,AGI(imap.pl);
```

with the DeadAGI command:

```
exten => 4627,n,DeadAGI(imap.pl);
```

It's that easy.

A word of warning, with DeadAGI, it is vitally important to make sure the script exits after a hang up, or else you may end up with processes waiting for a response that will never come, tying up server resources in the process. This can be an issue if you have a script that is used a lot.

EAGI

EAGI is identical to AGI, with the exception of an audio path on file descriptor 3. This can be useful if you want to record people interacting with your script for usability studies or to make sure the script is functioning properly.

For example, if we wanted to be nosey and listen to everyone using IMAP by Phone, we would replace the AGI command:

```
exten => 4627,n,AGI(imap.pl);
```

with the EAGI command:

```
exten => 4627,n,EAGI(imap.pl);
```

Then adjust the code to read the audio to file descriptor 3.

Checklist

- Make sure that if you are using the *System()* dial plan command, you have taken steps to mitigate the possible use of un-escaped data.
- Make sure that remote AGI scripts are coming from a trusted source.

Summary

One of Asterisk's greatest features is its ability to interact with other programs on the computer. This can be done in two main ways: through Asterisk's *System()* dial plan command and the Asterisk Gateway Interface.

Calling external applications from the dial plan is a quick and easy way to execute another application for Asterisk. The problem is that once this program is executed, Asterisk can no longer interact with the script. This severely limits both Asterisk's and the script's functionality, but is handy if you don't need to interact with either application once it is executed.

The Asterisk Gateway Interface is a powerful yet simple system that allows scripts to interact with callers through Asterisk. AGI is controlled via the standard Unix file descriptors STDIN and STDOUT, so almost any programming language can use AGI. AGIs can play audio files, get data from the caller via the telephone keypad, and do many other things. There are numerous ways to interact with the caller, both on the input side and the output side. Callers can interact with the AGI script via the telephone keypad or by speech recognition, and the AGI script can interact with the caller via recordings or text-to-speech programs.

AGI has become popular enough that numerous third-party libraries are available for use that automate most of the repetitive tasks of AGI programming. This is advantageous to the programmer since they can focus more on developing the application rather than having to interface with Asterisk. Libraries are available for almost every popular language. Two of the more popular libraries are Asterisk::AGI for Perl and phpAGI for PHP.

There are three "special" Asterisk AGI commands: FastAGI, DeadAGI, and EAGI. FastAGI lets you offload the AGI script onto a separate server and have Asterisk connect to it via a network connection. DeadAGI allows an AGI script to continue functioning after the channel is hung up. EAGI is identical to AGI, except that all audio is on a special file descriptor that the script can read from.

Solutions Fast Track

Calling Programs from within the Dial Plan

☑ Calling external programs from within the dial plan is the simplest way to execute a program using Asterisk.

☑ Once the program is forked, there is no way to control the execution of the program or interact with it in any other way.

☑ At one time, Asterisk had numerous add-ons that let you embed various programming languages directly in the dial plan; however, they do not support the newer versions of Asterisk.

Using the Asterisk Gateway Interface

☑ AGI lets Asterisk hand off the user to a script that will take control of playing prompts, listen for input, and do all the jobs the dial plan usually handles.

☑ AGI is supported by any programming language that can handle STDIN and STDOUT.

☑ In normal AGI operation, once a channel is hung up, the script will be terminated.

☑ You can get input from your caller to your script in two ways: Interactive Voice Response menus and Speech recognition. IVR menus are much easier to implement, but often have usability issues. Speech recognition is harder and will cost extra; however, it is generally easier to use from the user's standpoint.

☑ You can get output from your script to your users in two ways as well: recordings and text to speech. Recordings sound better, but are fairly limited as to what they can say. Text to speech can sound less life-like, but it can say text that is dynamic.

Using Third-Party AGI Libraries

☑ Third-party AGI libraries automate most of the repetitive tasks of AGI programming, allowing the programmer to focus more on the application rather than the interface with Asterisk.

☑ Libraries exist for almost every popular language today: C, Perl, PHP, Java, Python, C#, and shell scripting.

☑ Two of the more popular ones—Asterisk::AGI for Perl and phpAGI for PHP—are commonly used in Asterisk today.

Using Fast, Dead, and Extended AGIs

☑ FastAGI allows you to host AGIs on a separate server in order to save overhead in executing the scripts on the same server that is handling the calls.

☑ DeadAGIs let you continue to execute the script after the channel has gone into a hung up state.

☑ EAGIs allow you to record audio on the channel through a special file descriptor.

Links to Sites

■ www.perl.com/pub/a/2000/10/begperl1.html — perl.com's "Introduction to Perl." It's a bit old, but still on-topic.

■ http://user.it.uu.se/~matkin/documents/shell/ — A good guide on the basics of shell programming.

■ http://wls.wwco.com/ledsigns/ — Walt's LED Sign page, a great resource if you have a LED sign that you want to hook up to a computer.

■ www.digium.com/en/products/voice/ — Digium's IVR recording service.

■ http://asterisk.gnuinter.net/ — Asterisk::AGI homepage.

■ http://phpagi.sourceforge.net/ — phpAGI library homepage.

Frequently Asked Questions

The following Frequently Asked Questions, answered by the authors of this book, are designed to both measure your understanding of the concepts presented in this chapter and to assist you with real-life implementation of these concepts. To have your questions about this chapter answered by the author, browse to **www.syngress.com/solutions** and click on the **"Ask the Author"** form.

Q: What are my options for developing my own application with Asterisk?

A: Numerous options are available to you. You can use the Asterisk Gateway Interface, which allows you to interact with Asterisk and callers with an external application, or you can call an external application with the *System()* dial plan command, which will limit your ability to interact with the caller.

Q: What can I accomplish using the *System()* dial plan command?

A: Not much. Calling a program from the *System()* dial plan command allows you to fork a program from the dial plan. Other than that, it executes autonomously.

Q: What can I accomplish through the use of AGI?

A: AGI gives the ability to be fully interactive with the script. The caller can enter data, the script can act upon this data, and the script can be used to interact with external data.

Q: What programming languages does AGI support?

A: Almost anything that supports Unix file descriptors. AGI operates over STDIN, STDOUT, and STDERR. Any programming language that works on Unix/Linux should support these.

Q: How can I have my AGI call me back once I hang up?

A: Use the DeadAGI dial plan command rather than the AGI dial plan command. DeadAGI allows the script to continue executing past hang up.

Q: Are there any libraries for AGI?

A: Yes. Libraries are available for almost every popular language today: C, Perl, PHP, Java, Python, C#, and shell scripting. Everyone has their favorite pet language, so there is a choice here for most everyone.

## Chapter 5

Understanding and Taking Advantage of VoIP Protocols

Solutions in this chapter:

- **Understanding the Basic Core of VoIP Protocols**

- **How Compression in VoIP Works**

- **Signaling Protocols**

☑ Summary

☑ Solutions Fast Track

☑ Frequently Asked Questions

Introduction

Understanding how to install and configure Asterisk is important, but for the "hacking" side, it's also important to understand the "core" of how VoIP works. This doesn't only deal with Asterisk, but VoIP in general. Asterisk uses a standard set of protocols to communicate with remote systems—be it Asterisk or other types of VoIP systems and hardware.

Knowing how these VoIP protocols function will not only give you a clear picture of how Asterisk deals with VoIP, but show you how other systems work as well. Many VoIP systems deal with standardized protocols for interoperability.

Your Voice to Data

In order for your voice to travel across the wires, routers, and "tubes" of the Internet (as Senator Ted Stevens so amusingly put it), several conversions and protocols are used. The back-end protocol for SIP and H.323, the one where your voice is actually stored in data packets, is known as the Real Time Protocol, or RTP.

Other protocols are used to get your call from one side of the Internet to the other. These are known as "signaling" protocols. We'll discuss these protocols later, but it's important to understand how and why RTP is used to transfer your voice. RTP uses the User Datagram Protocol (UDP), which is part of the TCP/IP suite.

Upon first glance, UDP may sound like a terrible thing to use if you're not familiar with it. It is a stateless protocol, which means UDP doesn't offer any guarantee the packet will even make it to its destination. It also doesn't guarantee the order in which the packet will be received after it's sent. This reduces the size of the packet "headers," which describe how the packet should get to its destination. Within the UDP header, all that is sent is the length, source, destination, and port numbers. The actual data is stored in what is known as a UDP datagram. This is where the short snippets of your digitized voice or other data are stored.

Since UDP is stateless and can be broken down into small packets, the bandwidth and timing overhead is low—which is a good thing. Let's now compare this to using TCP for VoIP. TCP provides verification on packet delivery and the order it was received. If a TCP packet is "out of order," it simply reassembles it in the correct order. Though this sounds like a good idea, it actually causes some problems in real-time/time-sensitive applications like VoIP. For example, with TCP, if a packet is "dropped," the packet will be re-sent at the receiver's request. Considering that we are dealing with real-time VoIP, by the time the TCP packet with our voice snippet is retransmitted, it's too late to put it into our audio stream! Minor network issues could render a VoIP conversation useless due to retransmissions and the reordering of packets.

Since UDP doesn't ensure packet delivery or their order, if there's a minor network "hiccup," the VoIP stream can recover. Thus, you might notice a minor "skip" or "chop" in a conversation, but it may still be able to recover. Basically, if a UDP packet is sent and it makes it, it makes it. Otherwise, it might be discarded and the conversation will continue with minor interruptions. If TCP was used, however, your conversation might never recover since TCP attempts to resequence and resend packets.

RTP/UDP is only part of the overall picture of how VoIP works. It'll place snippets of your voice within a datagram and get it across the Internet, but it doesn't help you place a call to your intended target. That's where other "signaling" protocols, like SIP, come into play.

Making Your Voice Smaller

When the "audio" data of a VoIP call is placed into an RTP packet, a codec (enCOder/DECoder) is used. This is the method of how the "audio" data is placed within the UDP datagram. Information about what codec to use is between the systems and is negotiated during the call setup (signaling).

Some codecs use compression, while others do not. Compressed codecs will be able to cram more data into the RTP packet, but there is always a trade-off. With compressed codecs, your CPU will work harder at cramming data into the UDP datagram. You'll also lose a bit of quality in the audio. However, less network bandwidth will be used to send the information. With noncompressed codecs, the audio data will be placed in the UDP datagram in a more "raw"-like form. This requires much less CPU time, but necessitates more network bandwidth. There's always a trade-off of CPU power versus bandwidth when using compressed and noncompressed codecs.

Currently, Asterisk supports ADPCM (Adaptive Differential Pulse Code Modulation), G.711 (A-Law and μ-Law), G.723.1 (pass through), G.729, G.729, GSM, iLBC, Linear, LPC-10, and Speex. G.711 is a commonly used uncompressed codec. Within the United States, G.711 u-law (pronounced *mu-law*—the "u" is greek) is typically used. In Europe and elsewhere, G.711 a-law is used. G.711 creates a 64-kbit/second stream that is sampled at a fairly standard 8kHz. This means, the

CPU doesn't have to work very hard encoding/decoding the RTP packets, but for each channel/call, 64 kbit/second will be used. This could be a problem if you're limited on bandwidth by your provider and wish to make several calls simultaneously.

For example, some DSL providers will limit your upstream bandwidth. If you're making several concurrent calls at one time, you might run into problems. In these situations, increasing your bandwidth or using a codec that employs compression might be a good idea. G.729 does an excellent job at compressing the data. When using G.729, rather than creating a 64-kbit/second stream, utilizing compression will reduce bandwidth usage to 8 kbit/second. The trade-off is that your CPU will be working harder per channel to compress that data. The CPU usage might limit you to the number of calls you can place, and the call quality won't be as good since you're using a compressed codec. In some situations, the quality loss might not be a huge issue. Typical person-to-person conversations might be fine, but with applications like "music on hold," compression might introduce slight chops.

It should be noted that in order to use the G.729 commercial environment, proper licensing is required. It can be used without licensing in noncommercial environments. For noncommercial usage, check out www.readytechnology.co.uk/open/ipp-codecs-g729-g723.1.

A popular, more "open," compressed codec is GSM. While it doesn't accomplish the same compression as G.729, it does a good job in trading bandwidth for compression. It's also free to use in both commercial and noncommercial environments. Quality ranges with different codecs. For example, LPC10 makes you sound like a robot but tightly compresses the data. Plus, it's important to understand codecs since some providers only support certain kinds. It's also important to be knowledgeable in this area during certain types of attacks.

Session Initiation Protocol

At this time, Session Initiation Protocol (SIP) is probably the most commonly used VoIP signaling protocol. SIP does nothing more than set up, tear down, or modify connections in which RTP can transfer the audio data. SIP was designed by Henning Schulzrinne (Columbia University) and Mark Handley (University College of London) in 1996. Since that time, it's gone through several changes. SIP is a lightweight protocol and is similar in many ways to HTTP (Hyper-Text Transport Protocol). Like HTTP, SIP is completely text-based. This makes debugging easy and reduces the complexity of the protocol. To illustrate SIP's simplicity, let's use HTTP "conversation" as an example.

At your workstation, fire up your favorite Web browser. In the URL field, type **http://www.syngress.com/Help/Press/press.cfm**. Several things happen between your Web browser and the Syngress Web server. First off, your local machine does a DNS (Domain Name Service) lookup of www.syngress.com. This will return an IP address of the Syngress Web server. With this IP address, your browser and computer know how to "contact" the Syngress Web server. The browser then makes a connection on TCP port 80 to the Syngress Web server. Once the connection is made, your Web browser will send a request to "GET" the "/Help/Press/press.cfm" file. The Syngress Web server will respond with a "200 OK" and dump the HTML (Hyper-Text Markup Language) to your Web browser and it'll be displayed. However, let's assume for a moment that the "press.cfm" doesn't exist. In that case, the Syngress Web server will send to your browser a "404 Not Found" Message. Or, let's assume that Syngress decided to move the "press.cfm" to another location. In that case, your Web browser might receive a "301 Moved Permanently" message from Syngress's Web server, and then redirect you to the new location of that file.

The 200, 404, and 301 are known as "status codes" in the HTTP world. Granted, the HTTP example is a very basic breakdown, but this is exactly how SIP works. When you call someone via SIP, the commands sent are known as "SIP Methods." These SIP methods are similar to your browser sending the *GET* command to a remote Web server. Typically, these SIP methods are sent on TCP port 5060. See Table 5.1.

Table 5.1 SIP Methods

INVITE	Invite a person to a call.
ACK	Acknowledgment. These are used in conjunction with INVITE messages.
BYE	Terminates a request
CANCEL	Requests information about the remote server. For example, "what codecs do you support?"
OPTIONS	This "registers" you to the remote server. This is typically used if your connection is DHCP or dynamic. It's a method for the remote system to "follow you" as your IP address changes or you move from location to location.

Continued

Table 5.1 continued SIP Methods

REGISTER	This "registers" you to the remote server. This is typically used if your connection is DHCP or dynamic. It's a method for the remote system to "follow you" as your IP address changes or you move from location to location.
INFO	This gives information about the current call. For example, when "out-of-band" DTMF is used, the INFO method is used to transmit what keys where pressed. It can also be used to transmit other information (Images, for example).

As stated before, response codes are similar and extend the form of HTTP/1.1 response codes used by Web servers. A basic rundown of response codes is shown in Table 5.2.

Table 5.2 Response Codes

Code	Definition
100	Trying
180	Ringing
181	Call is being forwarded
182	Queued
183	Session in progress
200	OK
202	Accepted: Used for referrals
300	Multiple choices
301	Moved permanently
302	Moved temporarily
305	Use proxy
380	Alternate service
400	Bad request
401	Unauthorized: Used only by registrars. Proxies should use Proxy authorization 407.
402	Payment required (reserved for future use)
403	Forbidden
404	Not found (User not found)
405	Method not allowed

Continued

Table 5.2 continued Response Codes

Code	Definition
406	Not acceptable
407	Proxy authentication required
408	Request timeout (could not find the user in time)
410	Gone (the user existed once, but is not available here any more)
413	Request entity too large
414	Request-URI too long
415	Unsupported media type
416	Unsupported URI scheme
420	Bad extension (Bad SIP protocol extension used. Not understood by the server.)
421	Extension required
423	Interval too brief
480	Temporarily unavailable
481	Call/transaction does not exist
482	Loop detected
483	Too many hops
484	Address incomplete
485	Ambiguous
486	Busy here
487	Request terminated
488	Not acceptable here
491	Request pending
493	Undecipherable (could not decrypt S/MIME body part)
500	Server internal error
501	Not implemented (The SIP request method is not implemented here.)
502	Bad gateway
503	Service unavailable
504	Server timeout
505	Version not supported (The server does not support this version of the SIP protocol.)

Continued

Table 5.2 continued Response Codes

Code	Definition
513	Message too large
600	Busy everywhere
603	Decline
604	Does not exist anywhere
606	Not acceptable

Intra-Asterisk eXchange (IAX2)

Inter-Asterisk eXchange (IAX) is a peer-to-peer protocol developed by the lead Asterisk developer, Mark Spencer. Today, when people refer to IAX (pronounced *eeks*), they most likely mean IAX2, which is version 2 of the IAX protocol. The original IAX protocol has since been depreciated for IAX2. As the name implies, IAX2 is another means to transfer voice and other data from Asterisk to Asterisk. The protocol has gained some popularity, and now devices outside of Asterisk's software support the IAX2 protocol.

The idea behind IAX2 was simple: build from the ground up a protocol that is full featured and simple. Unlike SIP, IAX2 uses one UDP port for both signaling and media transfer. The default UDP port is 4569 and is used for both the destination port and the source port as well. This means signaling for call setup, tear down, and modification, along with the UDP datagrams, are all sent over the same port using a single protocol. It's sort of like two protocols combined into one! This also means that IAX2 has its own built-in means of transferring voice data, so RTP is not used.

When IAX was being designed, there where many problems with SIP in NAT (Network Address Translation) environments. With SIP, you had signaling happening on one port (typically TCP port 5060) and RTP being sent over any number of UDP ports. This confused NAT devices and firewalls, and SIP proxies had to be developed. Since all communications to and from the VoIP server or devices happen over one port, using one protocol for both signaling and voice data, IAX2 could easily work in just about any environment without confusing firewalls or NAT-enabled routers.

This alone is pretty nifty stuff, but it doesn't stop there! IAX2 also employs various ways to reduce the amount of bandwidth needed in order to operate. It uses a different approach when signaling for call setup, tear down, or modification. Unlike

SIP's easy-to-understand almost HTTP-like commands (methods) and responses, IAX2 uses a "binary" approach. Whereas SIP sends almost standard "text" type commands and response, IAX2 opted to use smaller binary "codes." This reduces the size of signaling.

To further reduce bandwidth usage, "trucking" was introduced into the protocol. When "trunking" is enabled (in the iax.conf, "trucking=yes"), multiple calls can be combined into single packets. What does this mean? Let's assume an office has four calls going on at one time. For each call, VoIP packets are sent across the network with the "header" information. Within this header is information about the source, destination, timing, and so on. With trunking, *one packet* can be used to transfer header information about all the concurrent calls. Since you don't need to send four packets with header information about the four calls, you're knocking down the transmission of header data from 4 to 1. This might not sound like much, but in VoIP networks that tend to have a large amount of concurrent calls, trunking can add up to big bandwidth savings.

IAX2 also supports built-in support for encryption. It uses an AES (Advanced Encryption Standard) 128-bit block cipher. The protocol is built upon a "shared secret" type of setup. That is, before any calls can be encrypted, the "shared secret" must be stored on each Asterisk server. IAX2's AES 128-bit encryption works on a call-by-call basis and only the data portion of the message is encrypted.

Getting in the Thick of IAX2

As mentioned before, IAX2 doesn't use RTP packets like SIP. Both the signaling and audio or video data is transferred via UDP packets on the default port 4569. In the iax.conf file, the port can be altered by changing the "bindport=4569" option; however, you'll probably never need to change this. In order to accomplish both signaling and stuffing packets with the audio data of a call, IAX2 uses two different "frame" types. Both frame types are UDP, but used for different purposes.

"Full Frames" are used for "reliable" information transfer. This means that when a full frame is sent, it expects an ACK (acknowledgment) back from the target. This is useful for things like call setup, tear down, and registration. For example, when a call is made with IAX2, a full frame requesting a "NEW" call is sent to the remote Asterisk server. The remote Asterisk server then sends an ACK, which tells the sending system the command was received.

With Wireshark, full frame/call setup looks like the following:

```
4.270389 10.220.0.50 -> 10.220.0.1 IAX2 IAX, source call# 2, timestamp 17ms NEW
4.320787 10.220.0.1 ->  10.220.0.50 IAX2 IAX, source call# 1, timestamp 17ms ACK
4.321155 10.220.0.1 ->  10.220.0.50 IAX2 IAX, source call# 1, timestamp 4ms ACCEPT
4.321864 10.220.0.50 -> 10.220.0.1 IAX2 IAX, source call# 2, timestamp 4ms ACK
```

Full frames are also used for sending other information such as caller ID, billing information, codec preferences, and other data. Basically, anything that requires an ACK after a command is sent will use full frames. The other frame type is known as a "Mini Frame." Unlike the Full Frame, the Mini Frame requires no acknowledgment. This is an unreliable means of data transport, and like RTP, either it gets there or it doesn't. Mini Frames are not used for control or signaling data, but are actually the UDP datagram that contains the audio packets of the call. Overall, it works similar to RTP, in that it is a low overhead UDP packet. A Mini Frame only contains an F bit to specify whether it's a Full or Mini Frame (F bit set to 0 == Mini Frame), the source call number, time stamp, and the actual data. The time stamps are used to reorder the packets in the correct order since they might be received out of order.

Capturing the VoIP Data

Now that you understand what's going on "behind the scenes" with VoIP, this information can be used to assist with debugging and capturing information.

Using Wireshark

Wireshark is a "free" piece of software that is used to help debug network issues. It's sometimes referred to as a "packet sniffer," but actually does much more than simple packet sniffing. It can be used to debug network issues, analyze network traffic, and assist with protocol development. It's a powerful piece of software that can be used in many different ways. Wireshark is released under the GNU General Public License.

In some ways, Wireshark is similar to the tcpdump program shipped with many different Unix-type operating systems. tcpdump is also used for protocol analysis, debugging, and sniffing the network. However, tcpdump gives only a text front-end display to your network traffic. Wireshark comes with not only the text front-end, but a GUI as well. The GUI layout can assist in sorting through different types of packet data and refining the way you look at that data going through your network.

While tcpdump is a powerful utility, Wireshark is a bit more refined on picking up "types" of traffic. For example, if a SIP-based VoIP call is made and analyzed with tcpdump, it'll simply show up as UDP traffic. Wireshark can see the same traffic and "understand" that it's SIP RTP/UDP traffic. This makes it a bit more powerful in seeing what the traffic is being used for.

Both tcpdump and Wireshark use the "pcap" library for capturing data. pcap is a standardized way to "capture" data off a network so it can be used between multiple applications. PCAP (libpcap) is a system library that allows developers not to worry about how to get the network packet information off the "wire," and allows them to make simple function calls to grab it.

We'll be using pcap files. These are essentially snapshots of the network traffic. They include all the data we'll need to reassemble what was going on in the network at the time. The nice thing with pcap dump files is that you can take a snapshot of what the network was doing at the time, and transfer it back to your local machine for later analysis. This is what we'll be focusing on. The reason is, while Wireshark might have a nice GUI for capturing traffic, this doesn't help you use it with remote systems.

Unfortunately, not all pcap files are the same. While Wireshark can read tcpdump-based pcap network files, characteristics of that traffic might be lost. For example, if you create a pcap file of SIP RTP traffic with tcpdump and then transfer that "dump" back to your computer for further analysis, tcpdump will have saved that traffic as standard UDP traffic. If created with Wireshark, the pcap files a "note" that the traffic is indeed UDP traffic, but that it's being used for VoIP (SIP/RTP).

As of this writing, Wireshark can only understand SIP-based traffic using the G.711 codec (both ulaw and alaw). The audio traffic of a VoIP call can be captured in two different ways. In order to capture it, you must be in the middle of the VoIP traffic, unless using arp poisoning. You can only capture data on your LAN (or WAN) if you are somehow in line with the flow of the VoIP traffic. For these examples, we'll be using the command-line interface of Wireshark to capture the traffic. The reason for this is that in some situations you might not have access to a GUI on a remote system. In cases like this, the text-only interface of Wireshark is ideal. You'll be able to fire up Wireshark (via the command *tethereal* or *twireshark*) and store all the data into a pcap file which you can then download to your local system for analysis.

To start off, let's create an example pcap file. In order to capture the traffic, log in to the system you wish to use that's in line with the VoIP connection. You'll need "root" access to the system, because we'll be "sniffing" the wire. We'll need more than normal user access to the machine to put the network interface in promiscuous mode. Only "root" has that ability.

Once the network device is in promiscuous mode, we can capture all the network traffic we want. Running Wireshark as "root" will automatically do this for us. To begin capturing, type

```
# tethereal -i {interface} -w {output file}
```

So, for example, you might type

```
# tethereal -i eth0 -w cisco-voip-traffic.pcap
```

Unfortunately, this won't only capture the VoIP traffic but everything else that might pass through the *eth0* interface. This could include ARP requests, HTTP, FTP, and whatever else might be on the network. Fortunately, the tethereal program with Wireshark works on the same concept as tcpdump. You can set up "filters" to grab only the traffic you want. So, if we know our VoIP phone has an IP address of 192.168.0.5, we can limit what we grab by doing the following:

```
# tetheral -i eth0 -w cisco-voip-traffic.cap host 192.168.0.5
```

Once fired up, you should then see *Capturing on eth0*. As packets are received, a counter is displayed with the number of packets recorded. You can further reduce the traffic by using tcpdump type filters. Depending on the amount of calls, we might need to let this run for a while.

Extracting the VoIP Data with Wireshark (Method # 1)

Once you've captured the data, you'll need to get it to a workstation so you can do further analysis on it. This may require you transferring it from the target system where you created the pcap file to your local workstation. Once you have the data in hand, start Wireshark and load the pcap. To do this, type

```
$ wireshark {pcap file name}
```

You'll no longer need to be "root" since you won't be messing with any network interfaces and will simply be reading from a file. Once started and past the Wireshark splash screen, you'll be greeted with a screen similar to Figure 5.1

Figure 5.1 pcap Wireshark

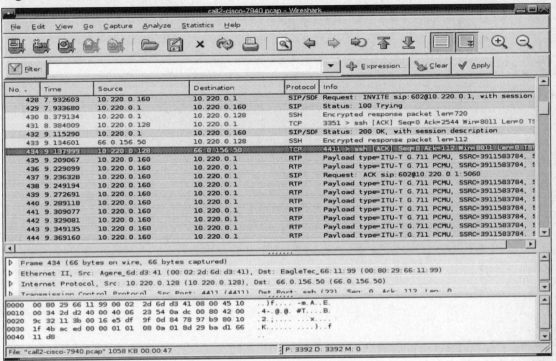

If you look closely at the example screen, you'll notice things like "SSH" traffic. We now need to filter out all the unwanted traffic, since we're only interested in UDP/RTP/VoIP traffic. So, the first thing we need to do is "Filter" the traffic. Note the "Filter" option at the top left-hand corner. This allows you to enter the criteria used to filter the packet dump. For example, you could enter "tcp" in this field and it'll only show you the TCP packets. In this case, we'll filter by **RTP**, as shown in Figure 5.2.

After entering **RTP** and clicking the **Apply** button, Wireshark removes all other TCP/IP packet types and only leaves you with RTP (UDP) type packets. In this case, the Source of 10.220.0.160 is my Cisco 7940 IP phone using the SIP. The Destination is my in in-house Asterisk server. Also notice the Info field. This tells us the payload type of the RTP packet. In this case, it's G.711.

Figure 5.2 Filter by RTP

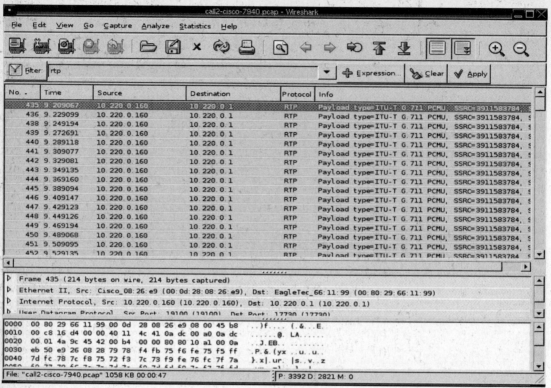

Now that we're only looking at RTP-type packets, this might be a good time to browse what's left in our filtered packet dump. Wireshark will also record what phone pad buttons (DTMF) were pressed during the VoIP session. This can lead to information like discovering what the voicemail passwords and other pass codes are that the target might be using.

To get the audio of the VoIP conversation, we can now use Wireshark's "RTP Stream Analysis." To do this, select **Statistics | RTP | Stream Analysis**.

Afterward, you should see a screen similar to Figure 5.3.

Figure 5.3 Wireshark RTP Stream Analysis

Packet .	Sequence	Delta (ms)	Jitter (ms)	IP BW (kbps)	Marker	Status
435	4257	0.00	0.00	1.60	SET	[Ok]
436	4258	20.03	0.00	3.20		[Ok]
438	4259	20.10	0.01	4.80		[Ok]
439	4260	23.50	0.23	6.40		[Ok]
440	4261	16.43	0.44	8.00		[Ok]
441	4262	19.96	0.41	9.60		[Ok]
442	4263	20.00	0.39	11.20		[Ok]
443	4264	20.05	0.36	12.80		[Ok]
444	4265	20.02	0.34	14.40		[Ok]
445	4266	19.93	0.33	16.00		[Ok]
446	4267	20.05	0.31	17.60		[Ok]
447	4268	19.98	0.29	19.20		[Ok]
448	4269	20.00	0.27	20.80		[Ok]
449	4270	20.07	0.26	22.40		[Ok]
450	4271	19.87	0.25	24.00		[Ok]

Analysing stream from 10.220.0.160 port 19100 to 10.220.0.1 port 17730 SSRC = 3911583784

Max delta = 0.023497 sec at packet no. 439
Total RTP packets = 1558 (expected 1554) Lost RTP packets = -4 (-0.26%) Sequence errors = 4

Save payload...   Save as CSV...   Refresh   Jump to   Graph   Next non-Ok   Close

From here, it's as simple as selecting **Save Payload**. You should then be greeted with a menu that looks similar to Figure 5.4.

Figure 5.4 Wireshark Save Payload

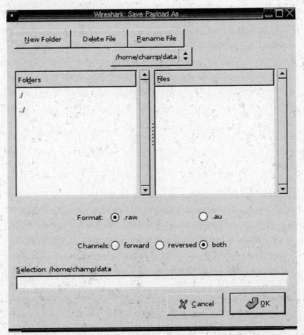

Before saving, look at the "Format" radio box. I typically change this from ."raw" (the default) to ."au." The reason is because it's an older audio format for Unix that was produced by Sun Microsystems. Conversion from the .au format to other formats is trivial and well supported. Under the "Channels" field, you'll probably want to leave this set to "both." With "both" enabled, you'll save the call as it was recorded with both sides of the conversation. The "forward" and "reversed" allows you to save particular channels of the conversation. This might be useful in certain situations, but most of the time you'll probably want the full conversation recorded to the .au file as it happened.

Once your .au file is recorded, conversion to other formats is trivial; using sound utilities like "sox" (http://sox.sourceforge.net/) is trivial. At the Unix command line with "sox" installed, you'd type: **sox {input}.au {output}.wav**.

Extracting the VoIP Data with Wireshark (Method # 2)

As of Wireshark version 0.99.5, VoIP support has improved a bit and will probably get even better. Wireshark versions before 0.99.5 do not contain this method of extracting and playing the VoIP packet dump contents.

To get started, we once again load Wireshark with our pcap file:

```
$ wireshark {pcap file name}
```

After the Wireshark splash screen, you'll be greeted with a screen similar to that from Figure 5.1. This time, the menu options we'll select are **Statistics | VoIP Call**.

Unlike before, we won't need to filter by "RTP" packet. Wireshark will go through the packet dump and pull out the VoIP-related packets we need. You should see a screen similar to Figure 5.5.

In this example, the packet dump contains only one VoIP call. Like the previous example, this packet dump is from my Cisco 7940 VoIP phone using SIP (10.220.0.160) to my Asterisk server (10.220.0.1). If multiple calls were present, this screen would show each call. Since there is only one call, we'll select that one. Once chosen, the Player button should become available. Upon selecting the Player button, you should see something similar to Figure 5.6.

Figure 5.5 VoIP Calls Packet

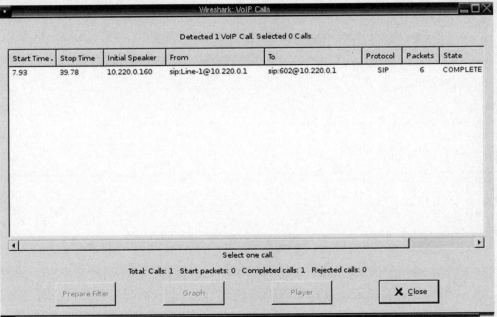

Figure 5.6 Wireshark RTP Player

Select the Decode button. The Wireshark RTP player should then appear, looking something like the one in Figure 5.7.

Figure 5.7 Decoded Wireshark RTP Player

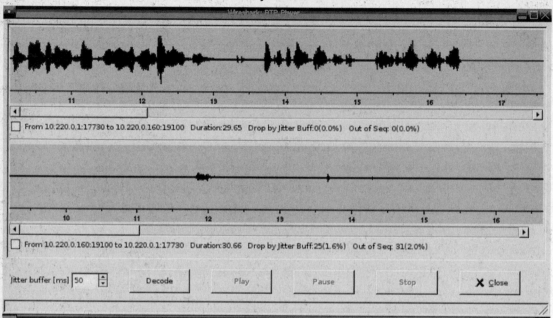

From here, select the stream to listen to and then click Play. It's as easy as that. The only disadvantage at this time is that you can't save the audio out to a file. That'll probably change as Wireshark supports more VoIP options.

This method also has a nice "Graph" feature, which breaks down the call into a nice, simple format. To use this, we perform the same steps to get to the Player button, but rather than selecting Player, we click Graph. Clicking the Graph button should generate a screen similar to that in Figure 5.8.

This breaks down the VoIP communication data. Note that timestamp 16.111 shows that the DTMF of "#" was sent. This type of information can be useful in determining what DTMF events happened. This can lead to revealing pass codes, voice-mail passwords, and other information.

Figure 5.8 VoIP Graph Analysis

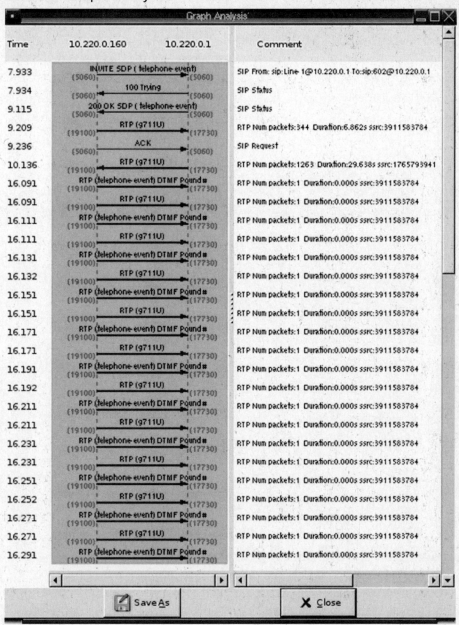

Getting VoIP Data by ARP Poisoning

ARP (Address Resolution Protocol) is used to located equipment within a LAN by the hardware MAC (Media Access Control). A MAC address is a preassigned to the network hardware. It uses a 48-bit address space, so there is plenty of room to grow. The 48-bit address space is expressed as 12 hexadecimal digits. The first six are assigned to the manufacture of the network device. For example, on my home Linux workstation, the Ethernet card MAC address is 00:04:61:9E:4A:56. Obtaining your MAC address depends on what operating system you're running. Under Linux, an *ifconfig -a* will display the various information, including your MAC address. On BSD-flavored systems, a *netstat -in* will usually do it. The output from my workstation is shown in Figure 5.9.

Figure 5.9 The MAC Address of the Author's Workstation

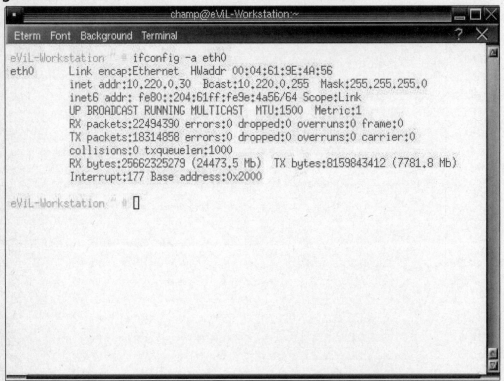

As you can see, the *HWaddr* field contains my MAC address. As stated earlier, the first six digits reveal the vendor of the hardware. So how do you determine the vendor, you ask? Well, it just so happens that the IEEE (Institute of Electrical and Electronics Engineers) maintains a list of vendors that is freely available at http://standards.ieee.org/regauth/oui/oui.txt. It is a flat ASCII text file of all vendors and their related MAC prefixes. So, looking up my MAC address in that list, we see that the 00:04:61 prefix belongs to:

```
0-04-61     (hex)                      EPOX Computer Co., Ltd.
000461      (base 16)        EPOX Computer Co., Ltd.
                             11F, #346, Chung San Rd.
                             Sec. 2, Chung Ho City, Taipei Hsien 235
                             TAIWAN TAIWAN R.O.C.
                             TAIWAN, REPUBLIC OF CHINA
```

This is the company that made my network card. While this is all interesting, you might wonder how it ties in to ARP address poisoning. Well, MAC addresses are unique among all networking hardware. With TCP/IP, the MAC address is directly associated with a TCP/IP network address. Without the association, TCP/IP packets have no way of determining how to get data from one network device to another. All computers on the network keep a listing of which MAC addresses are associated with which TCP/IP addresses. This is known as the systems ARP cache (or ARP table). To display your ARP cache, use *arp -an* in Linux, or *arp -en* in BSD-type systems. Both typically work under Linux, as shown in Figure 5.10.

Figure 5.10 Display of ARP Cache in Linux

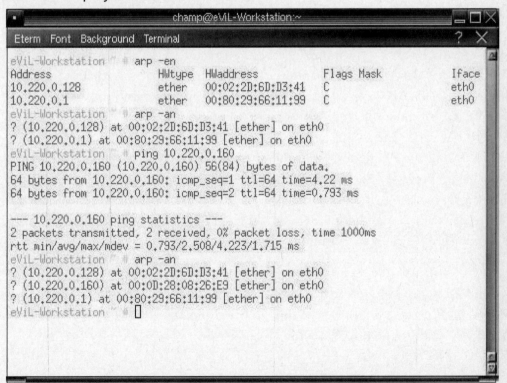

Notice that when I entered *arp -en* and *arp -an*, there were only two entries. Did you see what happened when I sent a *ping* request to my Cisco phone (10.220.0.160) and re-ran the *arp -an* command? It added the Cisco IP phone's MAC address into the ARP cache. To obtain this, my local workstation sent out what's known as an "ARP request." The ARP request is a network broadcast, meaning the request was sent networkwide. This is done because we don't know "where" 10.220.0.160 is. When an ARP request is sent, a packet is sent out saying "Who has 10.220.0.160?" networkwide. When 10.220.0.160 receives the ARP request, it replies "That's me. My MAC address is 00:0D:28:08:26:E9."

The following is a Wireshark dump of an ARP request and reply:

```
04:04:12.380388 arp who-has 10.220.0.160 tell 10.220.0.30
04:04:12.382889 arp reply 10.220.0.160 is-at 00:0d:28:08:26:e9
```

As you can see, this is literally what is happening! Now that both sides have their TCP/IP network addresses associated with the MAC, they can start transferring data. If 10.220.0.30 (my workstation) needs to talk to 10.220.0.160 (my Cisco IP phone), my workstation knows to send the data to the 00:0D:28:08:26:E9 MAC address, which is 10.220.0.160.

The underlying flaw with ARP is that in many cases it's very "trusting" and was never built with security in mind. The basic principle of ARP poisoning is to send an ARP reply to a target that never requested it. In most situations, the target will blindly update its ARP cache. Using my Cisco IP phone and Linux workstation as an example, I can send a spoofed ARP reply to a target with the Cisco IP phone's TCP/IP network address, but *with my workstation's MAC address*.

For this simple example, I'll use the arping2 utility (www.habets.pp.se/synscan/programs.php?prog=arping). This utility works much like the normal *ping* command but sends ARP requests and ARP replies. My target for this simple example will be my default route, which happens to be another Linux machine (10.220.0.1). The command I'll issue from my workstation (10.220.0.30) is

```
# arping2 -S 10.220.0.160 10.220.0.1
```

This -*S* option tells arping2 to spoof the address. So my Linux workstation will send an ARP request to 10.220.0.1 informing it that 10.220.0.160 is my workstation's MAC address. Figure 5.11 shows a screenshot from my Linux gateway.

Figure 5.11 Display from Author's Linux Gateway

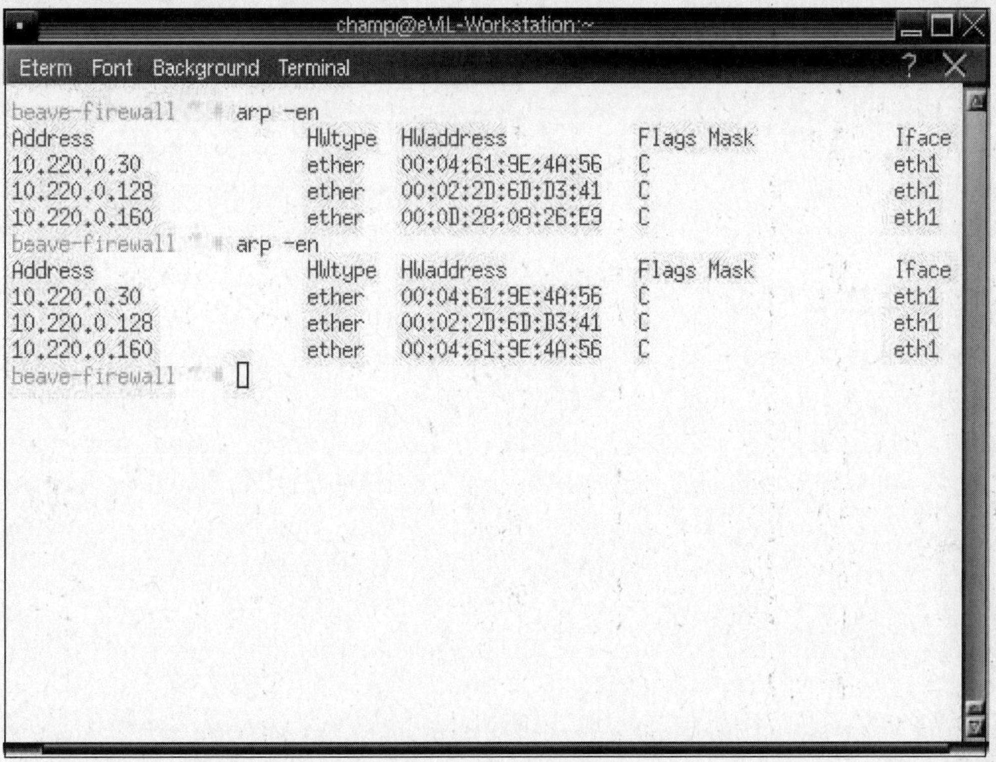

```
champi@eViL-Workstation:~                          _ □ X
 Eterm  Font  Background  Terminal                    ?  X
beave-firewall  # arp -en
Address          HWtype  HWaddress          Flags Mask      Iface
10.220.0.30      ether   00:04:61:9E:4A:56  C              eth1
10.220.0.128     ether   00:02:2D:6D:D3:41  C              eth1
10.220.0.160     ether   00:0D:28:08:26:E9  C              eth1
beave-firewall  # arp -en
Address          HWtype  HWaddress          Flags Mask      Iface
10.220.0.30      ether   00:04:61:9E:4A:56  C              eth1
10.220.0.128     ether   00:02:2D:6D:D3:41  C              eth1
10.220.0.160     ether   00:04:61:9E:4A:56  C              eth1
beave-firewall  # []
```

If you look closely at the first time I issue the *arp -en* command, the MAC
address is that of the Cisco IP phone (00:0D:28:08:26:E9). This is before the arping2
spoof command was issued. The second time *arp -en* is run is after I've spoofed with
arping2. You might have noticed that the *Hwaddress* has changed to my Linux workstation (00:04:61:9E:4A:56). Until the ARP tables get updated, whenever my Linux
gateway attempts to communicate with the Cisco phone, it'll actually be sending
packets to my workstation.

This basic example is not very useful other than in causing a very basic temporary DoS (Denial of Service). While I'll be receiving packets on behalf of the Cisco
IP phone, I won't be able to respond. This is where the Man-in-the-Middle attack
comes in.

Man in the Middle

A Man-in-the-Middle (MITM) attack is exactly what it sounds like. The idea is to
implement some attack by putting your computer directly in the flow of traffic. This
can be done in several ways, but we'll keep focused on ARP poisoning. To accom-

plish a MITM and capture all the VoIP traffic, we'll be ARP poisoning two hosts. The Cisco IP phone (10.220.0.160) and my gateway's ARP cache (10.220.0.1). I'll be doing the actual poisoning from my workstation (10.220.0.30), which is connected via a network switch and is not "in line" with the flow of VoIP traffic. Considering I have a network switch, I normally shouldn't see the actual flow of traffic between my Cisco phone and my gateway. Basically, my workstation should be "out of the loop." With a couple of nifty tools, we can change that.

Using Ettercap to ARP Poison

Ettercap is available at http://ettercap.sourceforge.net/. It primarily functions as a network sniffer (eavesdropper) and a MITM front end. It's a fairly simple utility that helps assist in grabbing traffic you shouldn't be seeing. Ettercap comes with a nice GTK (X Windows) interface, but we won't be focusing on that. We'll be looking more at the command line and ncurses interfaces. One nice thing about ettercap is that the curses interface is similar to the GUI, so moving from curses to GUI shouldn't be a hard transition.

I also don't want to focus on the GUI because many times your target might not be within your LAN. It's much easier to use the command line or curses menu when the network you're testing is remote. To kick things off, we'll look at the ncurses front end. In order to use Ettercap for sniffing and ARP poisoning purposes, you'll need to have "root" access. To start it up, type ***ettercap –curses***, and you should see something like Figure 5.12.

Figure 5.12 Ettercap Sniffing Startup

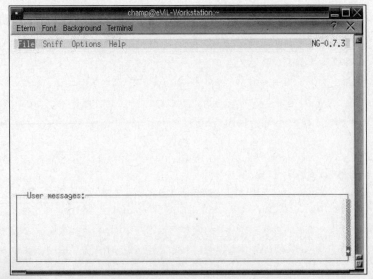

You'll want to store the data you've captured while sniffing, so you'll need to build a PCAP file you can later analyze. To do this, press **Shift + F**. Notice that the curses menu options are almost always the **Shift** key and the first letter of the menu option. To get more information about Ettercap's menu function, see the Help (**Shift + H**) options shown in Figure 5.13.

Figure 5.13 Help Option for Ettercap's Menu Function

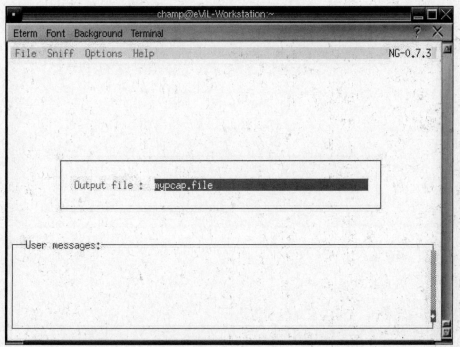

Type in the filename you wish to store the PCAP file as and press **Enter**. You'll now want to start sniffing the network. To do this, press **Shift + S** for the Sniff menu option, shown in Figure 5.14.

Figure 5.14 The Sniff Menu Option

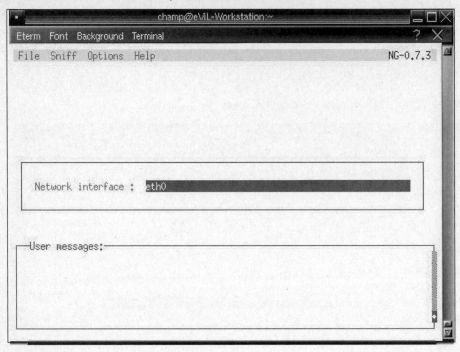

It'll now ask you which Ethernet device to use. Enter the device and press
Enter. The screen should change and look something like Figure 5.15.

Figure 5.15 Ethernet Device Selection

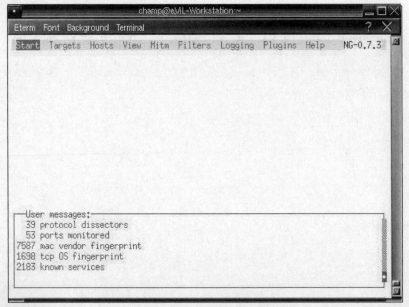

Press **Shift + H** to select the hosts in your network. The easiest way to populate this list is to choose **Scan for hosts**. So, select this option, as shown in Figure 5.16.

Figure 5.16 Selecting Network Hosts

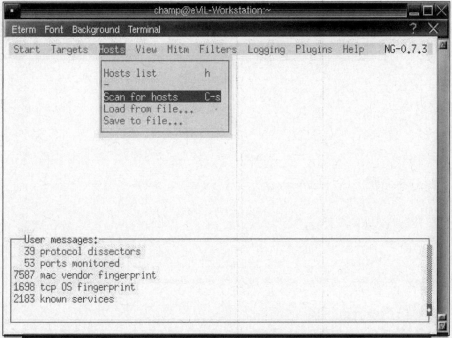

The way Ettercap scans for local network hosts is that it examines your network setup. In my case, I use a 10.220.0.0 network, with a netmask of 255.255.255.0. So, Ettercap sends out ARP requests for all hosts. In my case, 10.220.0.1 to 10.220.0.255. Ettercap stores all these responses in a "host list." My host list looks like Figure 5.17.

If you press the spacebar, it'll give you a little help, as shown in Figure 5.18.

Figure 5.17 Host List Displayed

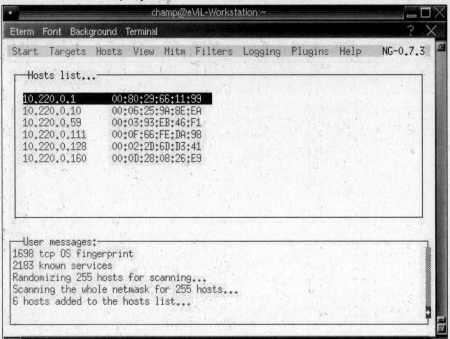

Figure 5.18 Help Shortcut List

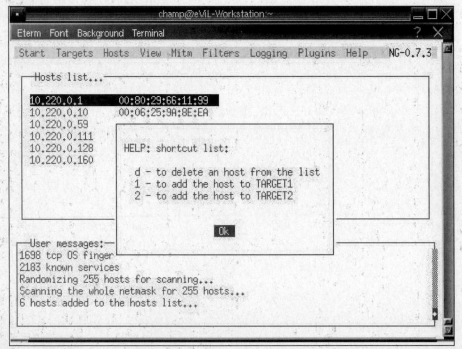

You can press enter to exit the Help screen. Once you exit the Help screen, you can use your up and down arrow keys to "mark" your target. To mark a target, use the **1** or **2** keys. In this example, I'm going to select 10.220.0.1 (my gateway) as Target 1, by pressing the numeric 1. I'll then add 10.220.0.160 (my Cisco IP phone) to the second target list by pressing the numeric 2, as shown in Figure 5.19

Figure 5.19 Target Selection

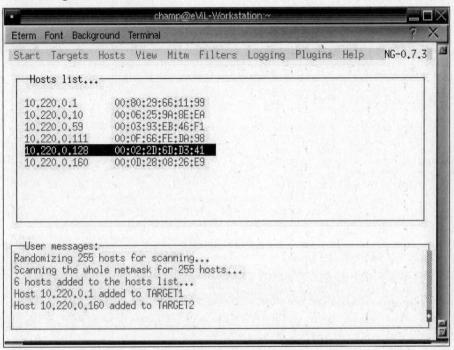

Note that when I select a target, in the User Messages section at the bottom of the screen it confirms my targets. Now that our targets are selected, you can double-check your target setup by pressing **Shift + T**, as shown in Figures 5.20A and 5.20B.

Figure 5.20A Target Setup Check

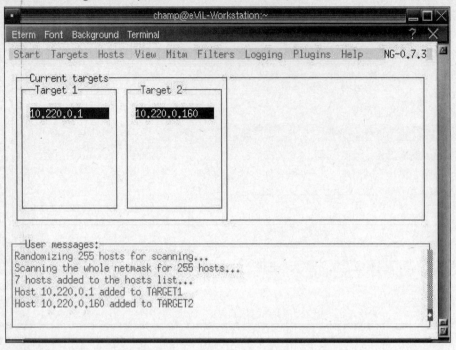

Figure 5.20B Target Setup Check

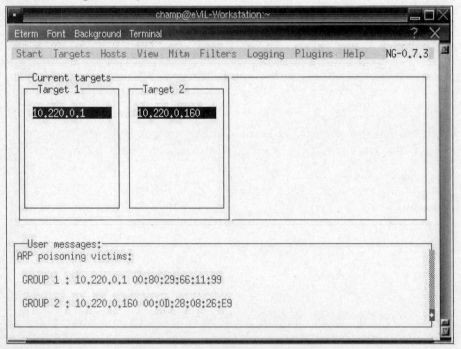

Now we're ready to set up the MITM attack. To do this, press **Shift + M** and select **ARP poisoning**, as shown in Figure 5.21.

Figure 5.21 MITM Attack Setup

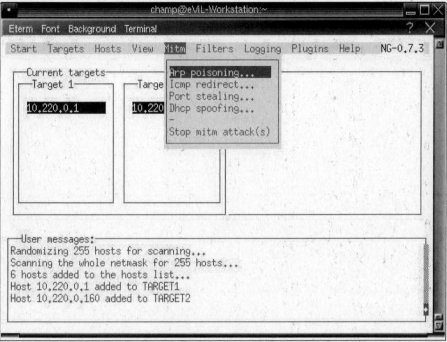

Once selected, it will prompt you for "Parameters." We want to do a full session sniffing MITM attack, so enter **remote** in this field. Now press **Enter**. You should see something like Figure 5.22.

Again, note the bottom of the screen. We are now ARP poisoning our targets and sniffing the traffic! Once you've let it run and feel that you've gotten the data you want, you can stop the MITM attack by pressing **Shift + M** (Stop MITM attack). This will re-ARP the targets back to what they originally were before the attack. You can then press **Shift + S** and select **Exit**. You should now have a PCAP file to analyze.

Figure 5.22 Parameters for Full Session Sniffing MITM Attack

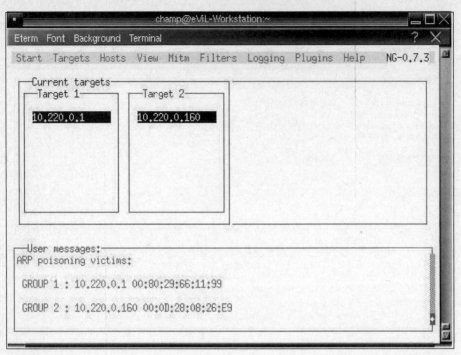

As you can see, the ncurses Ettercap interface is quite nice and powerful, but we can accomplish the exact same thing much easier! How could we possibly make it simpler? We can do all of the preceding in one simple command line. As "root," type

```
# ettercap -w my.pcap --text --mitm arp:remote /10.220.0.1/ /10.220.0.160/
```

That's all there is to it! The —*text* tells Ettercap we want to remain in a "text" mode.

We don't want anything fancy, just your basic good ol' text. The —*mitm* should be pretty obvious by now. The *arp:remote* option tells Ettercap we want to ARP poison the remote targets and we'd like to "sniff" the traffic. Once you capture the traffic, you can load it into something like Wireshark or Vomit and extract the SIP or H.323-based traffic.

Summary

Understanding how VoIP protocols function is important. This knowledge will help you debugging problems, assist in generating attacks in a security audit and help protect you against attacks targeting you Asterisk system. Like any other network protocols, there is no "magic" involved but a set of guidelines. These guidelines are covered in various RFC's (Request for Comments) and describe, in detail, how a protocol functions. Developers follow and use these RFC's to assist in development to help build applications. There are multiple RFCs covering various VoIP protocols. These describe how signaling works, how audio and video data is transferred and various other features. Reading and understanding these RFC's can help you unlock the "magic" of how VoIP works.

As shown in the chapter, two major functions with IAX2 and SIP is signaling and passing the audio/video data. Signaling handles the call build up, tear down and modification of the call. The two protocols handle passing the audio data and signaling differently. While SIP is a signaling protocol in itself and uses RTP to pass the audio/video data, IAX2 chose to build both into one protocol.

If you understand how the protocols work, building attacks becomes easier. For example, *fuzzing* or looking for flaws at the SIP level (typically TCP port 5060). If you know the SIP methods supported on a particular piece of SIP hardware, you can *probe* the target with bogus or invalid requests and see how it responds.

In conjunction with other hacking techniques, like ARP poisoning, you can perform man in the middle attacks. These types of attacks will not only let you grab the audio of a conversation, but other data as well. For example, authentication used between devices during the call build up or the DTMF used to authenticate with other devices. For example, voice mail.

Solutions Fast Track

Understanding the Core of VoIP Protocols

- ☑ VoIP data is transferred using small UDP packets.
- ☑ UDP is not time sensitive, which is good for VoIP.
- ☑ With SIP, these UDP packets are known as RTP packets. IAX2 uses a built in method known as mini-frames.

How Compression in VoIP Works

☑ Compression can further reduce the bandwidth needed for VoIP by compressing the UDP/VoIP packets.

☑ Compression uses more CPU time and less bandwidth. No compression uses more bandwidth but less CPU time.

☑ Compression codecs come in open and closed standards. For example, GSM and Speex is open, while G.729 requires licensing to use in corporate environments.

Signaling Protocols

☑ SIP is a signaling protocol used to setup/tear down/modification calls.

☑ SIP uses RTP (Real Time Protocol) packets for voice data. These are small UDP packets.

☑ The SIP protocol is similar to HTTP. This makes debugging easier, but requires a little bit more bandwidth.

☑ IAX2 has signaling and audio transfer built into one protocol. Unlike SIP, IAX2 does signaling via binary commands, which uses less bandwidth. VoIP audio is sent by mini-frames (small UDP packets).

Frequently Asked Questions

The following Frequently Asked Questions, answered by the authors of this book, are designed to both measure your understanding of the concepts presented in this chapter and to assist you with real-life implementation of these concepts. To have your questions about this chapter answered by the author, browse to **www.syngress.com/solutions** and click on the **"Ask the Author"** form.

Q: Since SIP is similar to the HTTP protocol, could similar methods be used to attack SIP and find weaknesses.

A: Yes. Fuzzing and probing equipment at the SIP level (typically port 5060) could possibly reveal programming flaws. The basic idea would be to build bogus SIP methods and see how the hardware responds. SIP responses to bogus or invalid methods could also help reveal flaws

Q: Could attacks, like brute forcing passwords, reveal password?

A: A good administrator would notice this, but it is possible. For example, brute forcing via the SIP REGISTER method would be trivial. Brute forcing is possible, but slow and might get noticed.

Q: Wouldn't encryption help prevent easy dropping?

A: Of course. However, many organizations don't bother to implement encryption on the LAN between the Asterisk server and the phone equipment. It is not always that the equipment cannot handle protocols like SRTP (Secure RTP); it is just rarely thought of. Between remote/satellite endpoints, using IPSec, OpenVPN, SRTP or IAX2's built in encryption is advised. Whatever type of VPN you chose to use, it'll need to be UDP based as TCP VPNs can wreak a VoIP network.

Q: Can't VLANs prevent ARP spoofing?

A: If properly setup, yes. The VoIP equipment should be setup on its own VLAN, away from the typical users. The idea is that the "users" VLAN won't be able to ARP poison the "voip" VLAN.

Chapter 6

Asterisk
Hardware Ninjutsu

Solutions in this chapter:

- Serial

- Motion

- Modems

- Legalities and Tips

☑ Summary

☑ Solutions Fast Track

☑ Frequently Asked Questions

Introduction

With Asterisk and the flexibility it offers, you can do some truly amazing things. With a "stock" configuration, only using what Asterisk has built in, you can build systems that do some really nifty stuff. If you throw the power of AGIs (Asterisk Gateway Interfaces) into the mix, you can write customized applications that might be difficult to accomplish with other VoIP systems.

Most AGI examples are typically written to take advantage of external resources that Asterisk itself might not have direct access to, or know how to deal with. For example, AGIs have been written to look up ISBNs (book numbers), ANACs (Automatic Number Announcement Circuits) that look up a telephone number information from external sources, text-based games, and IDSs (Intrusion Detection Systems) for monitoring.

We can take the power of AGIs a bit further to interface Asterisk with actual hardware. For example, security cameras, electronic door locks, and card readers to name a few. Creativity is the key.

If you can interface with the hardware externally and interact with it, odds are you can come up with some means to write an AGI to pass that information back.

Serial

To start off, we'll touch on serial communications—yes, that old communications method you used with a modem to connect to the Internet. Even though it's old, traditional serial is used to communicate with room monitoring equipment, magnetic card readers, robotics, environmental control systems, and various other things. It's used where high-speed bandwidth isn't important, but getting data and passing commands is.

These examples are only meant to stir your mind so you come up with creative ways to integrate hardware with Asterisk. While the code does function, the idea is to plant a seed on things you might be able to do with hardware and Asterisk.

Serial "One-Way" AGI

For the first example, we'll be using "one-way" communication via a serial port to the Asterisk server. "One way" means that we don't have to send commands to the device attached via serial. It'll send the information over the serial port automatically. For the generic example code, we use a magnetic stripe reader like the ones that read your credit card. The idea behind this simple code is that the user must "swipe" a card before they are allowed to place a call. If the card information matches, the call is placed. If it does not, the user is notified and the call is dropped. Before we jump into the code, we must place the AGI in line with outbound calls. That is, before the call is completed, it must run through our routine first. To our extensions.conf, we'd add something like:

```
[ serial-code-1 - extensions.conf ]

exten => _9.,1,agi,serial-code-1.agi
exten => _9.,2, Dial(.....)
```

This is a simple example, and depending on your environment and how you make outbound calls through your Asterisk server, you'll need to modify this. The idea is that, if the number starts with a 9, it'll go through this part of the extensions.conf. If it does, before Asterisk gets to step 2 and dials out, it'll have to pass the *serial-code-1.agi* tests first.

```
[ serial-code-1.agi perl routine ]


#!/usr/bin/perl -T
#
###########################################################################
# serial-code-1.agi                                                       #
#                                                                         #
# By Champ Clark - June 2007                                              #
# Description: This is a simple routine that'll take data from a serial
# port and respond to it. The example is something like a magstripe
# reader (credit card type). This only deals with one-way communication
# from the device to the AGI. We don't have to send commands to the
# device, so we'll simply listen and parse the data we get and act
# accordingly.                                                            #
###########################################################################

use strict;

use Asterisk::AGI;                      # Makes working with Asterisk AGI
                                        # a bit easier

use Device::SerialPort;                 # Used to connect/communicate with
                                        # the serial device.
```

```perl
# Following is the string we'll be searching for from the serial port. For
# this simple example, we'll hard-code in a fake driver's license
# to search for. The idea is that before anyone can make an outbound
# call, they must first swipe their licenses through the magstripe
# reader. Of course, this is just an example and could be used for
# anything.

my $searchfor =        "C000111223330";      # My fake driver's license number to
                                             # search for.


my $device    =        "/dev/ttyS1";         # Serial device used.


my $welcomefile = "welcome-serial"; #      This file is played at the
                                             # beginning of the call. It
                                             # explains that some form of
                                             # authentication is needed.


my $grantedfile =      "granted-serial";     # If authentication succeeds, we
                                             # play this and continue through
                                             # the extensions.conf.


my $deniedfile  =      "denied-serial";      # If the authentication fails,
                                             # we'll play this.


my $timeoutfile =  "timeout-serial";         # If we don't see any action on the
                                             # serial port for $timeout seconds,
                                             # we play this file and hang up.


my $errorfile   =      "error-serial";        # This is only played in the event
                                             # of a serial error.



my $serial    =        Device::SerialPort->new ($device) ||
                       die "Can't open serial port $device: $!";


# These are the settings for the serial port. You'll probably want to alter
# these to match whatever type of equipment you're using.

$serial->baudrate(9600) ||
die "Can't set baud rate";
```

```perl
$serial->parity("none") ||
die "Can't set parity";

$serial->databits(8) ||
die "Can't set data bits";

$serial->stopbits(1) ||
die "Can't set stop bits";

$serial->handshake("none") ||
die "Can't set handshaking";

$serial->write_settings ||
die "Can't write the terminal settings";

# After being prompted to "swipe their card," or do whatever you're trying
# to accomplish, we give the user 30 seconds to do so.
# If they don't, we play the $timeoutfile.

my $timeout="30";

# Various other variables are used to pull this together to make it work.

my $string;                        # From the serial port, concatenated.
my $serialin;                      # What we receive from the serial port.
my $i;                             # Counter (keeps track of seconds passed)
my $AGI = new Asterisk::AGI;

$serial->error_msg(1);             # Use built-in error messages from
$serial->user_msg(1);              # Device::SerialPort.

# Play the welcome file and inform the user that we'll need a serial-based
# authentication method (as in the example magstripe reader). Something
# like "Swipe your card after the tone..."

$AGI->exec('Background', $welcomefile );

# Enter the serial "terminal" loop. We now start watching the
# serial port and parsing the data we get.
```

```perl
while($i < $timeout )
      {

      # We sleep for a second so we don't hammer the CPU monitoring the
      # serial port. We also use it to increment $i, which keeps track
      # of how long we've been in the loop (for $timeout). To increase
      # polling, you might want to consider using Time::HiRes. I've
      # not run into any problems.

      sleep(1); $i++;

      # Do we have data?

      if (($serialin = $serial->input) ne "" )
        {
        # Append it to $string so we can search it.
        $string = $string . $serialin;


# Now, search for the magic string ( $seachfor ) that will let us continue.
      if ( $string =~ /$searchfor/i )
          {
          $AGI->exec('Background', $grantedfile );
           exit 0;
          }

      # If we receive an enter/carriage return, we'll assume the unit
      # has sent all the data. If that's the case, and we've not
      # matched anything in the above, we'll play $deniedfile and
      # hang up.

      if ( $string =~ /\cJ/ || $string =~ /\cM/ )
          {
          $string = "";
           $AGI->exec("Background", $deniedfile );
           $AGI->hangup();
           exit 0;
          }
    }
```

```
# If there is some sort of serial error, we'll play this file to let
# the user know that something isn't set up correctly on our side.

if ( $serial->reset_error)
    {
    $AGI->exec("Background", $errorfile );
    $AGI->hangup();
    exit 0;
    }
}


# If the user doesn't respond to our request within $timeout, we
# tell them and hang up.

$AGI->exec("Background", $timeoutfile );
$AGI->hangup();
exit 0;
```

Before this routine will function, you'll need to record a few prompt and response audio files.

welcome-serial	This tells the user that they'll need to "swipe" their card before the call is placed. Use something like "Please swipe your card after the tone (tone)."
granted-serial	Lets the user know that the card was read and accepted. For example, "Thank you. Your call is being placed."
denied-serial	Lets the user know the card was declined for the call. For example, "I'm sorry. Your card was not accepted." The call will automatically terminate.
timeout-serial	Informs the user that they didn't swipe their card within the allotted amount of time (via *$timeout*). For example, "I'm sorry. This session has timed out due to inactivity."
error-serial	Lets the user know that there has been some sort of communication error with the serial device. The call is not placed. For example, "There has been an error communicating with the card reader." The call is automatically hung up.

In the example, we are looking for a hard-coded string (*$searchfor*). You could easily make this routine search a file or database for "good" responses.

Dual Serial Communications

Unlike the first example, which relies on simple serial input from a remote device, this code "probes" (sends a command) to a serial device and parses the output for information we want. In the example code, we'll use an "environmental control" system. We want to know what the "temperature" is in a particular room. If the temperature goes above a certain level, we'll have Asterisk call us with a warning.

The interesting idea behind this AGI is that it works in a circular method that requires no addition to the extensions.conf. If the routine is called with a command-line option, it will probe the serial port. If nothing is wrong, it will simply exit. If something *is* wrong, it will create a call file that loops back to itself (without a command-line option) and notifies the administrators.

```
######################################################################
# serial-code-2.agi                                                 #
#
# By Champ Clark - June 2007                                        #
# Description: This is a simple routine that serves two roles. If
# called with a command-line option (any option), it will send a command to
# a serial device to dump/parse the information. In this example,
# it will send the command "show environment" to the serial device.
# What it looks for is the "Temperature" of the room. If it's under a set
# amount, nothing happens. If it's over the amount, it creates a call
# file (which loops back to serial-code-2.agi).
#
# That's where the second side of this routine kicks in. If _not_ called
# with a command-line argument, it acts as an AGI. This simply lets
# the administrator know the temperature is over a certain amount.     #
######################################################################

use Asterisk::AGI;              # Simply means to pass Asterisk AGI
                                # commands.
use Device::SerialPort;         # Access to the serial port.


# We check to see if the routine was called with a command-line argument.
# If it was (and it really doesn't matter what the argument was),
# we can safely assume we just need to check the serial port and
# parse the output (via cron). If the routine was called without
# a command-line argument, then the routine acts like a
```

```perl
# traditional perl AGI.

if ( $#ARGV eq "-1" ) { &agi();  exit 0; }


my $device        =        "/dev/ttyS1";      # Serial device to check

my $timeout    =       "10";                  # Timeout waiting of the
                                              # serial device to respond.
                                              # If it doesn't respond
                                              # within this amount of
                                              # seconds, we'll assume
                                              # something is broken.


# This is the command we'll send to the serial device to get information
# about what's going on.

my $serialcommand = "show environment\r\n ";

my $searchfor = "Temperature";        # The particular item from the
                                      # $serialcommand output we're
                                      # interested in.
my $hightemp = "80";                  # If the temp. is higher than this value,
                                      # we want to be notified!

my $overtemp="overtemp-serial";       # This is the audio file that's played
                                      # when the temperature gets out of range
                                      # (or whatever # you're looking for).

# This file is played if the serial device doesn't respond correctly or as
# predicted. The idea is that it might not be working properly, and so the
# system warns you.

my $timeoutfile="timeout-serial";


my $alarmfile="alarm.$$";                          # .$$ == PID
my $alarmdir="/var/spool/asterisk/outgoing";       # where to drop the call


# This tells Asterisk how to make the outbound call. You'll want to
```

```perl
# modify this for your environment.

my $channel="IAX2/myusername\@myprovider/18505551212";

my $callerid="911-911-0000";          # How to spoof the Caller
                                      # ID. Will only work over
                                      # VoIP networks that allow
                                      # you to spoof it.


# These should be pretty obvious...

my $maxretries="999";
my $retrytime="60";
my $waittime="30";

# This is how we'll communicate with the serial device in question. You
# will probably need to modify this to fit the device you're communicating
# with.

my $serial      =       Device::SerialPort->new ($device) ||
                        die "Can't open serial port $device: $!";

$serial->baudrate(9600) ||
die "Can't set baud rate";

$serial->parity("none") ||
die "Can't set parity";

$serial->databits(8) ||
die "Can't set data bits";

$serial->stopbits(1) ||
die "Can't set stop bits";

$serial->handshake("none") ||
die "Can't set handshaking";

$serial->write_settings ||
die "Can't write the terminal settings";
```

```perl
my $i;                              # Keeps track of the timer (in case of
                                    # serial failure).
my $stringin;                       # Concatenation of all data received on the
                                    # serial port. Used to search for our
                                    # string.


$serial->error_msg(1);              # Use built-in error messages from
$serial->user_msg(1);               # Device::SerialPort.


my $AGI = new Asterisk::AGI;


# Here we send a command (via the $serialcommand variable) to
# our device. After sending the command, we'll parse out what we need.


$serial->write( $serialcommand );


# We now enter the "terminal loop." The command has been sent,
# and we are looking for the data we are interested in. If we
# send the command but don't receive a response within $timeout seconds,
# we can assume the device isn't working and let the administrator know.


while ( $i < $timeout )
    {

        # We sleep for a second so we don't hammer the CPU monitoring the
        # serial port. We also use it to increment $i, which keeps track
        # of how long we've been in the loop (for $timeout). To increase
        # polling, you might want to consider using Time::HiRes. I've
        # not run into any problems.

        sleep(1); $i++;

        # Did we get any data from the serial port?

        if (($serialin = $serial->input) ne "" )
        {

            # We'll probably get multiple lines of data from our $serialcommand.
            # Every time we receive an "end of line" (carriage return or Enter)
            # we "clear" out the string variable and "new string" array.
```

```perl
 if ( $serialin =~ /\cJ/ || $serialin =~ /\cM/ )
        {
        $string = "";               # Clear the concatenated string.
        @newstring="";              # Clear our array used by "split."
        }


# If the preceding is not true, the routine concatenates $serialin
# to $string. Once $string + $serialin is concatenated, we look
# for the ":" delimiter. This means the serial port will return
# something like "Temperature: 75". We want the "Temperature"
# value and will strip out the rest.

 $string = $string . $serialin;
 @newstring=split /$searchfor:/, $string;


# In this example, we check to see if the devices return a higher
# temperature than what we expect. If so, we build a call file
# to "alert" the administrator that the A/C might not be working!
#
if ($newstring[1] > $hightemp )
        {
        if (!open (ALARM, "> $alarmdir/$alarmfile"))
            {
            die "Can't write $alarmdir/$alarmfile!\n";
            }

        print ALARM "Channel:  $channel\n";
        print ALARM "Callerid: Temp. Alert <$callerid>\n";
        print ALARM "MaxRetries: $maxretries\n";
        print ALARM "RetryTime: $retrytime\n";
        print ALARM "WaitTime: $waittime\n";
        print ALARM "Application: AGI\n";
        print ALARM "Data: serial-code-2.agi\n";
        print ALARM "Set: tempfile=$overtemp\n";
        close(ALARM);
        }
    }

}
```

```
# If for some reason communications with the serial device fails, we'll
# also let the administrator know.

if (!open (ALARM, "> $alarmdir/$alarmfile"))
   {
   die "Can't write $alarmdir/$alarmfile!\n";
   }

   print ALARM "Channel:  $channel\n";
   print ALARM "Callerid: Temp. Alert <$callerid>\n";
   print ALARM "MaxRetries: $maxretries\n";
   print ALARM "RetryTime: $retrytime\n";
   print ALARM "WaitTime: $waittime\n";
   print ALARM "Application: AGI\n";
   print ALARM "Data: serial-probe.agi\n";
   print ALARM "Set: tempfile=$timeoutfile";
   close(ALARM);

exit 0;

# end of routine.

# This subroutine acts as an AGI if the routine is called without a
# command-line argument.

sub agi
{

my $AGI = new Asterisk::AGI;
my %AGI;

#  If this subroutine is called, obviously something has gone seriously
#  wrong. The call (via a call file) has already been placed, this just
#  lets the administrator know "what" went wrong.

$AGI->answer();                      # Pick up! We need to tell the user
                                     # something!
$AGI->exec('Wait', '1');             # Give me a warm fuzzy...

# We grab the audio file we want to play from the "tempfile" variable in
```

```
# the call file and play it.

$tempfile=$AGI->get_variable('tempfile');
$AGI->exec('Background', $tempfile );
$AGI->hangup();

exit 0;
}
```

You will need to record a couple of prompt/audio files. They include the following:

overtemp-serial	This is the file that's played if the temperature (in our example) is over the *$hightemp*.
timeout-serial	This file is played if the serial device doesn't respond in *$timeout*. The idea is that the device might not be functioning.

Since the routine is operating as an AGI, it'll need to be copied to your AGI directory. This is typically done by using /var/lib/asterisk/agi-bin. This way, Asterisk will have access to the routine. To start monitoring the hardware, you'll want to create a cron job that would "test" every ten minutes or so. That cron entry would look something like this:

```
*/10  * * * * /var/lib/asterisk/agi-bin/serial-code-2.agi test 2>&1 > /dev/null
```

Motion

Motion is open-source software that uses video camera equipment to record "motion" in a room. It's primarily used for security purposes, and has many features. For example, you can take snapshots of an area every few seconds or create time-lapse movies.

Of course, to use Motion you'll need the proper hardware. "All weather cameras" and video capture cards have come down in price over the years. I like to keep things simple, so I use multiport BT848 (chipset) capture cards for my home security system. It's a generic chip set that works well with Linux. My particular card comes with four onboard built-in ports, but it can support up to eight ports with an external adapter. This means I can run up to eight cameras at a time. Considering I use this to monitor my home (front yard, back yard, inside my office, and so on), I'm not worried that the cameras and capture card chipset won't produce high-definition quality. I simply want a means to record events and watch my cameras over the Internet.

If you have spare camera equipment around, you might want to look into how well it's supported under Linux. Some USB cameras require proprietary drivers to work while others do not. The first step is to get the camera up and working under Linux, and then configure it to work with Motion.

To obtain Motion, simply go to http://motion.sourceforge.net. Once you've downloaded it, installation is typically at *./configure && make && make install*. Some Linux-based distributions have motion packages you might want to look into.

The motion.conf file can be quite daunting, but don't let it scare you. It'll probably take a bit of tweaking to get your configuration up and running and that will largely depend on the type of hardware you use. If you're using more than one camera, it's better to get one camera online first before trying to configure the rest of them. Motion uses a "threaded" system in monitoring multiple cameras, so you'll actually have multiple configuration files per camera.

[tail end of a default motion.conf file]

```
###############################################################
# Thread config files - one for each camera.
# However, if there's only one camera, you only need this config file.
# If you have more than one camera, you MUST define one thread
# config file for each camera in addition to this config file.
###############################################################

# Remember: If you have more than one camera, you must have one
# thread file for each camera. Thus, two cameras require three files:
# This motion.conf file AND thread1.conf and thread2.conf.
# Only put the options that are unique to each camera in the
# thread config files.
; thread /usr/local/etc/thread1.conf
; thread /usr/local/etc/thread2.conf
; thread /usr/local/etc/thread3.conf
; thread /usr/local/etc/thread4.conf
```

The option we'll be focusing on is a pre-thread configuration file, so once you have a working configuration:

```
# Command to be executed when a motion frame is detected (default: none)
; on_motion_detected value
```

The idea is that when motion is detected, we can have Motion (the program) run a routine. When I leave town for an extended period of time, I want to know if motion is detected within my home. I'm not as concerned about outside because false positives would drive me crazy. For example, cats or dogs that just happen to roam through my yard, I'm not interested in.

If an event happens inside the home and I know nobody is there, then I certainly want to know! So, with the cameras that are internal, we'll use the *on_motion_detect* option to run a routine that'll call my cell phone and alert me to something or someone in my house. We can do this on a per-thread configuration file basis. So, for cameras that are outside, we won't add the *on_motion_detect* option.

The Idea behind the Code

The idea behind this code is simple, but does two different jobs. One is to create the outgoing call file to let you know when "motion" has been detected. The other is to be an AGI so that once the call is made, Asterisk can "tell you" which camera saw the motion. Since this routine handles all the necessary functions, you can simply copy it to your Asterisk AGI directory (usually /var/lib/asterisk/agi-bin) and go! No modifications are needed to Asterisk configuration files (for example, extensions.conf).

When Motion "sees motion," it will call the routine via the *on_motion_detect* command. Within the Motion configuration files for each camera we wish to monitor, we'll pass the command to alert us if something is detected. It will look something like this:

```
on_motion_detect /var/lib/asterisk/agi-bin/alarm.agi 1
```

The number "1" is passed as a command-line argument. In this example, this represents *camera 1*. Since we are passing a command-line argument, the routine is programmed to know that this is coming from Motion. When called as an AGI from Asterisk, no command-line argument is passed. Let's run through the entire routine to pull everything together.

Motion is monitoring *camera 1*, which we'll say is your home office. Motion detects "motion" in the room and starts recording the action, firing off the /var/lib/asterisk/agi-bin/alarm.agi file with the command-line option of "1," which signifies the camera that was triggered. The alarm.agi creates a call file in the Asterisk outgoing call directory (typically, /var/spool/asterisk/outgoing). The contents of this file will look something like this:

```
Channel:  IAX2/myusername@myprovider/18505551212
Callerid: Security Camera <911-911-0001>
MaxRetries: 999
RetryTime: 60
WaitTime: 30
Application: AGI
Data: alarm.agi
Set: camera=1
```

The *Channel:* is an option in the outgoing call file that gives the method of "how" to make the outgoing call. In this example, I'm using a provider that supports IAX2. You could easily change this to use a Zap device or SIP. Note the *Applications: AGI* and *Data: alarm.agi*. When alarm.agi builds the call file, it creates it in such a way that it loops back on itself. The *Set: camera=1* passes the camera that recorded the event. We attempt to spoof the Caller ID to *911-911-0001*. This just shows that an emergency has occurred on camera 1 (0001 in the Caller ID field). It will only work if your provider allows you to modify your Caller ID (CID). On the PSTN, the *Security Camera* portion will be dropped completely, even if the number is spoofed. It'll work fine over VoIP networks, but the PSTN does a lookup of the number and fills in the Name section of the Caller ID field. On the PSTN, that's out of your control.

Once the call file is built and saved in the Asterisk "outgoing" directory, Asterisk will almost immediately grab this file and follow the instructions in it. Asterisk calls via the method in the Channel: field, and then waits for the call to supervise. *Supervision* is a term used to signify that something or someone has "picked up" the call.

Upon supervision by you answering your phone, Asterisk executes the AGI alarm.agi. Since Asterisk is calling the routine this time without command-line arguments, the routine is programmed to act as an AGI. Upon you answering, the AGI side of the alarm.agi kicks in and feeds Asterisk commands like the following:

```
ANSWER
EXEC  Wait 1
GET VARIABLE camera
EXEC Background camera-1
HANGUP
```

As you can see, it's pretty simple! Answer the call, and wait one second. Get the contents of the variable *camera*. Remember that variable? It holds the numeric value of what camera was triggered. Once that variable is obtained, we issue a "Background" (audio playback) of the file "camera{camera variable}". In this case, that'll be camera1.

This means you'll want to record a couple of audio files to use with this routine. In this example, we said that camera1 was our home office. So we'd want to pre-record some audio files that represent our cameras. In this case, we might have an audio file that says, "Warning! There appears to be motion in the home office." The following is the standalone routine we're using. Of course, you could take this simple routine and modify it to do a multitude of things.

[alarm.agi]

```perl
#!/usr/bin/perl -Tw
#
###########################################################################
#  alarm.agi                                                              #
#  By Da Beave (Champ Clark) - June 2007                                  #
#  Description: This routine actually serves two purposes. It acts as the
#  routine that creates the "call files" and that the AGI routine Asterisk
#  uses. When this routine is called by Motion, a command-line argument
#  is given to specify which camera saw the motion. If there is no
#  command-line argument, then the routine services as an Asterisk AGI
#                                                                         #
###########################################################################

use strict;
use Asterisk::AGI;                   # Makes working with Asterisk AGI a
                                     # little bit easier.


# This is the name of the "sound" file to call. For example, if Motion
# sees motion on camera #1, it'll send to this routine: /var/lib/asterisk/
# agi-bin/alarm.agi 1. So, the file (in the /var/lib/asterisk/sounds)
# "camera1" is called to inform that motion was caught on "camera1".
# This is the prefix of the file (that is, camera1, camera2, and so on).

my $camerafile="camera";


# We check to see if there is a command-line argument. If not, we assume
```

```
# the routine needs to act like an Asterisk AGI. If it _does_ have a
# command-line argument, then we assume Motion has called the
# routine and given the camera information via argv....

if ( $#ARGV eq "-1" ) { &agi();  exit 0; }

# $channel contains the information about how the call is to be placed.
# In this example, we'll be using IAX2. However, you could use Zap, SIP,
# or other methods Asterisk supports. Replace with your method of
# dialing/phone number.

my $channel="IAX2/myusername\@myprovider/18505551212";

# We spoof the Caller ID. This will only work if you're VoIP provider
# allows you to modify the Caller ID information. With my VoIP carrier,
# I have to supply a full ten-digit phone number. YMMV (you might be able
# to get away with something shorter). So, when Motion calls me, it will
# send "911-911-000" as the NPA/NXX. The last digit is the camera that has
# reported motion.

my $callerid="911-911-000";

# These should be fairly obvious...

my $maxretries="999";
my $retrytime="60";
my $waittime="30";

# To keep outgoing calls unique, we build call files based on their PID.

my $alarmfile="alarm.$$";                       # .$$ == PID
my $alarmdir="/var/spool/asterisk/outgoing";    # where to drop the call
                                                # file.

my $tmpfile;
my $setinfo;
my $camera;
```

```
#########################################################################
# This is where the actual call file is built. Remember, with Asterisk,
# any call files that show up in the outgoing queue (usually /var/spool/
# asterisk/outgoing) are used automatically.
#########################################################################

# Open the outgoing queue file and feed it the commands.

if (!open (ALARM, "> $alarmdir/$alarmfile"))
        {
        die "Can't write $alarmdir/$alarmfile!\n";
        }

print ALARM "Channel:  $channel\n";
print ALARM "Callerid: Security Camera <$callerid$ARGV[0]>\n";
print ALARM "MaxRetries: $maxretries\n";
print ALARM "RetryTime: $retrytime\n";
print ALARM "WaitTime: $waittime\n";
print ALARM "Application: AGI\n";
print ALARM "Data: alarm.agi\n";
print ALARM "Set: camera=$ARGV[0]\n";
close(ALARM);

#########################################################################
# AGI section: If no command-line arguments get passed, we can
# assume it's not Motion calling the routine (because Motion passes
# the camera on the command line). Asterisk calls alarm.agi without
# any command-line arguments, so we act as an AGI.
#########################################################################

sub agi
{

my $AGI = new Asterisk::AGI;
my %AGI;

# This pulls in our Asterisk variables. For example, $input{camera},
# which we are using to pass the camera number.

my %input = $AGI->ReadParse();
```

```
# Okay - now we do our song and dance for the user we called!

$AGI->answer();                           # Pick up! We need to tell the user
                                          # something!
$AGI->exec('Wait', '1');                  # Give me a warm fuzzy...

$camera=$AGI->get_variable('camera');     # Get the "camera" variable.
$tmpfile="$camerafile$camera";
$AGI->exec('Background', $tmpfile );
$AGI->hangup();
exit 0;
}
```

Modems

Traditional analog modems present a problem with VoIP. First off, you're probably asking "Why the heck would you even attempt to hook up a traditional modem via VoIP?" One practical reason is because many systems still use traditional analog modems for communications—for example, point-of-sales equipment, TiVo, and credit card equipment. Before attaching any devices like these to a VoIP network, security should be considered. Equipment of this type might transmit sensitive information. It's less than practical to play with the PSTN network via VoIP network, dial into old style BBS systems, use older networks that still require a dialup connection, or "scan" for modems and telephone equipment. Scanning for modems and telephone equipment is known as *war dialing*. The term comes from the 1984 film *War Games*, but the term and technique is actually older than the movie, and is sometimes referred to as *demon dialing*. The term war dialing, though, is the one that sort of stuck in the phreaking and hacking community. In the film, our hero (played by Matthew Broderick) dials every telephone number within an exchange searching for interesting telephone and computer equipment owned by a fictional company named Protovision, Inc. In real life, the idea of war dialing is the same as in the movie and can be useful during security audits. During a security audit, you're dialing numbers within a particular block around your target searching for things like modems, fax machines, environmental control systems, PBXs, and other equipment connected to the PSTN.

You'll need prior permission, and checking with your local laws is advised before war dialing!

So, why would you want this behind VoIP when you could hook up a modem on the traditional PSTN? With VoIP, you are able to mask "where" you are calling from. Unlike the PSTN, our ANI information, which cannot be easily spoofed, won't be passed. We can "spoof" things like our Caller ID. It makes it harder to track down where the calls are coming from.

For whatever reason you'll be using an analog modem with VoIP, several things must be considered. First off, you won't be able to make very high-speed connections. The top speed you'll be able to accomplish is about 4800 baud. This is due to how the modem MOdulates and DEModulates (hence the term *modem*) the signal and network latency. At very low speeds, like 300 baud, a simple means of encoding the data is used, known as frequency-shift keying (FSK). The originator of the call transmits at 1070Hz or 1270Hz. The answering side transmits at 2025Hz or 2225Hz. This is well within the range and type of encoding we can do over VoIP. A speed of 1200 baud is also achievable and stable. At that speed, a simple encoding scheme is used, known as Phase-Shift Keying (PSK). Once you step into higher speeds like 14.4k, 28.8k, 33.6k, and above, you get into very time-sensitive encoding techniques, like quadrature amplitude modulation (QAM), which don't respond well in a VoIP world.

To keep things stable, I generally keep my rates locked at 1200 baud. Not blinding fast, but it's good enough to detect and look at remote systems. You might be wondering, "Wait a minute! How come things like Fax over VoIP can handle such higher baud rates?!" Good question!

As VoIP became more and more popular, the ITU (International Telecommunications Union) created a protocol known as T.38, which is sometimes referred to as FoIP (Fax over IP). Asterisk and many VoIP adapters now support T.38. When you plug in your fax machine to a VoIP adapter, it may very well auto-detect and support the fax under T.38. What T.38 does is it takes the fax signal and converts it to more data-network-friendly SIP/SDP TCP/UDP packets that get transmitted over the Internet. Since the fax signal doesn't actually have to traverse the Internet, greater speeds can be achieved. If your adapter or provider does not support T.38 and the analog fax signal has to transverse the Internet, then you'll run into similar issues as you would with analog modem.

This might make you wonder why there isn't a Modem over IP protocol. Well… in truth, the ITU *has* created such a standard, known as V.150.1 (also known as

V.MOIP) in 2003. It operates much like T.38 in that it takes the analog signal and converts it to a UDP/TCP packet that can traverse the Internet easily. Unfortunately, even fewer VoIP providers and equipment support V.150.1. This might change as VoIP becomes more and more popular and people want to connect equipment that would traditionally connect to the PSTN. Until that time, though, we are stuck doing it the hard way.

It order to test Modem over IP, you'll obviously need an analog modem. You'll also need some sort of VoIP telephone adapter like the Linksys PAP2 or Cisco ATAs. These devices are normally used to connect normal telephones to a VoIP network. They typically have one or two RJ11 jacks on them to plug in your traditional telephone. They also have an RJ45 network jack that will connect to your LAN. Rather than plugging in a traditional telephone into the RJ11 jack, we'll use this port to attach our analog modem.

Configuration of the VoIP adapter largely depends on the hardware itself. Configuration on the Asterisk side is typically pretty straightforward and simple. I use a dual line Linksys PAP2, which employs SIP. Since it is a dual line (two RJ11s) configuration, I have a [linksys1] and [linksys2] section in my Asterisk sip.conf file. The following shows what mine looks like:

```
[linksys1]
type=friend                     # Accept inbound/outbound
username=linksys1
secret=mysecret
disallow=all
;allow=gsm                      # This will NOT work for a modem.
allow=ulaw                      # Works much better with a modem.
context=internal
host=dynamic
```

In order for Modem over IP to work, you must consider two important factors. First: the better your network connection, the better your modem connections will be. The lower the latency, the better. Second: you must *not* use any sort of compressed codec! Compressed codecs, like GSM or G.729, will alter the analog signal/encoding, which will cause connections to completely fail. You'll want to use the G.711 (u-law) codec. If you can accomplish these two requirements, you'll be much better off. If you are configuring point-of-sales equipment or some sort of consumer electronics (TiVo, and so on), you'll probably want to test a bit and play with baud rate settings to see what you can achieve.

Fun with Dialing

If the modem you wish to use is attached to a computer and is not PoS/consumer electronics gear, you can start up some terminal software and go! Under Linux, and other Unix-like operating systems, multiple terminal software programs can be used. Minicom is probably one of the more well known and useful terminal software programs around. It comes with most distributions, but if your system doesn't have it, the source can be downloaded from http://alioth.debian.org/projects/minicom/. If Minicom doesn't suit your tastes, check out Seyon for X Windows, which can be obtained from ftp://sunsite.unc.edu/pub/Linux/apps/serialcomm/dialout/ (look for the latest Seyon release). No matter which software you use, knowledge of the Hayes AT command set is a plus. Hayes AT commands instruct the modem in "what to do." For example, ATDT means *Dial Tone*. You'll probably want to read over your modem's manual to get a list of supported AT commands.

Okay, so your modem is hooked up. Now what? I like to use VoIP networks to dial in to remote countries and play with things that might not be accessible in the United States. For example, in France there is a public X.25 (packet-switched) network known as Transpac that I like to tinker with. I also use VoIP with my modem to call Russian BBSs and an X.25 network known as ROSNET. There's a lot of nifty stuff out there that's not connected to the Internet and this gives me a cheap, sometimes even free, way to call foreign countries.

War Dialing

Another useful feature for a modem connection via VoIP is security audits. Rogue modems and various telephone equipment are still a security problem in the corporate world. When hired to do a security audit for an organization, I'll suggest a "scan" of the telephone numbers around the company to search for such rogue equipment. It's not uncommon for a company to not even be aware it has equipment connected to the PSTN. The usefulness of scanning via VoIP is that I can mask where I'm coming from. That is, I can spoof my Caller ID and I know my ANI information on the PSTN will be incorrect—meaning I can hide better. One trick I do is to spoof my telephone number as the number from a known fax machine. This way, during my war dialing, if someone tries to call me back, they'll dial a fax machine. From there, they'll probably think the fax machine just misdialed their telephone number and forget about it.

Of course, spoofing Caller ID can be useful in other ways for security audits—such as with social engineering. Social engineering is nothing more than presenting yourself as someone you're not and requesting information you shouldn't have, or requesting someone do something they shouldn't—for example, spoofing the Caller ID of an Internet Service Provider (ISP) and requesting changes be done to the network (change proxies so you can monitor communications) or requesting a password. This is getting off the topic of war dialing, but it's still useful.

I don't particularly want to spoof every time I make a call through my Asterisk system, so I set up a prefix I can dial before the telephone number. In my case, if I want to call 850-555-1212 and I want to use caller ID spoofing, I'll dial 5-1-850-555-1212. The initial "5" directs Asterisk to make the outbound call using a VoIP provider with Caller ID spoofing enabled. My extensions.conf for this looks something like:

```
; Caller ID spoofing via my VoIP provider.
;

exten => _5.,1,Set,CALLERID(number)=904-555-7777
exten => _5.,2,monitor,wav|${EXTEN:1}
exten => _5.,3,Dial(IAX2/myusername@myprovider/${EXTEN:1})
```

You might be wondering why we don't do a *Set, CALLERID(name)*. There isn't really much point. Once the call hits the PSTN, the number is looked up at the telephone company database and the Name field is populated. This means, once the call hits the traditional PSTN, you can't modify the Name field anyways. One interesting thing you *can* do, if you're trying to figure out who owns a phone number is spoof the call as that phone number to yourself. Once the call reaches the PSTN and calls you, the telephone company will look up the spoofed number in its database and display the name of who owns it. This is known as backspoofing and isn't completely related to war dialing, but can be useful in identifying who owns particular numbers. The Monitor option lets you record the audio of the call, so you can listen later and see if anything was found that the war dialer might have missed. It's advised you check your local laws regarding recording telephone calls. If you don't wish to do this, the option can be removed.

With our adapter set up and Asterisk configured, we are ready to war dial! Now we just need the software to send the commands to our modem and then we can start dialing. Several programs are available, some commercial and some open source, that'll take over the dialing and analysis of what you find. One of the most popular is the MS-DOS–based ToneLoc. While an excellent war dialer, it requires the extra overhead of running a DOS emulator. Phonesweep is another option, but runs under Microsoft Windows and is commercial. For Linux, and Unix in general, I use the open-source (GPL) program iWar (Intelligent Wardialer). It was developed by Da Beave from the network security company Softwink, Inc. Many of its features compete with commercial products.

Some of the features iWar supports are random/sequential dialing, key stroke marking and logging, IAX2 VoIP support (which acts as an IAX2 VoIP client), Tone location (the same method ToneLoc uses), blacklist support, a nice "curses" console interface, auto-detection of remote system type, and much more. It will log the information to a standard ASCII file, over the Web via a CGI, MySQL, or PostgreSQL database. You probably noticed the IAX2 VoIP support. We'll touch more on this later.

To obtain iWar, go to www.softwink.com/iwar. You can download the "stable" version, but they suggest you check out the CVS (Conversion Version System). This is a development version that typically has more features. To download via CVS, you'll need the CVS client loaded on your machine. Many distributions have CVS preloaded or provide a package to install it. If your system doesn't have it, check out www.nongnu.org/cvs/ for more information about CVS.

To download iWar via CVS, type

```
$ CVSROOT=:pserver:anonymous@cvs.telephreak.org:/root; export CVSROOT
$ cvs login
```

When prompted for a password, simply press Enter (no password is required). This will log you in to the development CVS system. To download the source code, type

```
$ cvs -z9 co -A iwar    # -z9 is optional (for compression)
```

If you're using the CVS version of iWar, it's suggested you join the iWar mailing list. It's a low-volume mailing list (one or two e-mails per week) that contains information about updates, bug fixes, and new features.

After downloading the software, installation uses the typical ./configure && make && make install. For MySQL or PostgreSQL support, you'll need those libraries preloaded on your system before compiling iWar. If you wish to compile iWar with IAX2 support, you'll need to install IAXClient. You can locate and read about IAXClient at http://iaxclient.sourceforge.net/. This library allows iWar to become a full featured IAX2 VoIP client and war dialer. For proper installation of IAXClient, refer to their mailing list and Web page.

Of course, iWar will compile without MySQL, PostgreSQL, or IAXClient support and will work fine for our purposes with a standard analog modem. Once compiled, we are ready to fire it at our target!

To give you an idea of the options with everything built in (MySQL, PostgreSQL, and IAXClient), the following is the output of *iwar –help*.

```
iWar [Intelligent Wardialer] Version 0.08-CVS-05-24-2007 - By Da Beave
(beave@softwink.com)


[ iwar -help output]


usage: iwar [parameters] --range [dial range]


-h, --help           : Prints this screen
-E, --examples       : Examples of how to use iWar
-s, --speed          : Speed/Baud rate
                       [Serial default: 1200] [IAX2 mode disabled]
-S, --stopbit        : Stop bits [Serial Default: 1] [IAX2 mode disables]
-p, --parity         : Parity (None/Even/Odd)
                       [Serial default 'N'one] [IAX2 mode disabled]
-d, --databits       : Data bits [Serial default: 8] [IAX2 mode disabled]
-t, --device         : TTY to use (modem)
                       [Serial default /dev/ttyS0] [IAX2 mode disabled]
-c, --xonxoff        : Use software handshaking (XON/XOFF)
                       [Serial default is hardware flow control]
                       [IAX2 mode disabled]
-f, --logfile        : Output log file [Default: iwar.log]
-e, --predial        : Pre-dial string/NPA to scan [Optional]
-g, --postdial       : Post-dial string [Optional]
-a, --tonedetect     : Tone Location (Toneloc W; method)
                       [Serial default: disabled] [IAX2 mode disabled]
-n, --npa            : NPA (Area Code - ie 212)
-N, --nxx            : NXX (Exchange - ie - 555)
-A, --nonpa          : Log NPA, but don't dial it (Useful for local calls)
-r, --range          : Range to scan (ie - 5551212-5551313)
-x, --sequential     : Sequential dialing [Default: Random]
-F, --fulllog        : Full logging (BUSY, NO CARRIER, Timeouts, Skipped, etc)
-b, --nobannercheck  : Disable banners check
                       [Serial Default: enabled] [IAX2 mode disabled]
-o, --norecording    : Disable recording banner data
```

```
                              [Serial default: enabled] [IAX2 mode disabled].
-L, --loadfile        :  Load numbers to dial from file.
-l, --statefile       :  Load 'saved state' file (previously dialed numbers)
-H, --httplog         :  Log data via HTTP to a web server
-w, --httpdebug       :  Log HTTP traffic for CGI debugging
-C, --config          :  Configuration file to use [Default: iwar.conf]
-m, --mysql           :  Log to MySQL database [Optional]
-P, --postgresql      :  Log to PostgreSQL database [Optional]
-I, --iax2            :  Enabled VoIP/IAX2 for dialing without debugging
                         (See iwar.conf)
-i, --iax2withdebug   :  Enabled VoIP/IAX2 for dialing with debugging
                         (--iax2withdebug <filename>)
```

iWar also comes with a configuration file to set up things like your serial port, baud rate, and various logging options. The default iwar.conf is suited to work with most hardware, but it's advised to tweak it to your hardware.

```
[ default iwar.conf ]
###########################################################################
                                                                        ##
# iWar configuration file. Please see http://www.softwink.com/iwar for   ##
# more information.                                                       ##
###########################################################################


###########################################################################
# Traditional serial port information                                    #
###########################################################################


#
# Serial port information (prt, speed, data bits, parity). Command--
# line options override this, so you can use multiple modems.
#
port /dev/ttyS0
speed 1200
databits 8
parity N

#
# Modem INIT string. This can vary for modem manufacturers. Check your
# modem's manual for the best settings. Below is a very _basic_ init
# string. The main objective toward making things work better is DTR
```

```
# hangups and dial speed. Here's what is set in this string.
#
# E1   =  Echo on
# L3   =  Modem speaker on high
# M1   =  Modem speaker on until carrier detect
# Q0   =  Result codes sent
# &C1  =  Modem controls carrier detect
# &D2  =  DTE controls DTR (for DTR hangup!)
# S11  =  50 millisecond DTMF tones. On the PSTN in my area, 45ms DTMF
#           works fine, and might work for you. It's set to 50ms to be safe.
#           My ATAs can handle 40ms DTMF, which is as fast as my modem
#           can go. If you're having dial problems, slow down this
#           setting by increasing it. For faster dialing, decrease this.
# S06  =  How long to "wait" for a dial tone. Modems normally set this
#           to two seconds or so. This is terrible if you're trying to
#           detect tones! This is for Toneloc-type tone location
#           (ATDT5551212W;). You may need to adjust this.
# S07  =  Wait 255 seconds for something to happen. We set it high
#           because we want iWar to decide when to hang up. See
#           "serial_timeout."
#
# Extra things to add to the init string:
#
# +FCLASS=1 = Want to scan for fax machines (And only fax - however,
#              the Zylex modems might do data/fax)
#
# X4        = All modems support this. If you add "X4" to the init
#              string, your modem will detect "NO DIALTONE" and "BUSY".
# X6 or X7  = Certain modems (USR Couriers, for example) can
#              detect remote call progression. X7 is good because
#              it leaves everything on (RINGING, BUSY, NO CARRIER)
#              except "VOICE." "VOICE" is sometimes triggered by
#              interesting tones. X6 leaves everything on.
#              This is good when you're doing carrier detection!
# X0        = Set the modem to blind dialing. This is good if you're into
#              "hand scanning." The modem doesn't attempt to detect
#              anything like BUSY or VOICE (it will still detect carriers).
#              You can then use the manual keys to mark numbers.

init ATE1L3M1Q0&C1&D2S11=50S07=255
```

```
# If your modem is not capable of doing DTR hangups, then leave this
# enabled. This hangs up the modem by sending the old "+++ATH" to the
# modem on connections. If your _positive_ your modem is using DTR drops
# to hang up, you can save scan time by disabling this.
# If you enable this and DTR drops don't work, your line will NOT hang up
# on carrier detection!


plushangup 1
plushangupsleep 4

# "This only applies to modems that support remote call progression
# (for example, "RINGING"). Modems that can do this are the USR
# Couriers, Multitech and mccorma modems. If your modem doesn't
# support this, ignore it.

# remote_ring 5

# If remote ring is enabled and functional, then this is the max time
# we'll allow between rings with no result code (BUSY, CONNECT,
# VOICE). For example, if we receive two RINGING result codes,
# but for 30 seconds see nothing else, then something picked up on the
# remote side that the modem didn't register. It might be worth going back
# and checking.

# ring_timeout 20

# This is for modems that reliably detect remote "tones." This changes
# the dial string from the standard ATDT5551212 to ATDT5551212w;
# (See about the ATS06=255 - wait for dial tone). When the modem
# dials, it "waits" for another tone. If the iWar receives an "OK,"
# then we know the end supplied some sort of tone. Most modems can't
# do this. Leave this commented out if your modem doesn't support it.

# tone_detect 1

# Banner file. Banners are used to attempt to figure out what the remote
# system is.
```

```
banner_file /usr/local/etc/banners.txt

# Blacklist file. This file contains phone numbers that should
# never be called (for example, 911).

blacklistfile /usr/local/etc/iwar-blacklist.txt

# Serial connection timeout (in seconds). This is used to detect when
# the modem doesn't return a result code. In that event, we'll
# hang the modem up. See the ATS07 (S07) at the top of this config file.

serial_timeout 60

# When connected (carrier detected), this is the amount of time to
# wait (in seconds) for iWar to search for a  "login banner." If
# no data is received and/or there is no banner, we hang up
# when this amount of time is reached.

banner_timeout 20

    # On the off chance that we keep receiving data, and the banner_timeout is
# never reached, this is the amount of data we will receive before giving
# up (hang up). Data sent from the remote end reset the banner_timeout -
# without this safe guard, the system may never hang up because it keeps
# receiving data! Value is in bytes.

banner_maxcount 2048

# After connecting, this is how long we wait (in seconds) to send a
# return. Some systems won't reveal their banners until they
# receive several \r\r's. Value is in seconds.

banner_send_cr 10

# This is the number a carriage returns to send once banner_send_cr
# is reached.

banner_cr 3

# After connecting, wait this long until picking up and trying to
```

```
# redial out. Measured in seconds.

connect_re-dial 5

# How long to wait before redialing after a BUSY, NO CARRIER, or other type
# of event, in seconds. On PSTN environments, you need to wait a few
# seconds before dialing the next number (or it'll register as a
# "flash"). On VoIP hardware-based scans, you can probably lower
# this to decrease scan time. This does not affect IAX2 dialing.

redial 3

# DTR re-init. Some modems (USR Couriers), when DTR is dropped, have
# to re-init the modem (I assume the USR treats DTR drops like ATZ).
# This will re-init after DTR drops.

# dtrinit 1

# Amount of time to drop DTR. Some modems require longer DTR drops to
# hang up. Value is in seconds. (If possible, 0 is best!)

dtrsec 2

# You can log all your information into a MySQL database. These are the
# credentials to use
#

#############################################################################
## MySQL Authentication                                                    ##
#############################################################################

mysql_username iwar
mysql_password iwar
mysql_host 127.0.0.1
mysql_database iwar

#############################################################################
## PostgreSQL Authentiation                                                ##
#############################################################################
```

```
postgres_username iwar
postgres_password iwar
postgres_host 127.0.0.1
postress_database iwar


############################################################################
## HTTP Logging                                                          ##
############################################################################
#
# The following is an example URL that is based from iWar to a Web server
# during HTTP logging.
#
# http://www.example.com/cgi-bin/iWar-HTTP.cgi?NPA=850&NXX=555&Suffix=1225&
# Revision=0&NumberType=2&Description=Looks%20good%21%21&Username=myname&
# Password=mypassword
#
# If your CGI requires authentication (see earlier), then set these.
# Otherwise, just leave these values alone (the remote site will ignore
# the values)

http_username iwar
http_password iwar

# The web server you are logging to:

http_log_host www.example.com

# HTTP port the Web server is running on.

http_port 80

# The path of the application on the remote Web server doing the logging.
# For more information, see the example iWar-HTTP.cgi.

http_log_path /cgi-bin/iWar-HTTP.cgi

# The combination of http_log_host + http_log_path logging URL would
# look something like this:
#
# http://www.example.com/cgi-bin/iWar-HTTP.cgi
```

```
#
# The example "GET" string (at the top of "HTTP Logging") is automatically
# tacked to the end of this! Ta-da!

########################################################################
## Following are IAX2 values that have no affect when serial scanning ##
########################################################################

# IAX2 username of your VoIP provider/Asterisk server

iax2_username iwar

# IAX2 password of your VoIP provider/Asterisk server. This is _not_
# required if your Asterisk server doesn't use password authentication.

iax2_password iwar

# IAX2 provider/Asterisk server. Can be an IP address or host name.

iax2_host     192.168.0.1

# 99.9% of the time, it's not necessary  to "register" with your provider
# to make outbound calls! It's highly unlikely you need to enable this!
# In the event you have a strange provider that "requires" you to
# register before making outbound calls, enable this.

#iax2_register 1

# If you're using iWar directly with a IAX2 provider, then set this
# to your liking. If you're routing calls via an Asterisk server, you can
# callerid spoof there. With Asterisk, this will have no
# affect.

iax2_callerid_number 5551212

# iax2_millisleep is the amount of time to sleep (in milliseconds) between
# IAX2 call tear down and start up. Probably best to leave this alone.

iax2_millisleep 1000
```

As you can see, many options must be set and tweaked depending on your hardware. With Asterisk configured to spoof the caller ID and make the outbound call using G.711, your VoIP adapter set to communicate properly via G.711 to your Asterisk server, your modem set up through the VoIP adapter, and iWar configured to properly use your hardware, we are ready to launch the dialer! To do this, type

```
$ iwar -predial 5 -npa 904 -nxx 555 -range 1000-1100
```

The —*predial* option tells iWar to dial a "5" before the rest of the number. We do this to let our Asterisk server know we want to spoof Caller ID and to go out our VoIP provider. The —*npa* option (Numbering Plan Area) is another way to say what area code we wish to dial. The —*nxx* is the exchange within the NPA that we'll be dialing, and the —*range* lets iWar know we want to dial all numbers between 1000 and 1100. When iWar starts, it'll send your modem a command that looks like *ATDT51904555XXXX*. The *XXXX* will be a random number between and including 1000 to 1100. The output of iWar will be stored in a flat ASCII text file named iwar.log. By default, as iWar dials, it will record information about interesting numbers it has found and attempt to identify remote modem carriers it runs into. If you wanted to log the information into a MySQL database, you'd have to configure the iwar.conf with your MySQL authentication. Then, to start iWar with MySQL logging enabled, you'd simply add the —*mysql* flag.

iWar is highly configurable. Once started, your screen, after a bit of dialing, should look something like Figure 6.1.

Figure 6.1 The iWar Startup Screen

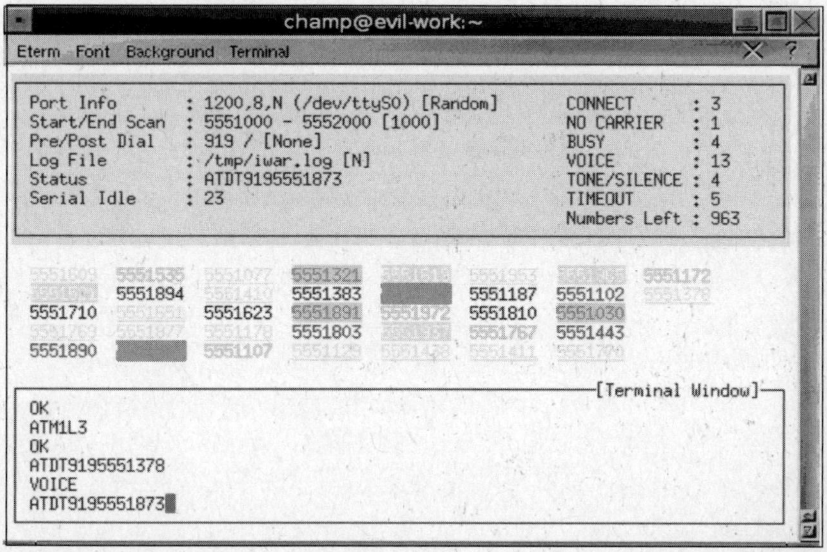

The top part of the screen gives you basic information like what serial port is being used, where the log file is being stored and statistics on what it has found (on the far top right). At the bottom is a "terminal window." This allows you to watch iWar interact with the modem. In the middle of the screen, with all the pretty colors, are the numbers that have been dialed. Those colors represent what iWar has found. By looking at those colors and the number highlighted, you can tell what numbers were busy, where modem carriers were found or numbers that gave no response. The color breakdown for iWar is shown in the following table.

Green / A_STANDOUT	Manually marked by the user
Yellow / A_BOLD	BUSY
Green / A_BLINK	CONNECT (modem found)!
Blue / A_UNDERLINE	VOICE
White / A_DIM	NO ANSWER
Magenta / A_NORMAL	Already scanned (loaded from a file)
Cyan / A_REVERSE	Blacklisted number (not called)
Red / A_NORMAL	Number skipped by the user (spacebar)
Blue / A_STANDOUT	Possible interesting number!
Cyan / A_UNDERLINE	Paused, then marked (IAX2 mode only)

The idea iWar uses behind the color coding is that, at a "glance," you can get an idea of what has been located.

iWar with VoIP

Up to now, we've talked about using iWar with physical hardware (a modem, a VoIP adapter, and Asterisk). iWar does contain some VoIP (IAX2) functionality. According to the projects Web page, it's the "first war dialer with VoIP functionality."

We used iWar via good old-fashioned serial because the VoIP detection engine is still under development. That is, in VoIP mode, iWar won't be able to detect modems, fax machines, and other equipment. It simply operates as a VoIP client. With a headset, you can let iWar do the dialing and even chat with people you call through it. According to the iWar mailing list, the addition of SIP and a detection engine is in the works. "Proof of concept" code has been chatted about on the mailing list for some time, but hasn't been included. While it is interesting to let iWar do your dialing and act as a VoIP client, you manually have to identify interesting numbers. Until the detection engine matures, the more practical way to war dial is to use a traditional modem. The detection engine should be added and released within the next couple of revisions of the code.

If you do wish to bypass the hardware way of scanning and have compiled iWar with IAX2 functionality, you can start iWar in IAX2 mode by passing the —*iax2* flag. For example:

```
$ iwar -npa 904 -nxx 555 -range 1000-1000 -iax2
```

Once started, the iWar curses screen will change a little bit since we are not using a traditional analog modem. It should look something like Figure 6.2.

Figure 6.2 The iWar Curses Screen

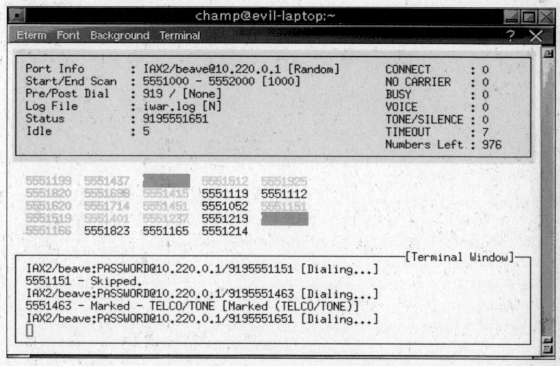

The color coding works the same as using iWar with a serial/analog modem, but the terminal window now shows the VoIP interaction with your provider or Asterisk. Future versions of iWar promise to be able to detect over VoIP the same things that a traditional modem can, and more.

All Modems Are Not Alike

Most people believe that all modems are created equally. This isn't the case. Some modems serve their basic function: to connect to another modem. Other modems are "smarter," and the smarter the modem, the better the results of your war dial. Most off-the-shelf modems will connect to other modems, but only detect things like BUSY, NO DIALTONE, and other trivial items, while smarter modems can detect remote RINGING and VOICE. Smarter modems will also speed up your scanning

If you're interested in scanning for fax machines and modem carriers, you'll probably have to make two sweeps: one to search for faxes, the other to search for modem carriers. Not many modems can do both within a single sweep.

If you are doing a serious security audit by war dialing, test the capabilities of your modem before throwing it into a war-dialing challenge.

The author of *iWar* (Da Beave) suggests the U.S. Robotics Courier (V.Everything) for the best results. You can typically find these types of modems on eBay for around $10 to $25. You can also use iWar with multiple modems to speed up your scanning.

Legalities and Tips

As stressed earlier in this chapter get *prior permission* before doing a war-dialing attack. Also, check your state's laws about war dialing. It might be that war dialing, even with prior permission, isn't legal within your state.

If you have prior consent to target a company and scan for rogue telephony-related devices and there are no legalities in your area regarding war dialing, it doesn't mean you can fire off a 10,000 number scan. Many VoIP providers have a clause against this in their terms of service. You'll want to scan "slow and low"—that is, instead of dialing 10,000 numbers at once, dial 50 numbers and wait for a while.

Timing is also an issue. Know the area you are dialing. For example, if the target is a state government agency, dialing in the evening will probably be better. Also, many state agencies have entire exchanges dedicated to them. This way, by dialing during the evening, you won't upset people at work. If it's a business that's located in a business district, the same applies. However, if the business is located in a suburban neighborhood, you'll probably want to war dial during the afternoon. The idea is that most people won't be at home, because they'll be at work. If you dialed the same exchange/area during the day, you'd likely upset many people.

War dialing is about looking for interesting things, not annoying people.

What You Can Find

There are literally thousands of different types of devices connected to the telephone network. RoguePCs with PC Anywhere installed, Xyplex terminal servers, OpenVMS clusters, SCO Unix machines, Linux machines, telco test equipment... The list goes on and on. Some require a type of authentication, while other hardware will let you in simply because you dialed the right number—that is, right into a network that might be guarded with thousands of dialers of firewalls and Intrusion Prevention Systems (IDSs).

Summary

Interfacing Asterisk with hardware can take some creativity. In these simple examples, we're using good old-fashioned serial communications. Serial is used quite a bit, but it's only one means to connect to hardware. The hardware you might want to connect to and write an interface for Asterisk might be connected by USB or something you probe over a TCP/IP network. The core ideas are still the same. Connect to the hardware, send a command if needed, and format the output so it can be used with Asterisk. Based on the information supplied by the device, an action can be taken, if needed. The AGI and functionality it will carry out is up to you. These examples use perl (Practical Extraction and Report Language) since it is a common and well-documented language. As the name implies, we are using it to "extract" information from the remote devices. perl also has some modules that assist in working with Asterisk (Asterisk::AGI), but just about any language can be used.

Solutions Fast Track

Serial

☑ Serial communications are simple and well documented. Many devices use serial to interface with hardware.

☑ One-way communications is data that is fed to us. Examples of this are things like magnetic card readers.

☑ Two-way communications require that a command be sent to the device beforewe can get a response. Examples of this are some environmental control systems and robotics.

☑ Serial is used as the basic example, but the same ideas apply with other communications protocols such as IR (Infrared). USB and TCP (for example, telnet) controlled equipment.

Motion

☑ Motion is a very powerful tool used to monitor camera(s) and detect events you might be concerned about. For example, if someone breaks into your home.

☑ One AGI/routine is used in conjunction with Motion and can be used to notify you if an event occurs.

☑ The routine will build the necessary "call files" and alert you by telephone if something was detected.

Modems

☑ Traditional analog modems are still used in point-of-sales equipment, TiVos, and other equipment that connect to the PSTN.

☑ Connections with a modem can be accomplished, but typically at lower speeds.

☑ You'll need to use noncompressed codecs, and the better the network connection, the better your modem connections will be.

☑ There are TCP/IP protocols for Fax over IP (T.38) and modems (ITU V.150.1). Unfortunately, ITU V.150.1 (also known as V.MOIP) isn't well supported.

☑ Using VoIP during security audits, you can mask where you are coming from. The data traditionally passed over the PSTN isn't passed over VoIP networks.

☑ VoIP can also be used to look up phone number information. This is known as backspoofing.

☑ When using a traditional modem and VoIP for scanning/war dialing, realize that not all modems are created equal. Some are better than others.

Legalities and Tips

☑ Before doing a security audit via VoIP and war dialing, check your local and state laws!

☑ Always get prior permission from the target before starting a security audit via war dialing.

Frequently Asked Questions

The following Frequently Asked Questions, answered by the authors of this book, are designed to both measure your understanding of the concepts presented in this chapter and to assist you with real-life implementation of these concepts. To have your questions about this chapter answered by the author, browse to **www.syngress.com/solutions** and click on the **"Ask the Author"** form.

Q: Using VoIP with point-of-sales modem equipment is sort of dangerous isn't it?

A: It can be. Before hooking up anything, you should first see what sort of data is being sent. Odds are, it's something you wouldn't want leaked out. Proper security measures should be in place before attaching such equipment to any VoIP network (encryption, VLANs, and so on).

Q: Seriously, what *can* you find via war dialing?

A: Many people and companies would be surprised. Often, the organization being targeted doesn't even know they have hardware that is connected to the PSTN. I've seen routers, dial-up servers, SCO machines, OpenVMS servers, rogue PC Anywhere installs, Linux machines, and much, much more. Some require authentications, while others simply let you into the network, bypassing thousands of dollars of network monitoring equipment.

Q: In the example with Motion, you use it in an environment where motion should not be detected. I'd like to use Motion outside to detect if people come to my front door, walk down my driveway, and so on. Can this be done?

A: Yes. With motion you can create "mask" files that will ignore motion from certain areas of the image. For instance, you can create a mask to ignore a tree in your front yard when the wind blows, but alert you when motion is detected on a walkway.

## Chapter 7

Threats to VoIP Communications Systems

Solutions in this chapter:

- Denial-of-Service or VoIP Service Disruption
- Call Hijacking and Interception
- H.323-Specific Attacks
- SIP-Specific Attacks

Introduction

Converging voice and data on the same wire, regardless of the protocols used, ups the ante for network security engineers and managers. One consequence of this convergence is that in the event of a major network attack, the organization's entire telecommunications infrastructure can be at risk. Securing the whole VoIP infrastructure requires planning, analysis, and detailed knowledge about the specifics of the implementation you choose to use.

Table 7.1 describes the general levels that can be attacked in a VoIP infrastructure.

Table 7.1 VoIP Vulnerabilities

Vulnerability	Description
IP infrastructure	Vulnerabilities on related non-VoIP systems can lead to compromise of VoIP infrastructure.
Underlying operating system	VoIP devices inherit the same vulnerabilities as the operating system or firmware they run on. Operating systems are Windows and Linux.
Configuration	In their default configuration most VoIP devices ship with a surfeit of open services. The default services running on the open ports may be vulnerable to DoS attacks, buffer overflows, or authentication bypass.
Application level	Immature technologies can be attacked to disrupt or manipulate service. Legacy applications (DNS, for example) have known problems.

Denial-of-Service or VoIP Service Disruption

Denial-of-service (DoS) attacks can affect any IP-based network service. The impact of a DoS attack can range from mild service degradation to complete loss of service. There are several classes of DoS attacks. One type of attack in which packets can simply be flooded into or at the target network from multiple external sources is called a distributed denial-of-service (DDoS) attack (see Figures 7.1 and 7.2).

In this figure, traffic flows normally between internal and external hosts and servers. In Figure 7.2, a network of computers (e.g., a botnet) directs IP traffic at the interface of the firewall.

Figure 7.1 Typical Internet Access

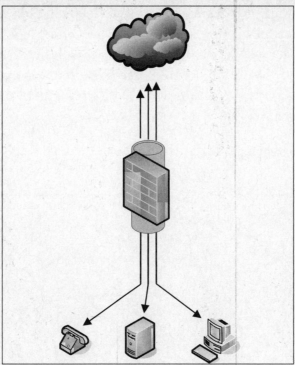

Figure 7.2 A Distributed Denial-of-Service Attack

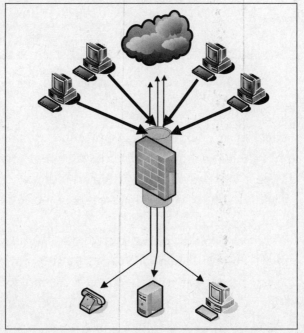

The second large class of Denial of Service (DoS) conditions occurs when devices within the internal network are targeted by a flood of packets so that they fail—taking out related parts of the infrastructure with them. As in the DdoS scenarios described earlier in this chapter, service disruption occurs to resource depletion—primarily bandwidth and CPU resource starvation (see Figure 7.3). For example, some IP telephones will stop working if they receive a UDP packet larger than 65534 bytes on port 5060.

Figure 7.3 An Internal Denial-of-Service Attack

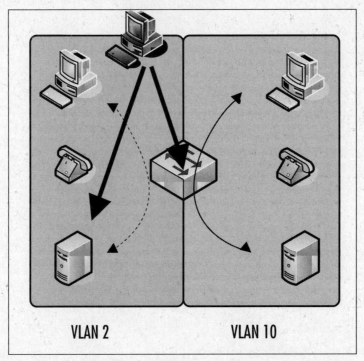

VLAN 2   VLAN 10

Neither integrity checks nor encryption can prevent these attacks. DoS or DDoS attacks are characterized simply by the volume of packets sent toward the victim computer; whether those packets are signed by a server, contain real or spoofed source IP addresses, or are encrypted with a fictitious key—none of these are relevant to the attack.

DoS attacks are difficult to defend against, and because VoIP is just another IP network service, it is just as susceptible to DoS attack as any other IP network services. Additionally, DoS attacks are particularly effective against services such as VoIP and other real-time services, because these services are most sensitive to adverse network status. Viruses and worms are included in this category as they often cause DoS or DDoS due to the increased network traffic that they generate as part of their efforts to replicate and propagate.

How do we defend against these DoS conditions (we won't use the term attack here because some DoS conditions are simply the unintended result of other unrelated actions)? Let's begin with internal DoS. Note in Figure 7.3 that VLAN 10 on the right is not affected by the service disruption on the left in VLAN 2. This illustrates one critical weapon the security administrator has in thwarting DoS conditions—logical segregation of network domains in separate compartments. Each compartment can be configured to be relatively immune to the results of DoS in the others. This is described in more detail in Chapter 8.

Point solutions will also be effective in limiting the consequences of DoS conditions. For example, because strong authentication is seldom used in VoIP environments, the message processing components must trust and process messages from possible attackers. The additional processing of bogus messages exhausts server resources and leads to a DoS. SIP or H.323 Registration Flooding is an example of this, described in the list of DoS threats, later. In that case, message processing servers can mitigate this specific threat by limiting the number of registrations it will accept per minute for a particular address (and/or from a specific IP address). An intrusion prevention system (IPS) may be useful in fending off certain types of DoS attacks. These devices sit on the datapath and monitor passing traffic. When anomalous traffic is detected (either by matching against a database of attack signatures or by matching the results of an anomaly-detection algorithm) the IPS blocks the suspicious traffic. One problem I have seen with these devices—particularly in environments with high availability requirements—is that they sometimes block normal traffic, thus creating their own type of DoS.

Additionally, security administrators can minimize the chances of DoS by ensuring that IP telephones and servers are updated to the latest stable version and release. Typically, when a DoS warning is announced by bugtraq, the vendor quickly responds by fixing the offending software.

NOTE

VoIP endpoints can be infected with new VoIP device or protocol-specific viruses. WinCE, PalmOS, SymbianOS, and POSIX-based softphones are especially vulnerable because they typically do not run antivirus software and have less robust operating systems. Several Symbian worms already have been detected in the wild. Infected VoIP devices then create a new "weak link" vector for attacking other network resources.

Compromised devices can be used to launch attacks against other systems in the same network, particularly if the compromised device is trusted (i.e., inside the firewall). Malicious programs installed by an attacker on compromised devices can capture user input, capture traffic, and relay user data over a "back channel" to the attacker. This is especially worrisome for softphone users.

VoIP systems must meet stringent service availability requirements. Following are some example DoS threats can cause the VoIP service to be partially or entirely unavailable by preventing successful call placement (including emergency/911), disconnecting existing calls, or preventing use of related services like voicemail. Note that this list is not exhaustive but illustrates some attack scenarios.

- **TLS Connection Reset** It's not hard to force a connection reset on a TLS connection (often used for signaling security between phones and gateways)—just send the right kind of junk packet and the TLS connection will be reset, interrupting the signaling channel between the phone and call server.

- **VoIP Packet Replay Attack** Capture and resend out-of-sequence VoIP packets (e.g., RTP SSRC—SSRC is an RTP header field that stands for Synchronization Source) to endpoints, adding delay to call in progress and degrading call quality.

- **Data Tunneling** Not exactly an attack; rather tunneling data through voice calls creates, essentially, a new form of unauthorized modem. By transporting modem signals through a packet network by using pulse code modulation (PCM) encoded packets or by residing within header information, VoIP can be used to support a modem call over an IP network. This technique may be used to bypass or undermine a desktop modem policy and hide the existence of unauthorized data connections. This is similar in concept to the so-called "IP over HTTP" threat (i.e., "Firewall Enhancement Protocol" RFC 3093)—a classic problem for any ports opened on a firewall from internal sources.

- **QoS Modification Attack** Modify non-VoIP-specific protocol control information fields in VoIP data packets to and from endpoints to degrade or deny voice service. For example, if an attacker were to change 802.1Q VLAN tag or IP packet ToS bits, either as a man-in-the-middle or by compromising endpoint device configuration, the attacker could disrupt the quality of service "engineered" for a VoIP network. By subordinating voice traffic to data traffic, for example, the attacker might substantially delay delivery of voice packets.

- **VoIP Packet Injection** Send forged VoIP packets to endpoints, injecting speech or noise or gaps into active call. For example, when RTP is used without authentication of RTCP packets (and without SSRC sampling), an attacker can inject RTCP packets into a multicast group, each with a different SSRC, which can grow the group size exponentially.

- **DoS against Supplementary Services** Initiate a DoS attack against other network services upon which the VoIP service depends (e.g., DHCP, DNS, BOOTP). For example, in networks where VoIP endpoints rely on DHCP-assigned addresses, disabling the DHCP server prevents endpoints (soft- and hardphones) from acquiring addressing and routing information they need to make use of the VoIP service.

- **Control Packet Flood** Flood VoIP servers or endpoints with unauthenticated call control packets, (e.g., H.323 GRQ, RRQ, URQ packets sent to UDP/1719). The attacker's intent is to deplete/exhaust device, system, or network resources to the extent that VoIP service is unusable. Any open administrative and maintenance port on call processing and VoIP-related servers can be a target for this DoS attack.

- **Wireless DoS** Initiate a DoS attack against wireless VoIP endpoints by sending 802.11 or 802.1X frames that cause network disconnection (e.g., 802.11 Deauthenticate flood, 802.1X EAP-Failure, WPA MIC attack, radio spectrum jamming). For example, a Message Integrity Code attack exploits a standard countermeasure whereby a wireless access point disassociates stations when it receives two invalid frames within 60 seconds, causing loss of network connectivity for 60 seconds. In a VoIP environment, a 60-second service interruption is rather extreme.

- **Bogus Message DoS** Send VoIP servers or endpoints valid-but-forged VoIP protocol packets to cause call disconnection or busy condition (e.g., RTP SSRC collision, forged RTCP BYE, forged CCMS, spoofed endpoint button push). Such attacks cause the phone to process a bogus message and incorrectly terminate a call, or mislead a calling party into believing the called party's line is busy.

- **Invalid Packet DoS** Send VoIP servers or endpoints invalid packets that exploit device OS and TCP/IP implementation denial-of-service CVEs. For example, the exploit described in CAN-2002-0880 crashes Cisco IP phones using jolt, jolt2, and other common fragmentation-based DoS attack methods. CAN-2002-0835 crashes certain VoIP phones by exploiting DHCP DoS CVEs. Avaya IP phones may be vulnerable to port zero attacks.

- **Immature Software DoS** PDA/handheld softphones and first generation VoIP hardphones are especially vulnerable because they are not as mature or intensely scrutinized. VoIP call servers and IP PBXs also run on OS platforms with many known CVEs. Any open administrative/maintenance port (e.g., HTTP, SNMP, Telnet) or vulnerable interface (e.g., XML, Java) can become an attack vector.

- **VoIP Protocol Implementation DoS** Send VoIP servers or endpoints invalid packets to exploit a VoIP protocol implementation vulnerability to a DoS attack. Several such exploits are identified in the MITRE CVE database (http://cve.mitre.org). For example, CVE-2001-00546 uses malformed H.323 packets to exploit Windows ISA memory leak and exhaust resources. CAN-2004-0056 uses malformed H.323 packets to exploit Nortel BCM DoS vulnerabilities. Lax software update practices (failure to install CVE patches) exacerbate risk.

- **Packet of Death DoS** Flood VoIP servers or endpoints with random TCP, UDP, or ICMP packets or fragments to exhaust device CPU, bandwidth, TCP sessions, and so on. For example, an attacker can initiate a TCP Out of Band DoS attack by sending a large volume of TCP packets marked "priority delivery" (the TCP Urgent flag). During any flood, increased processing load interferes with the receiving system's ability to process real traffic, initially delaying voice traffic processing but ultimately disrupting service entirely.

- **IP Phone Flood DoS** Send a very large volume of call data toward a single VoIP endpoint to exhaust that device's CPU, bandwidth, TCP sessions, and so on. Interactive voice response systems, telephony gateways, conferencing servers, and voicemail systems are able to generate more call data than a single endpoint can handle and so could be leveraged to flood an endpoint.

Call Hijacking and Interception

Call interception and eavesdropping are other major concerns on VoIP networks. The VOIPSA threat taxonomy (www.voipsa.org/Activities/taxonomy-wiki.php) defines eavesdropping as "a method by which an attacker is able to monitor the entire signaling and/or data stream between two or more VoIP endpoints, but cannot or does not alter the data itself." Successful call interception is akin to wiretapping in that conversations of others can be stolen, recorded, and replayed without their knowledge. Obviously, an attacker who can intercept and store these data can make use of the data in other ways as well.

Tools & Traps…

DNS Poisoning

A DNS A (or address) record is used for storing a domain or hostname mapping to an IP address. SIP makes extensive use of SRV records to locate SIP services such as SIP proxies and registrars. SRV (service) records normally begin with an underscore (_sip.tcpserver.udp.domain.com) and consist of information describing service, transport, host, and other information. SRV records allow administrators to use several servers for a single domain, to move services from host to host with little fuss, and to designate some hosts as primary servers for a service and others as backups.

An attacker's goal, when attempting a DNS Poisoning or spoofing attack, is to replace valid cached DNS A, SRV, or NS records with records that point to the attacker's server(s). This can be accomplished in a number of fairly trivial ways—the easiest being to initiate a zone transfer from the attacker's DNS server to the victim's misconfigured DNS server, by asking the victim's DNS server to resolve a networked device within the attacker's domain. The victim's

Continued

DNS server accepts not only the requested record from the attacker's server, but it also accepts and caches any other records that the attacker's server includes.

Thus, in addition to the A record for www.attacker.com, the victim DNS server may receive a bogus record for www.yourbank.com. The innocent victim will then be redirected to the attacker.com Web site anytime he or she attempts to browse to the yourbank.com Web site, as long as the bogus records are cached. Substitute a SIP URL for a Web site address, and the same scenario can be repeated in a VoIP environment.

This family of threats relies on the absence of cryptographic assurance of a request's originator. Attacks in this category seek to compromise the message integrity of a conversation. This threat demonstrates the need for security services that enable entities to authenticate the originators of requests and to verify that the contents of the message and control streams have not been altered in transit.

In the past several years, as host PCs have improved their processing power and their ability to process networked information, network administrators have instituted a hierarchical access structure that consists of a single, dedicated switched link for each host PC to distribution or backbone devices. Each networked user benefits from a more reliable, secure connection with guaranteed bandwidth. The use of a switched infrastructure limits the effectiveness of packet capture tools or protocol analyzers as a means to collect VoIP traffic streams. Networks that are switched to the desktop allow normal users' computers to monitor only broadcast and unicast traffic that is destined to their particular MAC address. A user's NIC (network interface card) literally does not see unicast traffic destined for other computers on the network.

The address resolution protocol (ARP) is a method used on IPv4 Ethernet networks to map the IP address (layer 3) to the hardware or MAC (Media Access Control) layer 2 address. (Note that ARP has been replaced in IPv6 by Neighbor Discovery [ND] protocol. The ND protocol is a hybrid of ARP and ICMP.) Two classes of hardware addresses exist: the broadcast address of all ones, and a unique 6 byte identifier that is burned into the PROM of every NIC (Network Interface Card).

Figure 7.4 illustrates a typical ARP address resolution scheme. A host PC (10.1.1.1) that wishes to contact another host (10.1.1.2) on the same subnet issues an ARP broadcast packet (ARPs for the host) containing its own hardware and IP addresses. NICs contain filters that allow them to drop all packets not destined for their unique hardware address or the broadcast address, so all NICs but the query target silently discard the ARP broadcast. The target NIC responds to the query request by unicasting its IP and hardware address, completing the physical to logical mapping, and allowing communications to proceed at layer 3.

Figure 7.4 Typical ARP Request/Reply

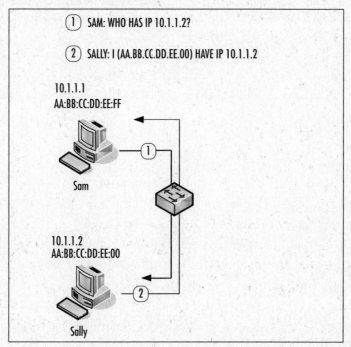

To minimize broadcast traffic, many devices cache ARP addresses for a varying amount of time: The default ARP cache timeout for Linux is one minute; for Windows NT, two minutes, and for Cisco routers, four hours. This value can be trivially modified in most systems. The ARP cache is a table structure that contains IP address, hardware address, and oftentimes, the name of the interface the MAC address is discovered on, the type of media, and the type of ARP response. Depending upon the operating system, the ARP cache may or may not contain an entry for its own addresses.

In Figure 7.4, Sam's ARP cache contains one entry prior to the ARP request/response:

Internet Address	Physical Address	
10.1.1.1	AA:BB:CC:DD:EE:FF	int0

After the ARP request/response completes, Sam's ARP cache now contains two entries:

Internet Address	Physical Address	
10.1.1.1	AA:BB:CC:DD:EE:FF	int0
10.1.1.2	AA:BB:CC:DD:EE:00	int0

Note that Sally's ARP cache, as a result of the request/response communications, is updated with the hardware:IP mappings for both workstations as well.

ARP Spoofing

ARP is a fundamental Ethernet protocol. Perhaps for this reason, manipulation of ARP packets is a potent and frequent attack mechanism on VoIP networks. Most network administrators assume that deploying a fully switched network to the desktop prevents the ability of network users to sniff network traffic and potentially capture sensitive information traversing the network. Unfortunately, several techniques and tools exist that allow any user to sniff traffic on a switched network because ARP has no provision for authenticating queries or query replies. Additionally, because ARP is a stateless protocol, most operating systems (Solaris is an exception) update their cache when receiving ARP reply, regardless of whether they have sent out an actual request.

Among these techniques, ARP redirection, ARP spoofing, ARP hijacking, and ARP cache poisoning are related methods for disrupting the normal ARP process. These terms frequently are interchanged and confused. For the purpose of this section, we'll refer to ARP cache poisoning and ARP spoofing as the same process. Using freely available tools such as ettercap, Cain, and dsniff, an evil IP device can spoof a normal IP device by sending unsolicited ARP replies to a target host. The bogus ARP reply contains the hardware address of the normal device and the IP address of the malicious device. This "poisons" the host's ARP cache (see Figure 7.5).

Figure 7.5 ARP Spoofing (Cache Poisoning)

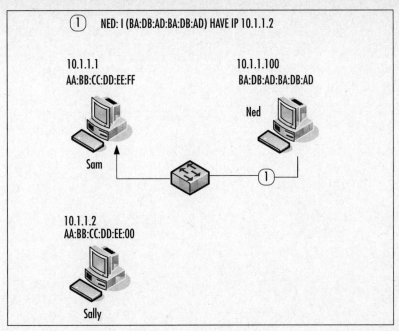

In Figure 7.5, Ned is the attacking computer. When SAM broadcasts an ARP query for Sally's IP address, Ned, the attacker, responds to the query stating that the IP address (10.1.1.2) belongs to Ned's MAC address, BA:DB:AD:BA:DB:AD. Packets sent from Sam supposedly to Sally will be sent to Ned instead. Sam will mistakenly assume that Ned's MAC address corresponds to Sally's IP address and will direct all traffic destined for that IP address to Ned's MAC. In fact, Ned can poison Sam's ARP cache without waiting for an ARP query since on Windows systems (9x/NT/2K), static ARP entries are overwritten whenever a query response is received regardless of whether or not a query was issued.

Sam's ARP cache now looks like this:

Internet Address	Physical Address	
10.1.1.1	AA:BB:CC:DD:EE:FF	int0
10.1.1.2	BA:DB:AD:BA:DB:AD	int0

This entry will remain until it ages out or a new entry replaces it.

ARP redirection can work bidirectionally, and a spoofing device can insert itself in the middle of a conversation between two IP devices on a switched network (see Figure 7.6). This is probably the most insidious ARP-related attack. By routing packets on to the devices that should truly be receiving the packets, this insertion (known as a Man/Monkey/Moron in the Middle attack) can remain undetected for some time. An attacker can route packets to /dev/null (nowhere) as well, resulting in a DoS attack.

Figure 7.6 An ARP MITM Attack

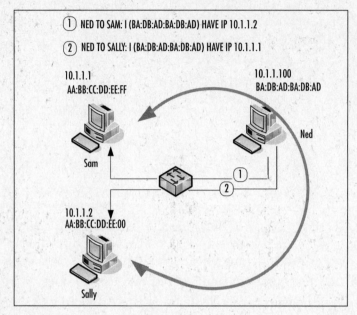

Sam's ARP cache:

Internet Address	Physical Address	
10.1.1.1	AA:BB:CC:DD:EE:FF	int0
10.1.1.2	BA:DB:AD:BA:DB:AD	int0

Sally's ARP cache:

Internet Address	Physical Address	
10.1.1.1	BA:DB:AD:BA:DB:AD	int0
10.1.1.2	AA:BB:CC:DD:EE:00	int0

As all IP traffic between the true sender and receiver now passes through the attacker's device, it is trivial for the attacker to sniff that traffic using freely available tools such as Ethereal or tcpdump. Any unencrypted information (including e-mails, usernames and passwords, and web traffic) can be intercepted and viewed.

This interception has potentially drastic implications for VoIP traffic. Freely available tools such as vomit and rtpsniff, as well as private tools such as VoipCrack, allow for the interception and decoding of VoIP traffic. Captured content can include speech, signaling and billing information, multimedia, and PIN numbers. Voice conversations traversing the internal IP network can be intercepted and recorded using this technique.

There are a number of variations of the aforementioned techniques. Instead of imitating a host, the attacker can emulate a gateway. This enables the attacker to intercept numerous packet streams. However, most ARP redirection techniques rely on stealth. The attacker in these scenarios hopes to remain undetected by the users being impersonated. Posing as a gateway may result in alerting users to the attacker's presence due to unanticipated glitches in the network, because frequently switches behave in unexpected ways when attackers manipulate ARP processes. One unintended (much of the time) consequence of these attacks, particularly when switches are heavily loaded, is that the switch CAM (Content-Addressable Memory) table—a finite-sized IP address to MAC address lookup table—becomes disrupted. This leads to the switch forwarding unicast packets out many ports in unpredictable fashion. Penetration testers may want to keep this in mind when using these techniques on production networks.

In order to limit damage due to ARP manipulation, administrators should implement software tools that monitor MAC to IP address mappings. The freeware tool, Arpwatch, monitors these pairings. At the network level, MAC/IP address mappings can be statically coded on the switch; however, this is often administratively untenable. Dynamic ARP Inspection (DAI) is available on newer Cisco Catalyst 6500 switches. DAI is part of Cisco's Integrated Security (CIS) functionality and is designed to prevent several layer two and layer three spoofing attacks, including ARP redirection attacks. Note that DAI and CIS are available only on Catalyst switches using native mode (Cisco IOS).

The potential risks of decoding intercepted VoIP traffic can be eliminated by implementing encryption. Avaya's Media Encryption feature is an example of this. Using Media Encryption, VoIP conversations between two IP endpoints are encrypted using AES encryption. In highly secure environments, organizations should ensure that Media Encryption is enabled on all IP codec sets in use.

DAI enforces authorized MAC-to-IP address mappings. Media Encryption renders traffic, even if intercepted, unintelligible to an attacker.

The following are some additional examples of call or signal interception and hijacking. This class of threats, though typically more difficult to accomplish than DoS, can result in significant loss or alteration of data. DoS attacks, whether caused by active methods or inadvertently, although important in terms of quality of service, are more often than not irritating to users and administrators. Interception and hijacking attacks, on the other hand, are almost always active attacks with theft of service, information, or money as the goal. Note that this list is not exhaustive but illustrates some attack scenarios.

- **Rogue VoIP Endpoint Attack** Rogue IP endpoint contacts VoIP server by leveraging stolen or guessed identities, credentials, and network access. For example, a rogue endpoint can use an unprotected wall jack and auto-registration of VOIP phones to get onto the network. RAS password guessing can be used to masquerade as a legitimate endpoint. Lax account maintenance (expired user accounts left active) increases risk of exploitation.

- **Registration Hijacking** Registration hijacking occurs when an attacker impersonates a valid UA to a registrar and replaces the registration with its own address. This attack causes all incoming calls to be sent to the attacker.

- **Proxy Impersonation** Proxy impersonation occurs when an attacker tricks a SIP UA or proxy into communicating with a rogue proxy. If an attacker successfully impersonates a proxy, he or she has access to all SIP messages.

- **Toll Fraud** Rogue or legitimate VoIP endpoint uses a VoIP server to place unauthorized toll calls over the PSTN. For example, inadequate access controls can let rogue devices place toll calls by sending VoIP requests to call processing applications. VoIP servers can be hacked into in order to make free calls to outside destinations. Social engineering can be used to obtain outside line prefixes.

- **Message Tampering** Capture, modify, and relay unauthenticated VoIP packets to/from endpoints. For example, a rogue 802.11 AP can exchange frames sent or received by wireless endpoints if no payload integrity check (e.g., WPA MIC, SRTP) is used. Alternatively, these attacks can occur through registration hijacking, proxy impersonation, or an attack on any component trusted to process SIP or H.323 messages, such as the proxy, registration servers, media gateways, or firewalls. These represent non-ARP-based MITM attacks.

- **VoIP Protocol Implementation Attacks** Send VoIP servers or endpoints invalid packets to exploit VoIP protocol implementation CVEs. Such attacks can lead to escalation of privileges, installation and operation of malicious programs, and system compromise. For example, CAN-2004-0054 exploits Cisco IOS H.323 implementation CVEs to execute arbitrary code. CSCed33037 uses unsecured IBM Director agent ports to gain administrative control over IBM servers running Cisco VoIP products.

Notes from the Underground…

ANI/Caller-ID Spoofing

Caller ID is a service provided by most telephone companies (for a monthly cost) that will tell you the name and number of an incoming call. Automatic Number Identification (ANI) is a system used by the telephone company to determine the number of the calling party. To spoof Caller-ID, an attacker sends modem tones over a POTS lines between rings 1 and 2. ANI spoofing is setting the ANI so as to send incorrect ANI information to the PSTN so that the resulting Caller-ID is misleading. Traditionally this has been a complicated process either requiring the assistance of a cooperative phone company operator or an expensive company PBX system.

In ANI/Caller-ID spoofing, an evildoer hijacks phone number and the identity of a trusted party, such as a bank or a government office. The identity appears on the caller ID box of an unsuspecting victim, with the caller hoping to co-opt valuable information, such as account numbers, or otherwise engage in malicious mischief. This is not a VoIP issue, per se. In fact, one of the big drawbacks about VoIP trunks is their inability to send ANI properly because of incomplete standards.

H.323-Specific Attacks

The only existing vulnerabilities that we are aware of at this time take advantage of ASN.1 parsing defects in the first phase of H.225 data exchange. More vulnerabilities can be expected for several reasons: the large number of differing vendor implementations, the complex nature of this collection of protocols, problems with the various implementations of ASN.1/PER encoding/decoding, and the fact that these protocols—alone and in concert—have not endured the same level of scrutiny that other more common protocols have been subjected to. For example, we have unpublished data that shows that flooding a gateway or media server with GRQ request packets (RAS registration request packets) results in a DoS against certain vendor gateway implementations—basically the phones deregister.

SIP-Specific Attacks

Multiple vendors have confirmed vulnerabilities in their respective SIP (Session Initiation Protocol) implementations. The vulnerabilities have been identified in the INVITE message used by two SIP endpoints during the initial call setup. The impact of successful exploitation of the vulnerabilities has not been disclosed but potentially could result in a compromise of a vulnerable device. (CERT: CA-2003-06.) In addition, many recent examples of SIP Denial of Service attacks have been reported.

Recent issues that affect Cisco SIP Proxy Server (SPS) [Bug ID CSCec31901] demonstrate the problems SIP implementers may experience due to the highly modular architecture or this protocol. The SSL implementation in SPS (used to secure SIP sessions) is vulnerable to an ASN.1 BER decoding error similar to the one described for H.323 and other protocols. This example illustrates a general concern with SIP: As the SIP protocol links existing protocols and services together, all the classic vulnerabilities in services such as SSL, HTTP, and SMTP may resurface in the VoIP environment.

Summary

DoS attacks, whether they are intentional or unintended, are the most difficult VoIP-related threat to defend against. The packet switching nature of data networks allows multiple connections to share the same transport medium. Therefore, unlike telephones in circuit-switched networks, an IP terminal endpoint can receive and potentially participate in multiple calls at once. Thus, an endpoint can be used to amplify attacks. On VoIP networks, resources such as bandwidth must be allocated efficiently and fairly to accommodate the maximum number of callers. This property can be violated by attackers who aggressively and abusively obtain an unnecessarily large amount of resources. Alternatively, the attacker simply can flood the network with large number of packets so that resources are unavailable to all other callers.

In addition, viruses and worms create DoS conditions due to the network traffic generated by these agents as they replicate and seek out other hosts to infect. These agents are proven to wreak havoc with even relatively well-secured data networks. VoIP networks, by their nature, are exquisitely sensitive to these types of attacks. Remedies for DoS include logical network partitioning at layers 2 and 3, stateful firewalls with application inspection capabilities, policy enforcement to limit flooded packets, and out-of-band management. Out-of-band management is required so that in the event of a DoS event, system administrators are still able to monitor the network and respond to additional events.

Theft of services and information is also problematic on VoIP networks. These threats are almost always due to active attack. Many of these attacks can be thwarted by implementing additional security controls at layer 2. This includes layer 2 security features such as DHCP Snooping, Dynamic ARP Inspection, IP Source Guard, Port Security, and VLAN ACLs. The fundamental basis for this class of attacks is that the identity of one or more of the devices that participate is not legitimate.

Endpoints must be authenticated, and end users must be validated in order to ensure legitimacy. Hijacking and call interception revolves around the concept of fooling and manipulating weak or nonexistent authentication measures. We are all familiar with different forms of authentication, from the password used to login to your computer to the key that unlocks the front door. The conceptual framework for authentication is made up of three factors: "something you have" (a key or token), "something you know" (a password or secret handshake), or "something you are" (fingerprint or iris pattern). Authentication mechanisms validate users by one or a combination of these. Any type of unauthenticated access, particularly to key infrastructure components such as the IP PBX or DNS server, for example, can result in disagreeable consequences for both users and administrators.

VoIP relies upon a number of ancillary services as part of the configuration process, as a means to locate users, manage servers and phones, and to ensure favorable transport, among others. DNS, DHCP, HTTP, HTTPS, SNMP, SSH, RSVP, and TFTP services all have been the subject of successful exploitation by attackers. Potential VoIP users may defer transitioning to IP Telephony if they believe it will reduce overall network security by creating new vulnerabilities that could be used to compromise non-VoIP systems and services within the same network. Effective mitigation of these threats to common data networks and services could be considered a security baseline upon which a successful VoIP deployment depends. Firewalls, network and system intrusion detection, authentication systems, anti-virus scanners, and other security controls, which should already be in place, are required to counter attacks that might debilitate any or all IP-based services (including VoIP services).

H.323 and SIP suffer security vulnerabilities based simply upon their encoding schemes, albeit for different reasons. Because SIP is an unstructured text-based protocol, it is impossibly to test all permutations of SIP messages during development for security vulnerabilities. It's fairly straightforward to construct a malformed SIP message or message sequence that results in a DoS for a particular SIP device. This may not be significant for a single UA endpoint, but if this "packet of death" can render all the carrier-class media gateway controllers in a network useless, then this becomes a significant problem. H.323 on the other hand is encoded according to ASN.1 PER encoding rules. The implementation of H.323 message parsers, rather than the encoding rules themselves, results in security vulnerabilities in the H.323 suite.

Index

GNU GENERAL PUBLIC LICENSE

Version 2, June 1991

Copyright (C) 1989, 1991 Free Software Foundation, Inc.

59 Temple Place - Suite 330, Boston, MA 02111-1307, USA

Everyone is permitted to copy and distribute verbatim copies
of this license document, but changing it is not allowed.

Preamble

The licenses for most software are designed to take away your freedom to share and change it. By contrast, the GNU General Public License is intended to guarantee your freedom to share and change free software—to make sure the software is free for all its users. This General Public License applies to most of the Free Software Foundation's software and to any other program whose authors commit to using it. (Some other Free Software Foundation software is covered by the GNU Library General Public License instead.) You can apply it to your programs, too.

When we speak of free software, we are referring to freedom, not price. Our General Public Licenses are designed to make sure that you have the freedom to distribute copies of free software (and charge for this service if you wish), that you receive source code or can get it if you want it, that you can change the software or use pieces of it in new free programs; and that you know you can do these things.

To protect your rights, we need to make restrictions that forbid anyone to deny you these rights or to ask you to surrender the rights. These restrictions translate to certain responsibilities for you if you distribute copies of the software, or if you modify it.

For example, if you distribute copies of such a program, whether gratis or for a fee, you must give the recipients all the rights that you have. You must make sure that they, too, receive or can get the source code. And you must show them these terms so they know their rights.

We protect your rights with two steps: (1) copyright the software, and (2) offer you this license which gives you legal permission to copy, distribute and/or modify the software.

Also, for each author's protection and ours, we want to make certain that everyone understands that there is no warranty for this free software. If the software is modified by someone else and passed on, we want its recipients to know that what they have is not the original, so that any problems introduced by others will not reflect on the original authors' reputations.

Finally, any free program is threatened constantly by software patents. We wish to avoid the danger that redistributors of a free program will individually obtain patent licenses, in effect making the program proprietary. To prevent this, we have made it clear that any patent must be licensed for everyone's free use or not licensed at all.

The precise terms and conditions for copying, distribution and modification follow.

TERMS AND CONDITIONS FOR COPYING, DISTRIBUTION AND MODIFICATION

0. This License applies to any program or other work which contains a notice placed by the copyright holder saying it may be distributed under the terms of this General Public License. The "Program", below, refers to any such program or work, and a "work based on the Program" means either the Program or any derivative work under copyright law: that is to say, a work containing the Program or a portion of it, either verbatim or with modifications and/or translated into another language. (Hereinafter, translation is included without limitation in the term "modification".) Each licensee is addressed as "you".

Activities other than copying, distribution and modification are not covered by this License; they are outside its scope. The act of running the Program is not restricted, and the output from the Program is covered only if its contents constitute a work based on the Program (independent of having been made by running the Program). Whether that is true depends on what the Program does.

1. You may copy and distribute verbatim copies of the Program's source code as you receive it, in any medium, provided that you conspicuously and appropriately publish on each copy an appropriate copyright notice and disclaimer of warranty; keep intact all the notices that refer to this License and to the absence of any warranty; and give any other recipients of the Program a copy of this License along with the Program.

You may charge a fee for the physical act of transferring a copy, and you may at your option offer warranty protection in exchange for a fee.

2. You may modify your copy or copies of the Program or any portion of it, thus forming a work based on the Program, and copy and distribute such modifications or work under the terms of Section 1 above, provided that you also meet all of these conditions:

a) You must cause the modified files to carry prominent notices stating that you changed the files and the date of any change.

b) You must cause any work that you distribute or publish, that in whole or in part contains or is derived from the Program or any part thereof, to be licensed as a whole at no charge to all third parties under the terms of this License.

c) If the modified program normally reads commands interactively when run, you must cause it, when started running for such interactive use in the most ordinary way, to print or display an announcement including an appropriate copyright notice and a notice that there is no warranty (or else, saying that you provide a warranty) and that users may redistribute the program under these conditions, and telling the user how to view a copy of this License. (Exception: if the Program itself is interactive but does not normally print such an announcement, your work based on the Program is not required to print an announcement.)

These requirements apply to the modified work as a whole. If identifiable sections of that work are not derived from the Program, and can be reasonably considered independent and separate works in themselves, then this License, and its terms, do not apply to those sections when you distribute them as separate works. But when you distribute the same sections as part of a whole which is a work based on the Program, the distribution of the whole must be on the terms of this License, whose permissions for other licensees extend to the entire whole, and thus to each and every part regardless of who wrote it.

Thus, it is not the intent of this section to claim rights or contest your rights to work written entirely by you; rather, the intent is to exercise the right to control the distribution of derivative or collective works based on the Program.

In addition, mere aggregation of another work not based on the Program with the Program (or with a work based on the Program) on a volume of a storage or distribution medium does not bring the other work under the scope of this License.

3. You may copy and distribute the Program (or a work based on it, under Section 2) in object code or executable form under the terms of Sections 1 and 2 above provided that you also do one of the following:

a) Accompany it with the complete corresponding machine-readable source code, which must be distributed under the terms of Sections 1 and 2 above on a medium customarily used for software interchange; or,

b) Accompany it with a written offer, valid for at least three years, to give any third party, for a charge no more than your cost of physically performing source distribution, a complete machine-readable copy of the corresponding source code, to be distributed under the terms of Sections 1 and 2 above on a medium customarily used for software interchange; or,

c) Accompany it with the information you received as to the offer to distribute corresponding source code. (This alternative is allowed only for noncommercial distribution and only if you received the program in object code or executable form with such an offer, in accord with Subsection b above.)

The source code for a work means the preferred form of the work for making modifications to it. For an executable work, complete source code means all the source code for all modules it contains, plus any associated interface definition files, plus the scripts used to control compilation and installation of the executable. However, as a special exception, the source code distributed need not include anything that is normally distributed (in either source or binary form) with the major components (compiler, kernel, and so on) of the operating system on which the executable runs, unless that component itself accompanies the executable.

If distribution of executable or object code is made by offering access to copy from a designated place, then offering equivalent access to copy the source code from the same place counts as distribution of the source code, even though third parties are not compelled to copy the source along with the object code.

4. You may not copy, modify, sublicense, or distribute the Program except as expressly provided under this License. Any attempt otherwise to copy, modify, sublicense or distribute the Program is void, and will automatically terminate your rights under this License. However, parties who have received copies, or rights, from you under this License will not have their licenses terminated so long as such parties remain in full compliance.

5. You are not required to accept this License, since you have not signed it. However, nothing else grants you permission to modify or distribute the Program or its derivative works. These actions are prohibited by law if you do not accept this License. Therefore, by modifying or distributing the Program (or any work based on the Program), you indicate your acceptance of this License to do so, and all its terms and conditions for copying, distributing or modifying the Program or works based on it.

6. Each time you redistribute the Program (or any work based on the Program), the recipient automatically receives a license from the original licensor to copy, distribute or modify the Program subject to these terms and conditions. You may not impose any further restrictions on the recipients' exercise of the rights granted herein. You are not responsible for enforcing compliance by third parties to this License.

7. If, as a consequence of a court judgment or allegation of patent infringement or for any other reason (not limited to patent issues), conditions are imposed on you (whether by court order, agreement or otherwise) that contradict the conditions of this License, they do not excuse you from the conditions of this License. If you cannot distribute so as to satisfy simultaneously your obligations under this License and any other pertinent obligations, then as a consequence you may not distribute the Program at all. For example, if a patent license would not permit royalty-free redistribution of the Program by all those who receive copies directly or indirectly through you, then the only way you could satisfy both it and this License would be to refrain entirely from distribution of the Program.

If any portion of this section is held invalid or unenforceable under any particular circumstance, the balance of the section is intended to apply and the section as a whole is intended to apply in other circumstances.

It is not the purpose of this section to induce you to infringe any patents or other property right claims or to contest validity of any such claims; this section has the sole purpose of protecting the integrity of the free software distribution system, which is implemented by public license practices. Many people have made generous contributions to the wide range of software distributed through that system in reliance on consistent application of that system; it is up to the author/donor to decide if he or she is willing to distribute software through any other system and a licensee cannot impose that choice.

This section is intended to make thoroughly clear what is believed to be a consequence of the rest of this License.

8. If the distribution and/or use of the Program is restricted in certain countries either by patents or by copyrighted interfaces, the original copyright holder who places the Program under this License may add an explicit geographical distribution limitation excluding those countries, so that distribution is permitted only in or among countries not thus excluded. In such case, this License incorporates the limitation as if written in the body of this License.

9. The Free Software Foundation may publish revised and/or new versions of the General Public License from time to time. Such new versions will be similar in spirit to the present version, but may differ in detail to address new problems or concerns.

Each version is given a distinguishing version number. If the Program specifies a version number of this License which applies to it and "any later version", you have the option of following the terms and conditions either of that version or of any later version published by the Free Software Foundation. If the Program does not specify a version number of this License, you may choose any version ever published by the Free Software Foundation.

10. If you wish to incorporate parts of the Program into other free programs whose distribution conditions are different, write to the author to ask for permission. For software which is copyrighted by the Free Software Foundation, write to the Free Software Foundation; we sometimes make exceptions for this. Our decision will be guided by the two goals of preserving the free status of all derivatives of our free software and of promoting the sharing and reuse of software generally.

NO WARRANTY

11. BECAUSE THE PROGRAM IS LICENSED FREE OF CHARGE, THERE IS NO WARRANTY FOR THE PROGRAM, TO THE EXTENT PERMITTED BY APPLICABLE LAW. EXCEPT WHEN OTHERWISE STATED IN WRITING THE COPYRIGHT HOLDERS AND/OR OTHER PARTIES PROVIDE THE PROGRAM "AS IS" WITHOUT WARRANTY OF ANY KIND, EITHER EXPRESSED OR IMPLIED, INCLUDING, BUT NOT LIMITED TO, THE IMPLIED WARRANTIES OF MERCHANTABILITY AND FITNESS FOR A PARTICULAR PURPOSE. THE ENTIRE RISK AS TO THE QUALITY AND PERFORMANCE OF THE PROGRAM IS WITH YOU. SHOULD THE PROGRAM PROVE DEFECTIVE, YOU ASSUME THE COST OF ALL NECESSARY SERVICING, REPAIR OR CORRECTION.

12. IN NO EVENT UNLESS REQUIRED BY APPLICABLE LAW OR AGREED TO IN WRITING WILL ANY COPYRIGHT HOLDER, OR ANY OTHER PARTY WHO MAY MODIFY AND/OR REDISTRIBUTE THE PROGRAM AS PERMITTED ABOVE, BE LIABLE TO YOU FOR DAMAGES, INCLUDING ANY GENERAL, SPECIAL, INCIDENTAL OR CONSEQUENTIAL DAMAGES ARISING OUT OF THE USE OR INABILITY TO USE THE PROGRAM (INCLUDING BUT NOT LIMITED TO LOSS OF DATA OR DATA BEING RENDERED INACCURATE OR LOSSES SUSTAINED BY YOU OR THIRD PARTIES OR A FAILURE OF THE PROGRAM TO OPERATE WITH ANY OTHER PROGRAMS), EVEN IF SUCH HOLDER OR OTHER PARTY HAS BEEN ADVISED OF THE POSSIBILITY OF SUCH DAMAGES.

END OF TERMS AND CONDITIONS

<u>How to Apply These Terms to Your New Programs</u>

If you develop a new program, and you want it to be of the greatest possible use to the public, the best way to achieve this is to make it free software which everyone can redistribute and change under these terms.

To do so, attach the following notices to the program. It is safest to attach them to the start of each source file to most effectively convey the exclusion of warranty; and each file should have at least the "copyright" line and a pointer to where the full notice is found.

one line to give the program's name and an idea of what it does.

Copyright (C) *yyyy name of author*

This program is free software; you can redistribute it and/or

modify it under the terms of the GNU General Public License

as published by the Free Software Foundation; either version 2

of the License, or (at your option) any later version.

This program is distributed in the hope that it will be useful,

but WITHOUT ANY WARRANTY; without even the implied warranty of

MERCHANTABILITY or FITNESS FOR A PARTICULAR PURPOSE. See the

GNU General Public License for more details.

You should have received a copy of the GNU General Public License

along with this program; if not, write to the Free Software

Foundation, Inc., 59 Temple Place - Suite 330, Boston, MA 02111-1307, USA.

Also add information on how to contact you by electronic and paper mail.

If the program is interactive, make it output a short notice like this when it starts in an interactive mode:

Gnomovision version 69, Copyright (C) *year name of author*

Gnomovision comes with ABSOLUTELY NO WARRANTY; for details

type `show w'. This is free software, and you are welcome

to redistribute it under certain conditions; type `show c'

for details.

The hypothetical commands 'show w' and 'show c' should show the appropriate parts of the General Public License. Of course, the commands you use may be called something other than 'show w' and 'show c'; they could even be mouse-clicks or menu items—whatever suits your program.

You should also get your employer (if you work as a programmer) or your school, if any, to sign a "copyright disclaimer" for the program, if necessary. Here is a sample; alter the names:

Yoyodyne, Inc., hereby disclaims all copyright

interest in the program `Gnomovision'

(which makes passes at compilers) written

by James Hacker.

signature of Ty Coon, 1 April 1989

Ty Coon, President of Vice

This General Public License does not permit incorporating your program into proprietary programs. If your program is a subroutine library, you may consider it more useful to permit linking proprietary applications with the library. If this is what you want to do, use the GNU Library General Public License instead of this License.

Syngress: *The Definition of a Serious Security Library*

Syn•gress (sin–gres): *noun*, *sing*. Freedom from risk or danger; safety. See *security*.

Cyber Spying: Tracking Your Family's (Sometimes) Secret Online Lives

Dr. Eric Cole, Michael Nordfelt,
Sandra Ring, and Ted Fair

Have you ever wondered about that friend your spouse e-mails, or who they spend hours chatting online with? Are you curious about what your children are doing online, whom they meet, and what they talk about? Do you worry about them finding drugs and other illegal items online, and wonder what they look at? This book shows you how to monitor and analyze your family's online behavior.

ISBN: 1-93183-641-8

Price: $39.95 US $57.95 CAN

Stealing the Network: How to Own an Identity

Timothy Mullen, Ryan Russell, Riley (Caezar) Eller,
Jeff Moss, Jay Beale, Johnny Long, Chris Hurley, Tom Parker, Brian Hatch

The first two books in this series "Stealing the Network: How to Own the Box" and "Stealing the Network: How to Own a Continent" have become classics in the Hacker and Infosec communities because of their chillingly realistic depictions of criminal hacking techniques. In this third installment, the all-star cast of authors tackle one of the fastest-growing crimes in the world: Identity Theft. Now, the criminal hackers readers have grown to both love and hate try to cover their tracks and vanish into thin air…

ISBN: 1-59749-006-7

Price: $39.95 US $55.95 CAN

Software Piracy Exposed

Paul Craig, Ron Honick

For every $2 worth of software purchased legally, $1 worth of software is pirated illegally. For the first time ever, the dark underground of how software is stolen and traded over the Internet is revealed. The technical detail provided will open the eyes of software users and manufacturers worldwide! This book is a tell-it-like-it-is exposé of how tens of billions of dollars worth of software is stolen every year.

ISBN: 1-93226-698-4

Price: $39.95 U.S. $55.95 CAN

Syngress: *The Definition of a Serious Security Library*

Syn•gress (sin‑gres): *noun, sing.* Freedom from risk or danger; safety. See *security*.

Buffer OverFlow Attacks: Detect, Exploit, Prevent

James C. Foster, Foreword by Dave Aitel

The SANS Institute maintains a list of the "Top 10 Software Vulnerabilities." At the current time, over half of these vulnerabilities are exploitable by Buffer Overflow attacks, making this class of attack one of the most common and most dangerous weapons used by malicious attackers. This is the first book specifically aimed at detecting, exploiting, and preventing the most common and dangerous attacks.

ISBN: 1-932266-67-4

Price: $34.95 US $50.95 CAN

Programmer's Ultimate Security DeskRef

James C. Foster

The Programmer's Ultimate Security DeskRef is the only complete desk reference covering multiple languages and their inherent security issues. It will serve as the programming encyclopedia for almost every major language in use.

While there are many books starting to address the broad subject of security best practices within the software development lifecycle, none has yet to address the overarching technical problems of incorrect function usage. Most books fail to draw the line from covering best practices security principles to actual code implementation. This book bridges that gap and covers the most popular programming languages such as Java, Perl, C++, C#, and Visual Basic.

ISBN: 1-932266-72-0

Price: $49.95 US $72.95 CAN

Hacking the Code: ASP.NET Web Application Security

Mark Burnett

This unique book walks you through the many threats to your Web application code, from managing and authorizing users and encrypting private data to filtering user input and securing XML. For every defined threat, it provides a menu of solutions and coding considerations. And, it offers coding examples and a set of security policies for each of the corresponding threats.

ISBN: 1-932266-65-8

Price: $49.95 U.S. $79.95 CAN

SYNGRESS®

SharePoint® 2007 and Office Development Expert Solutions

SharePoint® 2007 and Office Development

Expert Solutions

SharePoint® 2007 and Office Development

Expert Solutions

Randy Holloway

Andrej Kyselica

Steve Caravajal

Wiley Publishing, Inc.

SharePoint® 2007 and Office Development Expert Solutions

Published by
Wiley Publishing, Inc.
10475 Crosspoint Boulevard
Indianapolis, IN 46256
www.wiley.com

Copyright © 2007 by Wiley Publishing, Inc., Indianapolis, Indiana

Published simultaneously in Canada

ISBN: 978-0-470-09740-3

Manufactured in the United States of America

10 9 8 7 6 5 4 3 2 1

Library of Congress Cataloging-in-Publication Data

Holloway, Randy, 1974-
  SharePoint 2007 and Office development expert solutions / Randy Holloway, Andrej Kyselica, Steve Caravajal.
    p. cm.
  ISBN 978-0-470-09740-3 (paper/website)
  1. Intranets (Computer networks) 2. Web servers. I. Kyselica, Andrej, 1975- II. Caravajal, Steve, 1959- III. Title.
  TK5105.875.I6H64 2007
  004.6'82--dc22

                    2007023234

To my wife Donna and my two kids, Emily and Gavin. Thanks for your patience and support through many Saturdays and Sundays that were spent writing instead of spending time with family.

— Randy Holloway

To my wife Carol and my kids: Drew, Jennifer, Alison, Erik, and Josh.

— Andrej Kyselica

To my family, Rosemary, Stephen, and Christopher, and my parents, Ed and Frances, my inspiration in the past, now, and in the future.

— Steve Caravajal

About the Authors

Randy Holloway works as a technical advisor for Microsoft's enterprise software customers in St. Louis. Prior to joining Microsoft, Randy served as an IT director focused on software engineering and ERP systems support in addition to his writing and speaking engagements for a variety of industry publications and conferences. Randy holds a B.S. from Mississippi State University. He lives in St. Louis, Missouri, with his wife and two children. He also writes a weblog focused on technology and industry trends, located at `http://randyh.wordpress.com`.

Andrej Kyselica advises Microsoft's enterprise healthcare customers on technology strategy and architecture. Before joining Microsoft, Andrej worked in consulting for 9 years as a practice director and CTO focusing on .NET and portal solutions. Outside of work, Andrej enjoys cycling, running, and spending time outdoors with his wife and five children.

Steve Caravajal is a principal architect with the Microsoft Corporation. He has been architecting, deploying, and customizing SharePoint solutions since v1. Steve has over 25 years' experience in technology and product development. He holds a B.S. in chemistry and mathematics and a doctoral degree in chemistry and computer science. Steve is a software developer at heart, having written and managed the development of numerous enterprise software applications in .NET, Java, and C++. He currently lives in Cincinnati with his wife Rosemary. In his spare time he enjoys playing foosball, ping-pong, and blackjack.

Credits

Senior Acquisitions Editor
Jim Minatel

Senior Development Editor
Kevin Kent

Technical Editor
Todd O. Klindt

Production Editor
William A. Barton

Copy Editors
Luann Rouff
Ian Golder, Word One

Editorial Manager
Mary Beth Wakefield

Production Manager
Tim Tate

Vice President and Executive Group Publisher
Richard Swadley

Vice President and Executive Publisher
Joseph B. Wikert

Compositor
Laurie Stewart, Happenstance Type-O-Rama

Project Coordinator, Cover
Lynsey Osborn

Proofreading
Jen Larsen, Word One

Indexing
Johnna VanHoose Dinse

Anniversary Logo Design
Richard Pacifico

Acknowledgments

We'd like to thank our colleagues at Microsoft for their support and assistance during this project. Thanks to the Office Servers community alias and the Microsoft bloggers focused on SharePoint, enterprise content management, workflow, Word, and InfoPath for sharing technical information and providing a technical roadmap that helped to shape this book.

Thanks to Joe Wilson and Chris Reinhold from Microsoft for their support with the photography for this book and for not making us leave our offices to get the pictures done.

Contents

Contents

Contents

Contents

Introduction

The latest releases of Office and SharePoint, Microsoft Office 2007 and Microsoft Office SharePoint Server 2007, along with Windows SharePoint Services 3.0, provide a great deal of new functionality for application developers to build upon for their applications. At the same time, enterprise developers and independent software vendors are trying to reconcile the traditional approaches for developing in the enterprise with the Web 2.0 movement, which consists of many lightweight applications, loose aggregations of data and services, and mashups developed by users of the applications with little or no code.

The demand for agile application development based on standard frameworks is strong among both enterprise IT and their business users. This book aims to identify some of the key solutions and scenarios that can be supported with Office and SharePoint 2007. While the book is targeted at developers, equal weight is given to solutions that can be developed or configured with little or no code, providing a faster time to market for new applications and business solutions.

Who This Book Is For

The audience for this book includes software developers focused on enterprise IT scenarios involving Office and SharePoint, ISVs looking to integrate their software with Office and SharePoint 2007, and developers of Enterprise Content Management (ECM) software interested in the capabilities of the Microsoft ECM platform.

Those familiar with the new versions of Office and SharePoint will likely skip the Chapter 1 functional overview and dig in to the architectural components outlined in Chapter 2. For users primarily interested in Office development scenarios, you'll want to focus primarily on Chapters 4 through 7. For readers primarily interested in SharePoint content, you'll spend most of your time in Chapters 2, 3, 5 and 9.

What This Book Covers

This book is focused exclusively on developing solutions for Office 2007, Windows SharePoint Services 3.0, and Microsoft Office SharePoint Server 2007. The material included is focused on practical examples based on both configuration of SharePoint and Office 2007 applications along with development and customization through Visual Studio. Content related to previous versions of Office or SharePoint is included only to highlight new features of capabilities in the 2007 release.

How This Book Is Structured

This book consists of 10 chapters related to developing solutions with Office and SharePoint 2007. Unlike most books on these topics, we have included coverage of both Office and SharePoint to best highlight the integration between the client and server platforms and to help the reader understand how to build applications that leverage the Office client as a "smart client" user interface for processes that are based on a back-end SharePoint repository.

Introduction

The book begins with a technical overview and some detail on the architecture of the SharePoint 2007 technologies. It then transitions to chapters on the SharePoint object model, Office as a smart client for XML-based data, Web services integration, and a review of user interface development capabilities resident in SharePoint and Office. The book then moves through chapters on the various application services exposed through SharePoint supporting electronic forms, records management, and business intelligence (BI); a drill-down on Office and SharePoint BI features; a detailed review of workflow integration in SharePoint; and an examination of both Web Content Management and Enterprise Content Management scenarios in this platform. The chapters are organized to help the reader gain an understanding of the underlying technologies and then go through various solutions built on these technologies. In the end, the goal is for the reader to gain a thorough understanding of how these applications can be built and to think about how to apply these techniques in their own applications.

❑ **Chapter 1, "What's New for Developers in Microsoft Office SharePoint Server 2007?"** — This chapter includes an overview of the new features for developers in the SharePoint Server 2007 release, including both Windows SharePoint Services 3.0 (WSS) and Microsoft Office SharePoint Server 2007 (MOSS). This includes some of the new services such as InfoPath Forms Services, Excel Services, and other integration points with the Office 2007 client suite.

❑ **Chapter 2, "Understanding SharePoint 2007 Services and Architecture"** — In this chapter we break down the architecture of the SharePoint 2007 technologies, including WSS 3.0 and MOSS 2007, to learn more about new concepts including the Business Data Catalog, and integrated Web Content Management in WSS and MOSS, along with the architecture of enterprise search. This chapter provides the foundation for later chapters in terms of what technical features are supported within the SharePoint 2007 technologies.

❑ **Chapter 3, "Programming SharePoint Lists and Libraries"** — This chapter delves into the SharePoint object model and specifically focuses on lists and document libraries, which form the foundation of the SharePoint infrastructure for storing and retrieving data. The examples in this chapter include applications to add, update, and delete list items with both the SharePoint object model and supported web services. Concepts such as RSS in SharePoint are also covered here.

❑ **Chapter 4, "XML, Web Services, and Extensibility in Office 2007"** — Office as a smart client, including the format of the Office documents, the ability to bind documents to XML data sources and to integrate XML Web services into documents serve as the foundation of this chapter. This chapter serves as the focal point in the book for Office client development and document structure updates within the Office 2007 release.

❑ **Chapter 5, "User Interface Development in SharePoint and Office 2007"** — This chapter builds on concepts in Chapters 3 and 4 to include a variety of methods for creating customized user interfaces within Office and SharePoint 2007. The construction of custom Web Parts, development of master pages, extension of the Office Fluent Ribbon user interface, and the use of the Document Information Panel integrated with SharePoint content types make up this chapter.

❑ **Chapter 6, "SharePoint Application Services"** — This chapter examines the application services supported in SharePoint to help developers build electronic forms and workflow applications, business intelligence applications based on Excel Services, and records management solutions on the platform. This chapter provides an overview of several MOSS 2007 platform services that form the basis of several out-of-the-box site templates and applications within MOSS and includes detail on customizing these features.

❑ **Chapter 7, "Building Business Intelligence Applications with SharePoint and Office"** — Share-Point and Office 2007 have a significant set of new business intelligence (BI) features. This chapter provides an in-depth look at Excel 2007 BI capabilities, Excel Services features within MOSS, and support for key performance indicators and dashboards within the MOSS Report Center site template.

❑ **Chapter 8, "Creating Custom Workflows for Windows SharePoint Services and Office Share-Point Server"** — Workflow is integrated throughout SharePoint and Office 2007 as a system, from standard content routing and approval to advanced applications of the SharePoint Designer for content-based workflow activities and the use of Visual Studio for custom developed work-flow solutions. This chapter provides an in-depth look at the various workflow features within SharePoint 2007 and how those features are manifested within the Office client applications.

❑ **Chapter 9, "Web Content Management in Office SharePoint Server"** — The integration of con-tent management features into the WSS 3.0 and MOSS 2007 platforms marks one of the most sig-nificant architectural shifts for this platform since the previous release. This chapter examines the key Web Content Management features that developers need to understand when designing new web sites and content management processes based on WSS and MOSS.

❑ **Chapter 10, "Content Management and Workflow Scenario"** — This chapter puts it all together, combining elements of Office document features to support content management, workflow, the use of Features within SharePoint for managing delivery of changes to the server environment, along with document publishing features for web content and document content conversions within MOSS. In addition, this chapter focuses on scenarios for extending and customizing docu-ment and Web Content Management solutions based on key features within the SharePoint tech-nology platform.

What You Need to Use This Book

Readers of this book will need access to a server running Microsoft Office SharePoint Server 2007 along with an environment running Visual Studio 2005 and Microsoft Office 2007. All of the elements can be run on a single server, or the various elements can be broken out to include a dedicated Windows Server 2003 domain, a dedicated MOSS 2007 server, and a client running Office and Visual Studio. The client machines can be either Windows Vista or Windows XP SP2. Most of the code samples and examples included in this book were developed on a Windows Server 2003 image containing MOSS 2007 and Visual Studio 2005.

Conventions

To help you get the most from the text and keep track of what's happening, we've used a number of conventions throughout the book.

As for styles in the text:

❑ We *highlight* new terms and important words when we introduce them.

❑ We show keyboard strokes like this: Ctrl+A.

❑   We show file names, URLs, and code within the text like so: `persistence.properties`.

❑   We present code in two different ways:

```
In code examples we highlight new and important code with a gray background.
The gray highlighting is not used for code that's less important in the present
context, or has been shown before.
```

Source Code

As you work through the examples in this book, you may choose either to type in all the code manually or to use the source code files that accompany the book. All of the source code used in this book is available for download at `http://www.wrox.com`. Once at the site, simply locate the book's title (either by using the Search box or by using one of the title lists) and click the Download Code link on the book's detail page to obtain all the source code for the book.

Because many books have similar titles, you may find it easiest to search by ISBN; this book's ISBN is 978-0-470-09740-3.

The code samples for this book include both C# and VB.NET code where appropriate, along with supporting documentation and resource files.

Once you download the code, just decompress it with your favorite compression tool. Alternately, you can go to the main Wrox code download page at `http://www.wrox.com/dynamic/books/download.aspx` to see the code available for this book and all other Wrox books.

Errata

We make every effort to ensure that there are no errors in the text or in the code. However, no one is perfect, and mistakes do occur. If you find an error in one of our books, like a spelling mistake or faulty piece of code, we would be very grateful for your feedback. By sending in errata you may save another reader hours of frustration and at the same time you will be helping us provide even higher quality information.

To find the errata page for this book, go to `http://www.wrox.com` and locate the title using the Search box or one of the title lists. Then, on the book details page, click the Book Errata link. On this page you can view all errata that has been submitted for this book and posted by Wrox editors. A complete book list including links to each book's errata is also available at `www.wrox.com/misc-pages/booklist.shtml`.

If you don't spot "your" error on the Book Errata page, go to `www.wrox.com/contact/techsupport.shtml` and complete the form there to send us the error you have found. We'll check the information and, if appropriate, post a message to the book's errata page and fix the problem in subsequent editions of the book.

p2p.wrox.com

For author and peer discussion, join the P2P forums at p2p.wrox.com. The forums are a Web-based system for you to post messages relating to Wrox books and related technologies and interact with other readers and technology users. The forums offer a subscription feature to e-mail you topics of interest of your choosing when new posts are made to the forums. Wrox authors, editors, other industry experts, and your fellow readers are present on these forums.

At http://p2p.wrox.com you will find a number of different forums that will help you not only as you read this book, but also as you develop your own applications. To join the forums, just follow these steps:

1. Go to p2p.wrox.com and click the Register link.

2. Read the terms of use and click Agree.

3. Complete the required information to join as well as any optional information you wish to provide and click Submit.

4. You will receive an e-mail with information describing how to verify your account and complete the joining process.

You can read messages in the forums without joining P2P but in order to post your own messages, you must join.

Once you join, you can post new messages and respond to messages other users post. You can read messages at any time on the Web. If you would like to have new messages from a particular forum e-mailed to you, click the Subscribe To This Forum icon by the forum name in the forum listing.

For more information about how to use the Wrox P2P, be sure to read the P2P FAQs for answers to questions about how the forum software works as well as many common questions specific to P2P and Wrox books. To read the FAQs, click the FAQ link on any P2P page.

1

What's New for Developers in Microsoft Office SharePoint Server 2007?

Microsoft's SharePoint technologies provide the foundation for collaboration and communication for information workers. The first releases of SharePoint technologies in 2001 have undergone several substantial revisions, and the technology has moved from a productivity tool supporting a limited set of scenarios to a full-fledged platform for developers to build enterprise-class applications.

The core components of the SharePoint technology stack are Microsoft Windows SharePoint Services 3.0 and Microsoft Office SharePoint Server (MOSS) 2007. In this first chapter we cover what's new for developers in SharePoint and Office and focus on some key scenarios that developers are going to need to know about. This chapter will help to provide a roadmap for the rest of the book and help you envision some of the possible uses for the Office and SharePoint platforms as a developer. Because the integration between Microsoft Office 2007 and SharePoint is such a critical component of the technology, we discuss feature sets in Office that enable developers to build solutions on the platform. Most of these areas are new to the Office 2007 release.

To understand the SharePoint technology platform and the benefits to developers, this chapter covers the following key areas:

- ❏ Portal services in SharePoint
- ❏ Search
- ❏ Document and content management
- ❏ Workflow integration

In each of these areas we'll expand the topic to focus on integration with Office applications in some scenarios and cover other key aspects, including Visual Studio developer support, building Web parts for SharePoint, and leveraging the forms and business intelligence capabilities in SharePoint. Throughout the book we'll continue to examine each of these areas in greater depth and focus on scenarios to build out these applications and leverage the platform services along with custom code to create new solutions.

This chapter assumes that you have basic familiarity with Windows SharePoint Services 2.0 and SharePoint Portal Server 2003. We'll point out some of the differences between new and prior versions of the SharePoint technologies.

At the end of this chapter, developers should have a good understanding of the various technologies that comprise the Microsoft Office SharePoint Server and have a sense of how these can be applied to build solutions.

SharePoint and Office — Developer Platforms

The key components of Microsoft's information worker technologies include Office and SharePoint; and Microsoft's product development strategy hasn't shifted substantially in that regard. What has changed since the Office System 2003 time frame is that Microsoft has invested substantially more into the use of Office as a fully supported platform for developers. In the past, Office development was about creating macros in Excel or building tools with Visual Basic for Applications (VBA). Because of a lack of consistent APIs and platform-level services for Office, most developers have focused on writing applications that can work with Office document formats as transient data stores or output formats for data, but they have not tended to leverage the Office applications themselves as part of a business solution. Where developers have built on Office, those applications are often very client-centric and focus on plug-ins or extended functionality in just a single application such as Word or Excel.

Similarly, early versions of SharePoint and their support for Web Parts drove developers toward Web Part–centric development efforts on SharePoint. If the problem couldn't be solved with a Web Part, SharePoint was unlikely to be the tool selected for a particular application development project.

Now, in the new paradigm of SharePoint and Office development, Windows SharePoint Services and Microsoft Office SharePoint Server deliver fully featured web development and content management platforms that can leverage a rich set of services to provision, manage, and administer the sites both at a user level and from a technical architecture perspective. Office then becomes a smart client that can expose many of the business processes in SharePoint, and interact seamlessly with display, data exchange, workflow, and support for many other requirements.

Building Applications with Windows SharePoint Services

Windows SharePoint Services is a core feature of the Windows Server 2003 operating system and can be configured to be installed along with a Windows Server. An example of a basic Windows SharePoint Services site is shown in Figure 1-1. Windows SharePoint Services 3.0 provides a scalable platform to provision web sites of different types. In the overall SharePoint architecture, Windows Share-Point Services is a foundational component and is designed to support the creation of a substantial number of web sites, making them available to a large number of users. It is not uncommon for a single Windows SharePoint Services instance to host many sites, scaling to hundreds or even thousands in a single infrastructure and storing gigabytes to terabytes of data. Usage scenarios can range in the tens of thousands of users for these sites, even with only a single server. Like other web technologies, Windows SharePoint Services provides a web farm architecture, enabling you to scale the front-end web servers and leverage a centralized SQL Server for content storage and management of other site data.

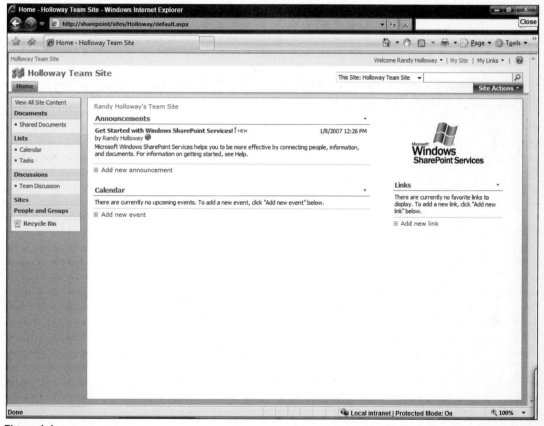

Figure 1-1

As in the previous versions of SharePoint, Windows SharePoint Services 3.0 provides collaboration and document management features that make it easy for end users and IT personnel to create and design web sites, with features including contact sharing, lists of links and resources, shared calendars, and document libraries that enable document collaboration and publishing (shown in Figure 1-2). Because of SharePoint Service's history as a collaboration tool primarily targeted to end users, developers may not see Windows SharePoint Services 3.0 as what it truly has become. In this most recent revision, the product is much more powerful than just a collaboration tool aimed at the end users, and it provides a strong development platform on top of ASP.NET. Nearly all SharePoint data is exposed in a series of lists, and a rich object model supports the manipulation of site data through the SPList class. This enables each SharePoint site to support the end user features while providing a store for developers to write code against.

A detailed review of some of the key classes supported through this object model is provided in Chapter 3.

The services provided by Windows SharePoint Services 3.0 on top of ASP.NET 2.0 include a provider model. This model supports provisioning and storage for pages, lists, and document libraries, and the provisioning can be driven through code or through user administration activities through the web browser. Without requiring user intervention or support from IT, Windows SharePoint Services 3.0 automatically determines where to store this content. Windows SharePoint Services 3.0 also eliminates many common development tasks required for a typical ASP.NET web site because of the availability of Web Parts and other UI components, enabling users to add, view, and modify content (see Figure 1-3).

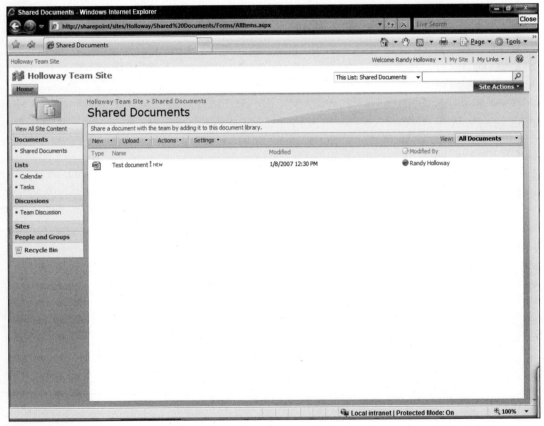

Figure 1-2

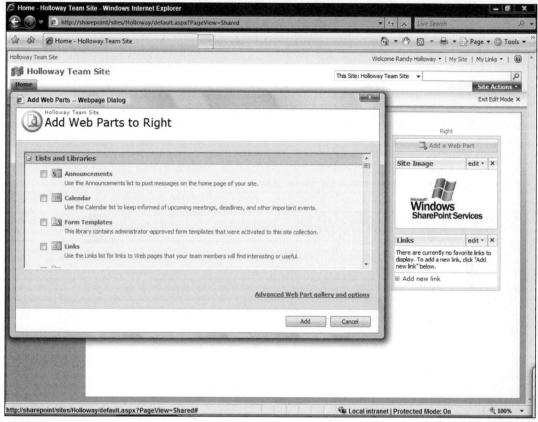

Figure 1-3

SharePoint and ASP.NET

A significant area of improvement in the architecture of SharePoint is the enhanced integration it offers with ASP.NET in the current product. In prior versions of the product, SharePoint was a "bolt on" technology that wasn't well integrated with other web functionality, and as a result it caused problems for developers trying to integrate ASP.NET code into their SharePoint infrastructure. In Windows SharePoint Services 3.0, provisioning starts at the level of the IIS web site, and before creating your first Windows SharePoint Services site, you must act as administrator to extend 3.0 functionality onto one or more IIS web sites. In the previous version, Windows SharePoint Services 2.0, the term *virtual server* was used to describe an IIS web site that had been extended with Windows SharePoint Services functionality. In the current version, Windows SharePoint Services 3.0 now refers to an IIS web site extended with Windows SharePoint Services functionality as a *web application*.

The previous version of Windows SharePoint Services was integrated with IIS 6.0 and ASP.NET 1.1 using an ISAPI DLL. The ISAPI integration technique results in IIS routing HTTP requests to Windows Share-Point Services before ASP.NET can process these requests. This routing has proven to be problematic in certain situations because Windows SharePoint Services takes control of an incoming HTTP request before it has a chance to be properly initialized with ASP.NET context. Windows SharePoint Services 3.0 integrates with ASP.NET in a different manner to help eliminate these problems. Windows SharePoint Services 3.0 is built upon ASP.NET 2.0, which provides enhancements over ASP.NET 1.1. In addition, the integration between Windows SharePoint Services 3.0 and ASP.NET 2.0 was changed to route incoming requests through the ASP.NET runtime before Windows SharePoint Services. This change was implemented by removing the ISAPI filter and adding an HttpModule and an HttpHandler that are registered with ASP.NET using standard Web.config entries. This works in the same manner as any developer extending the standard HTTP functionality would experience, meaning that incoming HTTP requests always enter the ASP.NET runtime environment and are initialized with ASP.NET context before they are forwarded to Windows SharePoint Services code to carry out SharePoint-related processing. This enables develop-ers to run a SharePoint server and extend it with custom web applications without being concerned about a complicated or convoluted approach to administering the web server environment.

When you extend an IIS web site to become a Windows SharePoint Services Web application, Windows SharePoint Services 3.0 adds a wildcard application map to the IIS metabase. This wildcard application map serves to route all incoming HTTP requests to the ASP.NET runtime regardless of their extension. This wildcard application map is necessary to forward a request for any type of document or file for ASP.NET, which then forwards the request to Windows SharePoint Services for processing.

Another relevant feature of the new architecture is related to how .aspx pages are parsed and compiled. The .aspx page parser used by ASP.NET 1.1 works only with .aspx pages that reside on the local file sys-tem. However, Windows SharePoint Services architecture relies on storing .aspx pages inside SQL Server. Because Windows SharePoint Services 2.0 relies on ASP.NET 1.1, the Windows SharePoint Services team had to create a custom .aspx page parser. Unfortunately, the .aspx parser of Windows SharePoint Services 2.0 does not support many of the richer features offered by the ASP.NET .aspx page parser. ASP.NET 2.0 has also introduced a new pluggable component type known as a *virtual path provider*. A developer can write a custom component that retrieves .aspx pages for any location, including a database such as SQL Server. Once a custom virtual path provider retrieves an .aspx page, it can then hand it off to ASP.NET to conduct the required parsing and compilation. ASP.NET also gives the virtual path provider a good deal of control over how .aspx pages are parsed and whether they are compiled or run in a noncompile mode.

The key takeaway for developers is that this new platform for SharePoint application development is sub-stantial and feature-rich, and provides many capabilities above and beyond previous product versions.

New SharePoint Developer Features

Microsoft Office SharePoint Server 2007 (along with Windows SharePoint Services) provides a number of capabilities that developers can leverage in their existing applications or use as a platform for build-ing new solutions. Among those features is support in SharePoint for building portal applications with concepts of site membership, content targeting, and flexibility in content organization in management as part of the core features. More advanced features in the areas of search, document and content management, and workflow are also significant capabilities that developers can build upon in their applications. The following sections provide an overview of some of these new key features.

Portal Services

Developers and IT professionals will find all the portal features in Microsoft Office SharePoint Server 2007 that they have become accustomed to in SharePoint Portal Server 2003, including user profiles, audience targeting, My Sites for private and public sites designed for the individual, enterprise search, and single sign-on capabilities. However, if you're experienced in building portal solutions on top of SharePoint Portal Server 2003, you will find that Office SharePoint Server 2007 has many differences as well. For instance, Office SharePoint Server 2007 also includes several new portal features, such as the Business Data Catalog, which extends the search capabilities of the platform. In addition, SharePoint Portal Server provides its own separate administrative web application for provisioning and configuring portal sites. Office SharePoint Server 2007 doesn't require its own separate administrative application and instead integrates its administrative features and configuration links into the Windows SharePoint Services Central Administration application.

Office SharePoint Server 2007 is also quite different from SharePoint Portal Server under the hood from an architectural perspective. SharePoint Portal Server builds its portal site infrastructure around the concepts of "areas" and "listings." Areas in SharePoint Portal Server 2003 represent a relatively complex layer on top of Windows SharePoint Services 2.0 that is hard to extend using standard Windows SharePoint Services development techniques. As a result, many developers were not effectively able to leverage this functionality. The concepts of areas and listings from SharePoint Portal Server 2003 have been eliminated in Office SharePoint Server 2007 and replaced with a portal infrastructure that has been designed and implemented much more in line with Windows SharePoint Services best practices.

An Office SharePoint Server 2007 portal site is a Windows SharePoint Services site collection containing a top-level site along with several child sites below it. Unlike SharePoint Portal Server 2003, the portal site does not have to be created at the root of an IIS web site. This provides more flexibility because you can host hundreds of portal sites inside a single IIS web site. This is important for many organizations that need a number of portals but don't want to support a complex or sprawling infrastructure for applications that shouldn't require it due to minimal load or complexity. This has also made it possible for application service providers and hosting companies to provide more functionality with SharePoint hosting.

You create a new Office SharePoint Server 2007 portal site the same way that you create any other new site collection — through the Windows SharePoint Services Central Administration application, the Stsadm.exe command-line utility, or through custom code. When creating a new Office SharePoint Server 2007 portal site, you can use one of the portal site templates that ship with the product. Examples of site templates that can be used to create a new Office SharePoint Server 2007 portal site include a template for an Internet-facing web site, a content publishing site, and a standard corporate intranet (see Figure 1-4).

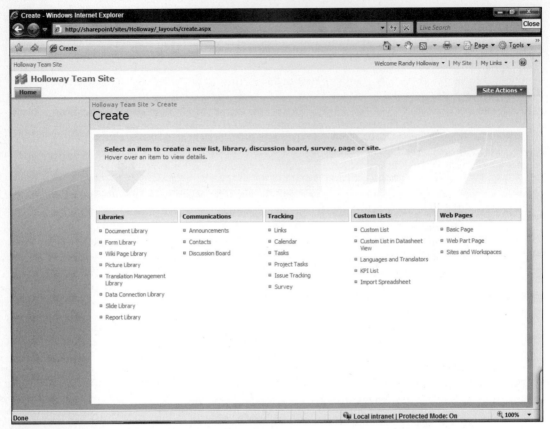

Figure 1-4

Enterprise Search

The enterprise search features have long been considered one of the most important features of the Share-Point product stack. SharePoint Portal Server 2003 Search was designed to enable searching through not only content and documents within Office SharePoint Portal Server portal sites and Windows SharePoint Services team sites, but also through external content such as Windows file shares, public Microsoft Exchange Server folders, standard web sites and other non-Microsoft products, including Lotus Notes or EMC's Documentum. Office SharePoint Server 2007 Search has been designed to give you these same features in a manner that is more performance-oriented and easier to configure. With the previous version of SharePoint technologies, Windows SharePoint Services 2.0 and SharePoint Portal Server 2003 each used a different underlying infrastructure to support indexing and searching. This created problems as companies upgraded from Windows SharePoint Services 2.0 to SharePoint Portal Server 2003. These upgrade problems have been addressed, as the Windows SharePoint Services Search service and the Office SharePoint Server Search service are now based on the same underlying indexing and search infrastructure, which is an improved version of what was provided by SharePoint Portal Server 2003.

What's New with Search?

In terms of significant differences in the new version of the product, consider the following:

❑ Microsoft Office SharePoint Server 2007 Search makes it possible to search external content across the network, such as Windows file shares and business data, whereas Windows SharePoint Services 3.0 Search is limited to searching through content and documents within the current site collection.

❑ Microsoft Office SharePoint Server 2007 search can be configured to run the indexing service and search service on different servers within a farm to increase scalability and throughput. Windows SharePoint Services 3.0 search is limited to running the indexing service and search service on the same physical server.

❑ Configuring Office SharePoint Server 2007 search is an administrative exercise in creating and configuring content sources within the scope of a particular Shared Service Provider (SSP). A content source defines a set of searchable content. When you create a new SSP, Office SharePoint Server 2007 automatically creates a content source to search through user profile data as well as the content and documents within Office SharePoint Server 2007 portal sites and Windows Share-Point Services sites within the web applications associated with the current SSP. However, the SSP administrator must explicitly create and configure additional content sources to support building indexes and searching through external content such as documents in a Windows file share or content from an intranet or public web site.

❑ Windows SharePoint Services 3.0 and Microsoft Office SharePoint Server 2007 provide a rich user interface for searching by adding search boxes and search result pages to each Windows SharePoint Services site and each Office SharePoint Server 2007 portal site. Microsoft Office SharePoint Server 2007 goes even further, supplying a dedicated child site named Search Center within a portal site collection that provides a specialized user interface for searching, as shown in Figure 1-5.

With all these new features, you might conclude that developers can take full advantage of Office Share-Point Server 2007 Search facilities without writing any custom code, as the product provides so many features out of the box. However, Office SharePoint Server 2007 makes it possible for you customize how search results look by modifying the Extensible Stylesheet Language Transformations (XSLT) it uses to display search results. Office SharePoint Server 2007 also exposes its search engine through a programmable API that enables developers to extend either the Windows SharePoint Services 3.0 Search service or the Office SharePoint Server 2007 Search service using custom code. This enables you to write a server-side component such as a custom workflow that queries the Office SharePoint Server 2007 Search service through code and renders the search results in a customized site.

Additionally, you can easily integrate SharePoint Search features into a third-party web site and preserve that site's look and feel by leveraging SharePoint web services for searching as part of the implementation, returning the results to that site as XML and then applying a transformation to render the results as you'd like them to appear.

To further look at the benefits of Search in SharePoint, we want to take a deeper look at the Business Data Catalog and features to enable the integration of Search into structured application data.

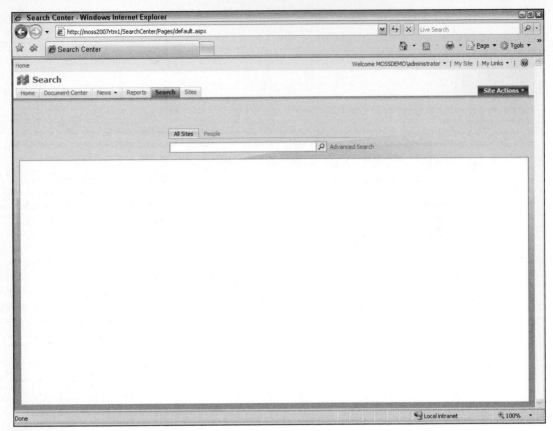

Figure 1-5

The Business Data Catalog

The Business Data Catalog (BDC) is a new framework that provides Office SharePoint Server 2007 portal sites and standard Windows SharePoint Services 3.0 sites with integration into systems such as those created by SAP, Siebel, and PeopleSoft or your custom applications for your business. The BDC additionally provides the means to integrate data directly with database systems such as SQL Server and Oracle. While SharePoint Portal Server 2003 makes it possible to integrate portal sites with back-end systems, it requires you to write custom code to manage connections to those systems and retrieve the data you need to display. Furthermore, the code you must write changes significantly as you switch between back-end systems from vendors such as SAP and PeopleSoft. The BDC is designed to make things much easier by providing a generalized layer of abstraction and some value-added functionality such as out-of-the-box Web Parts and search services for BDC data.

There are several key scenarios for extending BDC data to custom applications as well. For example, you can write a custom Web Part or a console application to display your business data. You can use the Runtime object model in a Web Part or console application to query the metadata database and display data from a back-end application. When using the Runtime object model, you can invoke methods and display business data on a page. The BDC also offers an Administration object model that you can use to create, edit, and delete metadata. The main goal of the Administration object model is to enable you to write tools that simplify the creation and administration of metadata. You can then use the Administration object model to support a Windows Form or a web application for managing metadata. In addition, the BDC enables you to bring data from business applications to user profiles. For example, you can add your contact information, such as an address field, from an SAP or Siebel application registered in the Business Data Catalog to the user profiles.

Integrating Back-End System Data with No Code

Fundamentally, the BDC enables you to integrate data from back-end systems without requiring you to write custom code for managing connections and retrieving data. The design of the BDC is based on standardized metadata that describes the location and format of a back-end system and the data entities defined inside. The BDC also provides an execution component capable of reading BDC metadata that is able to retrieve external data from any back-end system and return it to Office SharePoint Server 2007 in a standard format. Connectivity between the BDC and traditional line-of-business systems is achieved using standard web services. Connectivity between the BDC and database systems is achieved using ADO.NET providers.

The first step in using the BDC is to author an XML file containing the metadata to connect to a back-end system. When you author metadata for the BDC, you define the data you want to retrieve in terms of entities. For example, you might define a customer as one entity and an invoice as another entity. The BDC metadata format also enables you to define associations between entities in scenarios for which there is a one-to-many relationship, such as one that might exist between customers and invoices. The definition of a BDC entity contains identifiers, properties, and methods.

The methods define how the BDC interacts with entry points exposed by the back-end system. For a back-end system accessible through web services, methods define the names of the web service operations and the parameters required to call them. For a back-end system that is a database such as SQL Server or Oracle, methods define the names of stored procedures and SQL statements.

Entities can also define actions. A BDC action is used to dynamically parse together the URL behind a hyperlink that enables a user to navigate from a page in an Office SharePoint Server 2007 portal site to another location. For example, an action defined on a BDC customer entity could be written to redirect users to a web page in an SAP application that supports updates to customer information. Actions were designed to support scenarios in which the BDC is used to display read-only data and to bootstrap the user into another application when updating or some other type of external operation is required. The overview for this architecture is shown in Figure 1-6.

More details on the architecture of Search and the BDC are covered in Chapter 2.

Once you have authored or acquired the XML file with the required BDC metadata for a back-end system, you must import it into the BDC within the scope of a particular SSP to create a BDC application. You can accomplish this import process using the SSP administrative web pages. You can alternatively import an XML file with BDC metadata using custom code written against the BDC Administration object model. Once you have imported the required metadata to create a BDC application, there are several out-of-the-box techniques for leveraging and displaying its data within a portal site. Microsoft Office SharePoint Server 2007 ships with a set of Business Data Web Parts that can be quickly added to pages to query and display BDC data. You can also add new columns to lists and document libraries based on an entity defined in a BDC application. A user editing a column based on a BDC entity is automatically presented with a user interface, making it possible to query the back-end system. The BDC has been designed to integrate with the Office SharePoint Server 2007 Search Service as well.

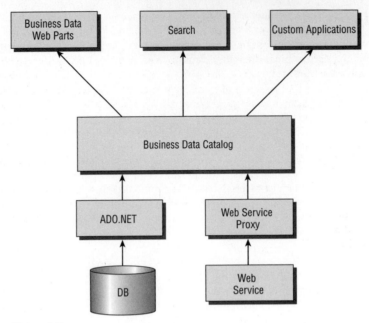

Figure 1-6

For example, a back-end system and its entities can be defined as a content source so that the Office Share-Point Server 2007 indexing service will crawl through its data and build indexes for the Office SharePoint Server 2007 Search engine. This becomes a powerful feature because it enables users to discover data from back-end systems about things such as customers and invoices when running standard search queries through both Office SharePoint Server 2007 portal sites and standard Windows SharePoint Services 3.0 sites.

The BDC provides convenient features to map data from a BDC entity to properties in a user profile and to synchronize this data at periodic intervals. For example, if your company has an SAP system that contains employee data you would like to include in user profiles, such as phone numbers or social security numbers, you can configure this type of data importing without writing any custom code.

Integrating Back-End System Data Using Custom Code

Finally, BDC entities can also be accessed programmatically using custom code written against the BDC object model. This makes it possible to write custom Web Parts as well as other server-side components and services that run their own BDC queries. One nice aspect of writing code to query BDC entities is that you don't have to worry about managing connections or whether you are accessing the back-end system through web services or ADO.NET. All those details are abstracted away by BDC metadata and the BDC execution engine.

> More details about the BDC architecture and the implementation of business data services in SharePoint is provided in Chapter 2.

The main value of the Business Data Catalog is that it enables a developer to integrate existing business data through the SharePoint Search infrastructure and then use a single set of APIs to query the search engine and render results within an application.

Document and Content Management

Developers familiar with previous SharePoint versions have offered the criticism that some key features supported with document libraries do not extend to lists. Features such as version control extended to document libraries, but lists did not have the same level of support. In Windows SharePoint Services 3.0, the functionality of lists has been extended, and they are now on a par with document libraries, supporting many of the same features, such as versioning, events, and folders. Additionally, some new features in Windows SharePoint Services 3.0 are supported by both lists and document libraries, such as exposing data through automatic RSS feeds.

Improved Performance with Large Lists and Document Libraries

Performance with large lists and document libraries has also been another concern with previous Share-Point versions. For example, lists often started showing degraded performance when the number of items exceeded several thousand. Document libraries have had similar performance concerns in the past, causing implementers and developers to focus on creative workarounds. Since the rule of thumb for not exceeding a few thousand documents applied to each folder within a document library, as opposed to the document library itself, schemes to partition documents across multiple folders within a document library became a popular workaround in Windows SharePoint Services 2.0 for dealing with a large number of documents. Windows SharePoint Services 3.0 introduces a new column indexing feature to help mitigate these performance issues. From a list settings page or a document library settings page, you can add an index to any column. Doing this does not actually create a physical index in SQL Server. Instead, it creates a table with the integer ID of the list item or document and the value of the indexed column. Windows SharePoint Services then uses this table to improve the performance of data returned from views, especially a view with a filter based on the indexed column.

Extensible Field Types

Many developers have expressed a desire to work with Windows SharePoint Services fields at a lower level to obtain more control over field rendering and validation, features that are important when building business applications on top of the SharePoint infrastructure. As a result, the product has been enhanced with the addition of extensible field types in Windows SharePoint Services 3.0. You can create an extensible field type by writing a class in C# or Visual Basic .NET that inherits from one of the built-in Windows SharePoint Services field types, such as `SPFieldText`, `SPFieldNumber`, or `SPFieldUser`. An extensible field type can also utilize an ASP.NET user control that contains your favorite web controls. This enables you to use the same techniques for control initialization and validation that you have used in ASP.NET applications.

Custom Site Columns

Another feature added to Windows SharePoint Services 3.0 is custom site columns. A *site column* is a reusable definition that can be used across multiple lists. A site column defines the name for a column, its underlying field type, and other characteristics such as its default value, formatting, and validation. Once you have defined a site column, you can then use it as you define the structure of your user-defined lists. An obvious advantage is that you can update the site column in a single place and have that update affect all the lists where the site column has been used.

A site column is defined within the scope of a single site, yet it is visible to all child sites below the site in which it has been defined. Therefore, you can create a site column that is usable across an entire site collection by defining it in the top-level site.

One convenient technique made available with the introduction of site columns is the ability to perform field lookups across sites. For example, you can create a site column in a top-level site that performs a lookup on a list in the same site. Then you can create other lists within child sites that use this site column to perform lookups on the list in the top-level site. In Windows SharePoint Services 2.0, you had no way to accomplish this task short of writing custom code.

Content Types

Imagine you want to store several different types of documents in the same document library — for example, if you needed to store presentations, sales proposals, and customer activity reports, and each of these document types had its own unique set of custom columns and its own unique event handlers. In Windows SharePoint Services 2.0, you can add extra columns and event handlers only to the document library itself, which always affects every document in that document library. In addition, a Windows SharePoint Services 2.0 document library can have only one associated document template. Windows SharePoint Services 3.0 introduces a powerful new storage mechanism known as *content types* to solve this problem. A content type is a flexible and reusable Windows SharePoint Services type that defines the shape and behavior for an item in a list or a document in a document library.

For example, you can create a content type for a presentation document with a unique set of columns, an event handler, and its own document template. You can create a second content type for a sales proposal document with a different set of columns, a workflow, and a different document template. Then you can create a new document library and configure it to support both of these content types. The introduction of content types is significant to Windows SharePoint Services 3.0 because it provides a capability that did not exist in Windows SharePoint Services 2.0 — the capability to deal with different types of content in lists and document libraries.

Web Content Management

In the past, Microsoft customers have been forced to choose between Content Management Server (CMS) and SharePoint Portal Server 2003 for Web Content Management solutions based on web site design requirements, authoring and editing processes for a web site, and other factors. While there is a connector that provides a degree of integration between CMS and SharePoint Portal Server 2003, these two products are built on very different architectures. This has resulted in frustration because you cannot build a site that fully benefits from both the CMS Web Content Management features and the SharePoint Portal Server 2003 portal features.

To address this problem of a lack of integration in these key products, the Microsoft product teams supporting SharePoint and CMS ultimately merged the functionality of the products and built the new Web Content Management infrastructure on top of Office SharePoint Server 2007. While this impacts customers who have already become familiar with CMS development, the good news is that Microsoft's Web Content Management strategy is now built on Windows SharePoint Services3.0 and Microsoft Office SharePoint Server 2007. Now you can mix Web Content Management features with the portal features in an Office SharePoint Server 2007 portal site to provide a richer web experience as a web site developer, with much more functionality available through out-of-the-box templates than in previous versions.

For those familiar with the previous CMS product, the infrastructure has been designed using basic Windows SharePoint Services 3.0 building blocks such as child sites, page templates, content types, document libraries, and security groups. This approach lends itself to building custom solutions that extend the basic Web Content Management infrastructure using standard Windows SharePoint Services components such as custom event handlers and workflows. When you need to brand an Office SharePoint Server 2007 portal site, you can modify a single ASP.NET master page to customize the basic look and feel of the entire web site, just as you would in a standard Windows SharePoint Services 3.0 site collection. However, Office SharePoint Server 2007 extends Windows SharePoint Services 3.0 by introducing a publishing scheme based on page layouts. A *page layout* provides a structured approach to collecting content from content authors and displaying it on a page within a portal site, as shown in Figure 1-7. Page layouts are designed to make it fairly straightforward to add and modify content from within the browser.

Microsoft Office SharePoint Server 2007 provides a toolbar within the browser to give content authors and approvers a convenient way to move content pages through the editing and approval processes. Each page layout is based on a Windows SharePoint Services 3.0 content type and an associated `.aspx` page template. By layering page layouts on top of content types, Office SharePoint Server 2007 makes it possible to add custom fields for storing different types of structured content such as HTML, links, and images. Once a custom field is defined inside the content type associated with a page layout, it can be data-bound to the associated `.aspx` page template using another new Office SharePoint Server 2007 component known as a *field control*.

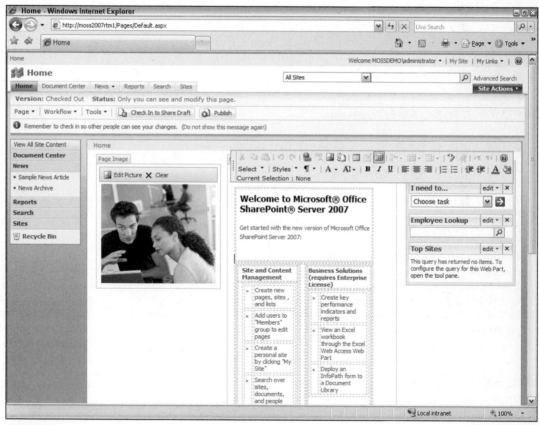

Figure 1-7

Combining these page layout and content delivery features with the workflow-driven approval (see Figure 1-8), the ability to automatically convert Office document formats into web-based content, multilingual and site synchronization support, and the ability to leverage Web Parts makes the Web Content Management features of Microsoft Office SharePoint Server 2007 compelling for many web site development efforts, both internal to organizations (for example, corporate portal) and for Internet-facing web sites.

Additional information about the SharePoint Web Content Management features and integration with the Office client applications is provided in Chapters 9 and 10.

Figure 1-8

Workflow Services

Workflow applications are some of the most challenging for developers, who often need to work across systems and provide business users with functionality without a lot of constant application and code maintenance. Microsoft's approach to solving workflow has been to develop and deploy workflow as a platform-level service that can be used across applications as part of the .NET Framework 3.0. The Microsoft workflow capabilities, named the Windows Workflow Foundation (WF), add a complete infrastructure for building workflow-enabled applications.

The WF infrastructure includes a workflow engine, pluggable components to persist workflow state, and a Visual Studio designer that makes it easy to create custom workflows by dragging components known as *activities* onto a workflow design surface. Windows SharePoint Services 3.0 builds on WF to provide a foundation for attaching business logic to list items and documents. Windows SharePoint Services 3.0 extends the basic workflow model of .NET 3.0 by associating a task list and history list with each workflow. Windows SharePoint Services 3.0's extensions add a degree of responsibility and accountability to workflows that are human-oriented in nature, such as a workflow for reviewing or approving a document. This provides the basis for a significant amount of functionality for workflow between Office applications and SharePoint.

Both Windows SharePoint Services 3.0 and Office SharePoint Server 2007 ship with workflows that are installed and ready to use out of the box. Windows SharePoint Services 3.0 includes some simple routing workflows for tasks such as moderation and approval. Office SharePoint Server 2007 supplies workflows that are more complex and are used to support features such as its Web Content Management approval process.

The creation of custom workflows represents an obvious extensibility feature for developers creating business solutions with Windows SharePoint Services 3.0 and Office SharePoint Server 2007. In addition to the standard support of the Visual Studio Extensions for WF, Microsoft provides a Windows SharePoint Services workflow SDK and a workflow starter kit, including Visual Studio project templates for creating custom workflows targeted at Windows SharePoint Services 3.0 sites. Office SharePoint Designer 2007 also provides support for creating custom workflows in SharePoint sites. This support is designed more for power users than developers because it provides a wizard to attach ad hoc business logic to list items and documents in a production Windows SharePoint Services 3.0 site.

For more in-depth coverage of workflow in SharePoint, see Chapter 8.

Integration with Office 2007

One of the major advantages of SharePoint and Office is the integration of the Office client tools with the back-end SharePoint services. Going forward, developers have the opportunity to build applications both in SharePoint and in Office that seamlessly work across environments. For example, you can build a custom workflow that is hosted in SharePoint and surfaced through Office documents to enable users to use their browser or just work within the Office client application to move the process along to the next step. Throughout this book, we'll point to some key examples where SharePoint and Office working together provide a great set of capabilities, and demonstrate how Office can be an integral part of the development platform.

Document Formats

In past versions of Microsoft Office Word, Excel, and PowerPoint, Microsoft has relied on a default file structure that is based on binary files written in a proprietary format. These formats have been very hard to read and modify unless you go through the object model of the hosting Office application such as Word or Excel. As a result, companies have tried to run Office desktop applications on the server, which poses serious problems with scalability and robustness. In many cases, these applications simply do not work due to concurrency issues. Office 2000 and Office 2003 added some limited capabilities for creating Excel spreadsheets and Word documents using XML. In the 2007 Microsoft Office system, Microsoft has taken this idea much further by adopting the Open XML Formats for Word, Excel, and PowerPoint documents.

The Open XML Formats is a new file standard for creating composite documents containing multiple inner XML files that factor out content from other aspects of the document, such as formatting instructions, data, and code.

The top-level file in the Open XML Formats is known as a *package*, and it is structured using standard ZIP file technology (see Figure 1-9). The internal files contained within a package are known as *parts*. Many parts within Word, Excel, and PowerPoint files contain XML structured in accordance with published XML schemas. Other parts within a package can consist of binary files for items such as graphics, audio clips, and videos. The Open XML Formats provide a standard approach for reading, manipulating, and generating documents in server-side scenarios where the automation of a desktop application such as Word or Excel isn't a viable option. Consider a scenario in an Office SharePoint Server 2007 portal site when you have created and configured an event handler, triggered whenever someone uploads a new Word document. The new Open XML Formats make it significantly easier to extract data or to perform hygiene such as removing comments and personal information from the document.

You can also leverage Open XML Formats to develop server-side components that generate Office documents using data pulled from content sources such as a Windows SharePoint Services list or the Business Data Catalog. To get started working with the Open XML Formats, you need to learn about the `Package` class under the `System.IO.Packaging` namespace in .NET Framework 3.0. This is the API to use when opening and creating packages. By leveraging these open formats, you can more effectively build applications to integrate Office content into your business processes.

For more detail on the Open XML Formats and how to apply them when building your applications to bind to data formats and extend the user interface in Office applications, see Chapters 4 and 5.

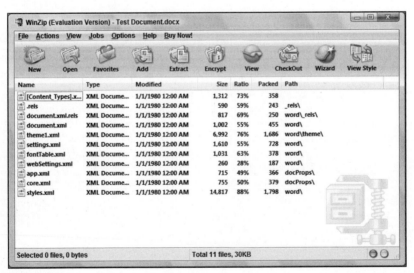

Figure 1-9

InfoPath Forms Services

Since Microsoft Office InfoPath's introduction with Microsoft Office 2003, many companies have found that it provides a quick and efficient solution for creating electronic forms to collect data from business users. The InfoPath 2003 forms designer provides a high level of productivity because it significantly reduces or eliminates the need to write custom code when developing forms, but can be extended to connect with web services or databases fairly easily. InfoPath also creates a very reliable way to collect data because InfoPath forms are built on top of XML schemas, enabling them to automatically validate the user's input and the structure of the document for processing.

InfoPath 2003 provides a convenient integration point with Windows SharePoint Services 2.0 whereby a form designer can publish an InfoPath form on a Windows SharePoint Services site, creating what is known as a forms library. A *forms library* is a hybrid Windows SharePoint Services document library that uses an InfoPath form template as its underlying document template and acts as a repository for XML documents containing form data entered by users.

A major limitation of InfoPath 2003 and its integration with previous versions of SharePoint is that it requires a full version of InfoPath to be installed on the user's desktop. This is true not only for those designing forms, but also for any users who need to read or modify data in an InfoPath form. The problem is that the features of InfoPath 2003 do not extend to users for whom deploying Office 2003 on their desktop isn't an available option. This is a common occurrence when forms need to be extended to users outside of your company or when the forms applications need to be accessed from computers that do not have Microsoft Office installed. To solve this issue, Microsoft Office SharePoint Server 2007 introduces InfoPath Forms Services. This functionality is also available in a standalone server product named Microsoft Office Forms Server 2007.

InfoPath Forms Services has been designed to render InfoPath forms within the browser to reach users who are not running InfoPath. In fact, InfoPath Forms Services doesn't even require users to be running a version of Microsoft Internet Explorer or the Windows operating system. Microsoft has tested InfoPath Forms Services for compatibility with browsers such as Firefox, Safari, and Netscape as well as several other HTML-enabled mobile devices.

You can use the Office InfoPath 2007 forms designer to create web-enabled forms, which can then be deployed to Office Forms Server 2007. This forms designer provides a compatibility checker to ensure that your forms contain only controls and elements that are compatible with what Office Forms Server 2007 can render to the browser. While it is a bit tricky to do, it's also possible to create and deploy InfoPath forms intended for dual use. Such a form is downloaded to the desktop and loaded into the rich client environment when InfoPath is available on the desktop or otherwise rendered through the browser when necessary. This provides a solid, rapid application tool to help quickly build and deploy data collection applications or forms to front-end existing web services or other exposed business APIs.

Examples of InfoPath Forms Services applications are included in Chapter 6.

Enterprise Content Management

Government agencies and other regulatory organizations have placed increasingly restrictive requirements on companies that generate and store large numbers of documents. In industries that aren't heavily regulated, a company's policies related to dealing with intellectual property and managing information assets can be a significant driver for advanced information management requirements. This inevitably impacts developers and IT professionals who build and manage the business systems that retain the company's information. The Sarbanes-Oxley Act and emerging privacy laws within the U.S. have made it much harder for companies to stay within acceptable levels of compliance, and these concerns are growing as the amount of information that companies are managing is increasing at a staggering pace. Microsoft Office SharePoint Server 2007 provides several enterprise content management features to address these challenges.

Many of the Office SharePoint Server 2007 features for enterprise content management are built on top of the new information management policy features. An *information management policy* is a SharePoint component that can be enabled and configured within the scope of a list or document library. Examples of information policies that ship with Office SharePoint Server 2007 include those for document expiration, auditing, and the automatic generation of bar code labels to identify physical documents and associate them with electronic copies maintained in a document library. The bar code and labeling policies can employ either descriptive labels based on underlying metadata or unique identifiers that are automatically generated for each document.

The SharePoint framework for information management policies was designed with extensibility in mind. For example, you can create a custom policy that checks the integrity of a digital signature on every document within a document library. You can create another policy that promotes document hygiene and privacy by removing all the comments and personal information from documents as they are uploaded to a document library. Other scenarios might include limitations of document printing to "high security" printers or print queues to help prevent dissemination of document content that must be printed out as part of its life cycle.

Another important feature of content management is the integration of the document authoring tools with the content management system. Often in document management systems the end users are required to enter metadata related to their documents. In Office 2007, the concept of a *Document Information Panel* has been introduced to facilitate this integration between Office and SharePoint. The Document Information Panel is designed to enable users to specify all the properties on a document at once, in one place, at any point when they are working with that document. The Document Information Panel is essentially an InfoPath form displayed within the client application that contains fields for the document metadata. The Document Information Panel enables users to enter metadata related to a file directly from the Office application. For files stored in Windows SharePoint Services 3.0 document libraries, the document information maps to the columns of the content type assigned to that particular document. The Document Information Panel displays a field for each content type property, or column, the user can edit.

Users can edit document column values either at the SharePoint document library level or from the Office client application. The metadata values are stored in the document itself as well as in SharePoint. If the user updates the document metadata in the document, the new values are saved in SharePoint when the document is saved back to its document library. Conversely, if the user updates the content type column values in the SharePoint site user interface, those values are saved in the document.

For more information about how documents can leverage the Document Information Panel to support advanced content management, see a document and web content management scenario described in Chapter 10.

Microsoft Office SharePoint Server 2007 provides a dedicated site template named the Records Center to assist with records management, as shown in Figure 1-10. A Records Center site provides archiving support for companies that are required to keep certain types of business documents as official records of the company's activities. While archiving requirements vary across different regulated industries, keeping records is required to provide evidence of a company's activities in the event of a litigation dispute or an audit. While these features are not necessarily compelling to developers directly, they do provide a foundation on which applications that are subject to these requirements can be based.

Numerous reports can be generated automatically from a Records Center site to help information managers track the status of various types of content being retained in a company. For an example of how the Records Center site can be used in an application, see Chapter 6.

Figure 1-10

SharePoint and Office Development in Visual Studio

Of greatest importance to developers when considering using SharePoint and Office platforms in their applications is whether or not the tools that they use to develop software can effectively be leveraged with these platforms. For developers who have been doing SharePoint development for a while, many have recognized the limitations of the developer tools and have been frustrated by those limitations. For those looking to build applications in Office, the situation was even worse. In the last couple of years, however, Microsoft has committed to making Office applications "first class" from a developer perspective. Part of this effort includes the revamped design application for SharePoint, now the SharePoint Designer instead of FrontPage (see Figure 1-11). The next sections take a look at a few of the key developer features that make Office and SharePoint 2007 viable for custom applications.

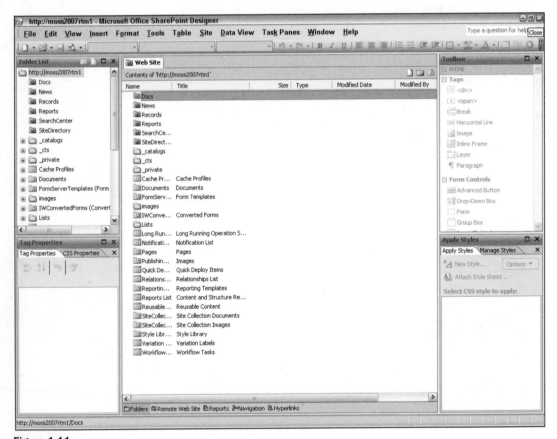

Figure 1-11

Web Part Development

One of the most popular ways for developers to extend Windows SharePoint Services 2.0 sites has been to create custom Web Parts. Web Parts are great because they add the extra dimensions of user customization and personalization. As a consequence, many teams at Microsoft, and third-party companies, have built custom Windows SharePoint Services 2.0 solutions using Web Parts. Because of the popularity of Web Parts in Windows SharePoint Services 2.0, Microsoft decided to add support for custom Web Part development to ASP.NET 2.0. The goal to reach a wider audience of developers was reached by creating a new Web Part infrastructure for ASP.NET 2.0 that is similar to the Web Part infrastructure created for Windows SharePoint Services 2.0. As a result, there are now two different styles of Web Parts. The older WSS-style Web Parts depend on `Microsoft.SharePoint.dll` and must inherit from the `WebPart` base class defined by Windows SharePoint Services 2.0 in the `Microsoft.SharePoint.WebPartPages` namespace. The newer ASP-style Web Parts depend on `System.Web.dll` and must inherit from a different base class, also named `WebPart`, defined by the ASP.NET 2.0 team in the `System.Web.UI.WebControls.WebParts` namespace. In this book, we focus on the development of Web Parts based on the `System.Web.UI.WebControls.WebParts` namespace.

It was important for Windows SharePoint Services 3.0 to run both the older WSS-style Web Parts as well as the newer ASP-style Web Parts because not all Web Parts will be upgraded by the original developers. This was accomplished by building the Windows SharePoint Services 3.0 support for Web Parts on top of the ASP.NET Web Part infrastructure, and then making changes to `Microsoft.SharePoint.dll` so that WSS-style Web Parts written for the Windows SharePoint Services 2.0 environment would be forwardly compatible with the Windows SharePoint Services 3.0 runtime environment.

In addition to the Web Parts compatibility across SharePoint environments, Web Parts targeted toward the ASP.NET 2.0 architecture can also be supported by SharePoint. The development of Web Parts is a key requirement to enable users to have flexible and dynamic web sites that are composed of the key business data elements and system interaction features that they need.

We go into much greater detail about Web Part development in Chapter 5.

Business Intelligence Features

Microsoft provided developers and IT professionals with business intelligence (BI) features in SharePoint Portal Server 2003, the Office 2003 Web Parts and Components Add-in, and Microsoft Office Business Scorecard Manager 2005. While these products had a wide range of capabilities, the mix of products and technologies was often hard to grasp and effectively implement. Over the past few years, many developers have used these BI products as a platform for building dashboard-style applications that provide business managers with current data that reflects the health of a business and flags potential problems in a timely matter.

Even products such as Microsoft's BizTalk Server have offered extended functionality to support this kind of business activity monitoring capability, delivering customer portals and web tools to provide visibility to the end users and IT administrators. Leveraging the previous experience with these earlier BI components, Microsoft has designed Office SharePoint Server 2007 to include a platform for building dashboards and integrating with other technologies in the broader Microsoft BI landscape, such as Microsoft Office Excel 2007, Microsoft SQL Server Reporting Services, and Microsoft SQL Server Analysis Services. As with all other aspects of Microsoft Office SharePoint Server 2007, its BI platform builds on top of ASP.NET and Windows SharePoint Services 3.0 and provides many opportunities for extending what is provided out of the box.

Excel Services

Microsoft has received extensive customer feedback telling them that a large percentage of corporations maintain a significant amount of business logic in Excel spreadsheets and that this business logic has been hard to leverage and reuse across a large organization. This feedback led Microsoft to create Excel Services for Office SharePoint Server 2007.

Excel Services represents a server-side version of the traditional Excel calculation engine that has been rewritten for SharePoint. Excel Services doesn't suffer from the same types of scalability problems that occur when you run the desktop version of Microsoft Excel on the server. Excel Services also provides a server-side rendering engine that can display spreadsheets in the browser as HTML. That means a company can store all its Excel spreadsheets in a centralized document library and make them viewable by users who don't even have Excel installed on their desktop. In addition, consumers of this data can see the numbers displayed by a spreadsheet within the browser without having any access to the business logic behind it that represents a company's intellectual property. This prevents users from gaining access to proprietary formulas or calculations, and it can also prevent users from accidentally modifying key data elements or formulas in a spreadsheet, which often leads to "multiple versions of the truth" within a company that relies heavily on Excel reporting for key business metrics.

A key observation is that the Microsoft Office 2007 system introduces a new paradigm, one that recognizes that companies maintain business logic within Excel spreadsheets just as they maintain business logic within managed code inside compiled assemblies and within stored procedures in a SQL Server database. To support this new paradigm, Microsoft has added many new features to Office 2007 products designed to expose and update this business logic as well as to protect its intellectual property from users that should not have access to sensitive information but need access to the data rendered by Excel to do their jobs.

The new desktop version of Office Excel 2007 has been enhanced to allow information workers with Excel expertise to publish and update their spreadsheets in a document library on an Office SharePoint Server 2007 portal site or a Windows SharePoint Services team site (see Figure 1-12). Users running a version of Excel can view these spreadsheets through a rich client experience, while other users can rely on Excel Services to view the same spreadsheet inside the browser. Note that this new spreadsheet publishing metaphor enables a company to maintain a single master copy of its critical spreadsheets. It also enables the spreadsheet author to post updates without the need to involve the development staff or the IT staff.

It's important to note that the use of Excel Services isn't restricted to the browser. You can create a Windows Forms application that leverages the server-side Excel calculation engine but doesn't use the rendering engine. For example, a Windows Forms application can use standard web services from Excel Services to load a spreadsheet on the server, enter input data, perform calculations, and return a result. This example extends the analogy that Excel Services exposes the business logic defined in a spreadsheet just as SQL Server exposes the business logic defined in a stored procedure. This enables developers to truly leverage these assets as an enterprise-level application infrastructure component and to build systems around the data and calculations stored in Excel Services.

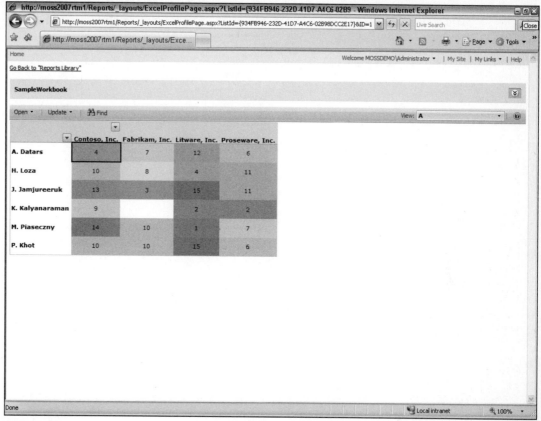

Figure 1-12

Report Center

Office SharePoint Server 2007 provides a special site template named Report Center, shown in Figure 1-13, for companies that want to build dashboard-style applications. Report Center was designed to make the new BI features of Office SharePoint Server 2007 easy to discover and use in the SharePoint paradigm. A Report Center site contains a document library named Reports Library that is designed to store and display BI reports such as Excel spreadsheets and reports built for SQL Reporting Services.

Another important aspect of Report Center is the built-in support it offers for creating and importing Key Performance Indicators (KPIs). A KPI is a common term used in the BI world to describe a visual indicator that tells a manager how some aspect of the business is doing. For example, the KPI for a product inventory level might display a green light when there is enough inventory to supply all the orders for the coming week. The light might turn from green to yellow when the inventory level drops below some predefined threshold, such as the amount of inventory required to supply orders for the next four days. The light then might change from yellow to red when the inventory level drops to a point where it will run out within the next 48 hours. In other words, a KPI flags aspects of a business that require immediate attention, alerting the appropriate users to take some action. Providing the capabilities to enable users to perform the banding and assignment of KPIs to the particular dashboards they want to monitor is critical to extending BI features to the broadest set of users and not just a few senior managers in an organization.

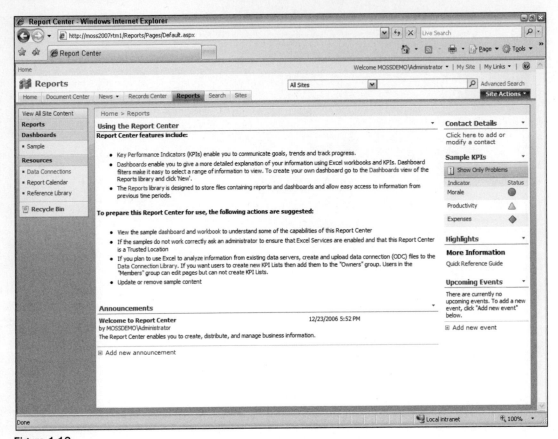

Figure 1-13

Office SharePoint Server 2007 provides support for creating several different types of KPIs. For example, you can create a KPI whose indicator changes automatically depending on data it dynamically reads from a Windows SharePoint Services list or an Excel spreadsheet (see Figure 1-14).

Office SharePoint Server 2007 also provides integration support for KPI in SQL Server 2005. That is, if you have already created KPIs with SQL Server Analysis Services, you can import and display them on a Report Center site along with other types of supported KPIs.

Another key feature of Report Center is the built-in framework it provides for filtering data before it is shown to the user. This is a key component of the Office SharePoint Server 2007 dashboard framework because it makes dashboard pages more relevant to the user. When a manager visits a Report Center site, the experience is enriched if the dashboard views have been customized with data that is relevant for that user. For example, a sales manager for the Eastern region of the United States can be presented with a different view of sales figures than the sales manager for the Western region of the United States might see.

Furthermore, managers like to be able to see high-level data at first and then be able to drill down into more specific categories on demand. Filtering support is built into Office SharePoint Server 2007 dashboards at the page level using Web Part connections. Office SharePoint Server 2007 supplies Web Parts

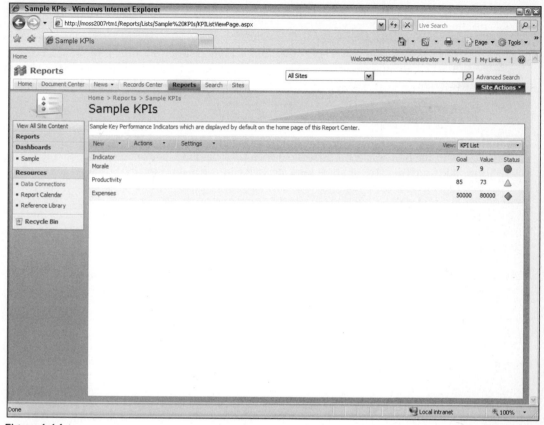

Figure 1-14

out of the box that enable page designers and users alike to specify criteria such as the name of the current user, a date range, or a product category. You also have many out-of-the-box Web Parts that can be configured to consume the filtering criteria supplied by other Web Parts, such as the standard Windows SharePoint Services List View Web Part and the Web Parts designed for use with the Business Data Catalog, Excel Services, SQL Server Reporting Services, and SQL Server Analysis Services.

Integrating the Desktop with Enterprise Systems

The key paradigm shift developers are facing is how to integrate rich desktop applications with enterprise systems, including Enterprise Resource Planning (ERP), manufacturing, Customer Relationship Management (CRM) and other repositories of business rules and critical business data. The trend toward service-oriented applications has been an important one, but one of the missing components has been integration of the standard desktop software that the average end users spend a lot of time working with.

Until Office 2003 there was no simple way to integrate a web service into an Office application. Most people needing to leverage Office would have to go back and forth between applications or systems, or perhaps some clever software developers would write code to generate the Office documentation required for a business process. Once Office 2003 implemented support for the Information Bridge Framework, this kind of integration became a bit easier, but even then, end-to-end scenarios involving workflow, document routing, and approvals across applications, along with the ability to integrate with third-party systems, just wasn't there.

In the most recent wave of Office and SharePoint releases, the scenarios for enterprise integration are becoming much richer. In the Office client, the new file formats and support for web services mappings and extensibility are driving the development of increasingly rich UI applications tied to business applications based on Office products such as Excel, Word, or InfoPath. In the case of SharePoint, the Business Data Catalog (BDC) provides a metadata mapping layer to enable the integration of business application data into enterprise search, SharePoint Web Parts, and other custom applications. As developers start leveraging these features in their applications, the convergence of Office client functionality and SharePoint functionality can lead to fully featured business applications, integrated with the back-end ERP systems, all driven by user interaction with their desktop installation of Office.

The Office 2007 system of applications and SharePoint technologies are designed to provide a rich set of integrated features for document-centric work as well as business processes that involve the exchange of documents, workflow that integrates with external systems, and the capability to search across the company's information assets; and the technologies provide all of these features in a very small number of tools and corresponding user interfaces. As you read this book, consider the possibilities of better integrating applications with standard desktop tools so that end users can more easily access information and get their work done; and so that as a developer you can focus on solving hard technical problems, not rebuilding and repurposing a significant amount of code to support basic business processes. This represents a significant shift in paradigm for many developers, but one that can lead to strong productivity improvement and faster time to market for both end users and developers focused on building new business processes and automating them to support efficient business operations.

Summary

This chapter has provided a broad survey of the most significant services and features targeted at software developers in Microsoft Office 2007, Microsoft Office SharePoint Server 2007, and Windows SharePoint Services 3.0. In the subsequent chapters, we'll take a deeper look at the architecture of the SharePoint product to understand what level of customization is possible within the product and which APIs and development techniques can be harnessed to extend the product. The focus in these chapters is the developer's ability to leverage APIs for Office and SharePoint features that support both automation and integration with other systems, along with the developer's ability to customize and extend these platforms for use in business applications.

In addition, new technical features that are supported out of the box, including new file formats for Office, content types within SharePoint to support document management, new site templates in SharePoint for Search, Business Intelligence, and Records Management among other product features, will be reviewed in detail.

2

Understanding SharePoint 2007 Services and Architecture

Microsoft's SharePoint technology architecture provides a foundation for developers to create web sites, collaboration tools and enterprise services for web content management, search, records management, and other document management and collaboration functions. One significant improvement in the SharePoint 2007 architecture is the improved integration of the Windows SharePoint Services components with the SharePoint Server technology, which helps developers to leverage their code for small-scale applications that don't require the SharePoint server functions along with larger SharePoint Server application deployments that include the additional services available in Microsoft Office SharePoint Server. Additional improvements include the areas of search services, including search support for data residing in business applications or third-party databases, and mobile services to enable greater reach for accessing SharePoint applications on mobile devices.

The core components of the SharePoint technology stack are Microsoft Windows SharePoint Services 3.0 and Microsoft Office SharePoint Server 2007. In this chapter we'll cover architectural elements of the SharePoint Server and Windows SharePoint Services products that are relevant to developers, and highlight the key technologies that developers need to understand to build SharePoint applications or to integrate SharePoint features into their existing code.

To understand SharePoint technologies and the overall architecture of the system, we'll look at the following components in the SharePoint product stack:

- ❑ Content services, including ASP.NET integration, business data services, and web content management features
- ❑ Search services, including content-based search and business data search
- ❑ SharePoint APIs, including web services and SharePoint's object model
- ❑ Mobile services features

In each of these areas we'll look at the architectural components of the SharePoint products and technologies that enable developers to build out these services and leverage them in their applications. In addition, we'll take a deeper look at the extensibility of these components and classify the various developer feature areas in both the Windows SharePoint Services and SharePoint Server products.

At the end of this chapter, you should have a good understanding of the architectural components of the Microsoft Office SharePoint Server and should have a good idea of which developer feature areas and technologies in the product should be deployed in various scenarios and how they can be utilized. This chapter also serves as the technical foundation for Chapter 3 and Chapter 5, which demonstrate Windows SharePoint Services and Microsoft Office SharePoint Server custom applications.

SharePoint Content Services

Microsoft's content services features are based on the integration of ASP.NET into SharePoint and the support for web content management, portal, and enterprise content management features. Microsoft Office SharePoint Server (MOSS) 2007 provides a set of document services to help users manage and work with the growing volumes of content generated by business process automation tools and the increased collaboration between users of tools such as Office. Features including the Records Center help users store and manage documents and apply policies and business rules to ensure the correct handling of content.

Another important content service is SharePoint's Web Content Management, which is new to SharePoint 2007. The web content management features of MOSS 2007 enable a Publishing feature on a site so you can brand the site to provide your company's look and feel and provide specific user interface elements. As a result, you can provide your users with a method to edit the corporate site within a web browser and can deploy content through staging servers into production. On a site with the Publishing functionality turned on, you can create a multilingual site by creating a source site and then have that site translated into other languages, which are then published as separate sites. The underlying functionality for web content management is based on ASP.NET and is easily extensible via code.

Related to ASP.NET integration is the support for master pages within SharePoint. Windows SharePoint Services pages that end users can customize, including list view pages, list form pages, and Web Part pages, are *content pages* that contain content for display. When a user requests a content page, it is merged with a *master page* to produce output that combines the layout of the master page with the content from the content page. All content pages share the same page structure. This is a valuable feature for ensuring that common elements appear consistently across pages, and sites can be more easily maintained and developed against using this technique.

For more information on how these content services function in a Web Content Management application, see Chapter 9.

Another key feature for providing SharePoint content is the *Business Data Catalog (BDC)*. This feature in SharePoint provides a metadata layer and an object model that enables developers to expose data from systems outside of SharePoint within SharePoint Web Parts and lists. In addition, this model can be used to develop metadata definitions that provide a generalized method of access to a set of APIs or data sources from multiple applications, both within and outside of SharePoint. To better understand how the Business Data Catalog supports the provision of content within SharePoint, the following sections go into detail regarding the use of this feature.

Business Data Catalog

The Business Data Catalog feature in Microsoft Office SharePoint Server is designed to enable access to business data and services in applications and data sources that are external to SharePoint. The Business Data Catalog is composed of object models for administration, security, and runtime access; a metadata model to describe the access methods to external sources; and a metadata database to store the descriptors of external data sources and services. The architecture is shown in Figure 2-1. This infrastructure is then consumed by other lists, Web Parts, and services in SharePoint, including Search. You can also build custom applications and SharePoint features to deliver this functionality to your users.

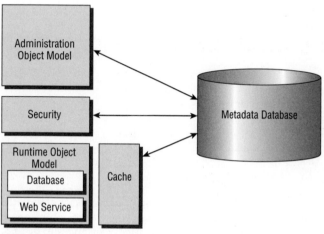

Figure 2-1

Business Data Catalog Metadata

The key element of the Business Data Catalog is the descriptive metadata for data sources and APIs. As an example, a data source such as a SQL Server database has a series of tables or other database objects that can be exposed through SQL queries. In the Business Data Catalog, the SQL queries have associated metadata that enables real-time access from a SharePoint service through this layer to the underlying database. In this sense, a Business Data Catalog service that's defined in SharePoint promotes reuse of that business data entity (API or underlying data source). The benefit to developers is that this model for using metadata to describe APIs and data sources provides a consistent way to access this business data across many services within SharePoint, and from external services and applications as well.

If you want to define a Business Data Catalog service, the metadata must be authored for each data source or API to enable interaction with the underlying data or API methods. In the case of a web service that exposes customer data from a Customer Relationship Management (CRM) system, you define metadata for the individual methods that return a customer's order history, credit status, or other customer profile information, including address and key contacts or company officers. You also define the metadata for the entities exposed by the services, which should be returned by calling a method of that service. Using the CRM system example again, entities will include Customer and Address and will have associated fields with descriptive names and related data elements such as Street Address, City, State, and Postal Code.

The metadata definitions are based on predefined relationships that specify how the data that's returned from the APIs or sources is related. To help you understand how this works, we'll examine the metadata associated with a web service that returns an order with line items.

The key data elements in the XML that define a Business Data Catalog are shown in Figure 2-2. The parent element is `LobSystem`, and child elements are `Properties`, `LobSystemInstances`, `Entities`, and `Associations`. `Properties` and `LobSystemInstances` are used to map the service name and location for use through the API and supporting administration and configuration tools.

```xml
<?xml version="1.0" encoding="utf-8" standalone="yes" ?>
+ <!--    -->
+ <!--    -->
- <LobSystem xmlns:xsi="http://www.w3.org/2001/XMLSchema-instance"
    xsi:schemaLocation="http://schemas.microsoft.com/office/2006/03/BusinessDataCatalog BDCMetadata.xsd" Type="WebService"
    Version="1.0.0.0" Name="SampleWebService" xmlns="http://schemas.microsoft.com/office/2006/03/BusinessDataCatalog">
  + <Properties>
  + <LobSystemInstances>
  + <Entities>
  + <Associations>
  </LobSystem>
```

Figure 2-2

We will next take a look at a sample entity definition for `Order` associated with a sample web service, shown in the listing that follows. For this example, we'll take a look at the order and line item related features of this web service, including the capability to retrieve an order by Order ID and to associate the order with its individual line items using Order ID as a foreign key relationship:

```xml
<Entity EstimatedInstanceCount="10000" Name="Order">
+ <Properties>
+ <Identifiers>
+ <Methods>
</Entity>
```

Order contains an `EstimatedInstanceCount` attribute and a name attribute, `Order`. The `EstimatedInstanceCount` is used to enable Business Data Catalog client applications (for example, Web Parts, lists, or custom applications) determine whether or not they'll display the data associated with this entity. If the entity contains too many instances to display, then the application can opt not to display the detail or can prompt the user to perform another action to drill into greater detail.

The child elements include `Properties`, `Identifiers`, and `Methods`. The `Properties` define a Property Name for the entity association, such as Order ID, used to describe the object, while `Identifiers` have a deeper functional implication. `Identifiers` act as a sort of primary key for entity instances that enable you to associate entities' data elements with other items. For example, if you have an Order ID as part of an order as an identifier, this enables you to associate the top-level order with its line item children.

The methods of the entity map to the functions that the back-end API or data store can perform to provide data to the Business Data Catalog. In the next listing we'll take a look at the metadata for a method from a sample web service and dig into the semantics of the metadata definitions a little deeper. The metadata for the `GetMyLineItems` method is shown in this listing:

```
<Method Name="GetMyLineItems">
 <Parameters>
  <Parameter Direction="Return" Name="LineItems">
  <TypeDescriptor TypeName="MyWebServiceProxy.LineItem[], MyWebService"
IsCollection="true" Name="ArrayOfLineItem">
    <TypeDescriptors>
     <TypeDescriptor TypeName="MyWebServiceProxy.LineItem, MyWebService"
Name="LineItem">
      <TypeDescriptors>
       <TypeDescriptor TypeName="System.String" IdentifierName="OrderID"
Name="OrderID" />
       <TypeDescriptor TypeName="System.String" IdentifierName="ProductID"
Name="ProductID" />
       <TypeDescriptor TypeName="System.String" Name="ProductName" />
       <TypeDescriptor TypeName="System.Int32" Name="OrderQty" />
       <TypeDescriptor TypeName="System.Decimal" Name="UnitPrice" />
       <TypeDescriptor TypeName="System.Decimal" Name="LineTotal" />
      </TypeDescriptors>
     </TypeDescriptor>
    </TypeDescriptors>
   </TypeDescriptor>
  </Parameter>
 </Parameters>
 <MethodInstances>
  <MethodInstance Type="Finder" ReturnParameterName="LineItems"
ReturnTypeDescriptorName="ArrayOfLineItem" ReturnTypeDescriptorLevel="0"
Name="FindLineItemInstances" />
 </MethodInstances>
</Method>
```

The `Method` element has an attribute to define method name. Child elements of `Method` include `Parameters` and `MethodInstances`. `Parameters` contains a description of the parameter type and `TypeDescriptors` to define the kind of data that is associated with the method, and the specific data types for that data. For example, for the method `GetMyLineItems`, the web service method returns an array of line items.

In order to map this to the Business Data Catalog metadata, we must define the `TypeDescriptor` for the line item that is defined by the instance of `MyWebServiceProxy.LineItem`, which is the web service proxy class instance of line items returned as part of the array. From there, we define the children of each line item, which include a `TypeDescriptor` for each Product ID, Product Name, Quantity Ordered, Unit Price, and Line Total for that line item, based on quantity. Note that the `TypeDescriptor` for `ProductName` is associated with `System.String`, while the type associated with `UnitPrice` is associated with `System.Decimal`.

Another important element to the use of Business Data Catalog metadata is the definitions of associations between entities. Associations define the relationships between entities. In the example we've been looking at, there are associations that define the relationship between an order and its line items. If an association is used to maintain a reference between an instance of an entity, such as an order, then it is possible to directly call another method defined in the metadata layer to return the entity's children or to traverse the other entity relationships that are defined. The following listing shows the association between orders and line items based on the metadata defined for our sample web service:

```
<Association Name="OrderToLineItems" AssociationMethodEntityName="LineItem"
AssociationMethodName="GetMyLineItems"
AssociationMethodReturnParameterName="LineItems"
AssociationMethodReturnTypeDescriptorName="ArrayOfLineItem"
AssociationMethodReturnTypeDescriptorLevel="0">
  <SourceEntity Name="Order" />
  <DestinationEntity Name="LineItem" />
</Association>
```

In this listing, the association of orders to line items uses the entity name `LineItem` and the method name `GetMyLineItems` to associate these line items with a source entity of `Order`. When an instance of the `Entity` class is created, a method named `FindAssociated` can retrieve the associated entities for that particular instance.

To understand how this can be useful, consider an example for a Business Data Catalog metadata definition supporting access to a reseller's CRM system that includes data on Customers, Orders, a customer's preferred stock items, and the suppliers that can provide those stock items. If a customer has orders and preferred stock defined in the CRM system, then you can start at the customer layer and determine which products they prefer and then determine which suppliers can provide those stock items. In short, these definitions provide a platform to leverage existing business services, APIs, and data sources to provide an application with a simple way to drill down on key business data, enabling end-users to view a business function very easily across multiple applications.

Object Model Support for Business Data Catalog

In addition to the metadata services, the other key area of functionality in the Business Data Catalog is provided through the Runtime and Administration object models. These are the programming interfaces that are used to access data from the Business Data Catalog, to connect Business Data lists, Web Parts, and other services interfacing with the back-end applications, and to manage and control the Business Data Catalog services.

The Runtime object model provides the developer with the primary mode of access to Business Data Catalog services. This object model provides a layer of abstraction for external Line of Business applications and data sources. This enables you to create a consistent way of accessing data across a disparate set of applications, web services, and databases. The Runtime object model doesn't directly access the data sources or applications outside of the Business Data Catalog, but rather uses an ADO.NET data provider for the database being accessed, or accesses a web service proxy that in turn calls the web service of the third-party application. The Runtime object model also provides the mode of access for the metadata model of the Business Data Catalog, eliminating the need to directly access the Shared Services database.

The Administration object model supports the management of Business Data Catalog metadata, including creating, updating, and deleting from that data store. The object model provides an alternative to the creation of XML files that define metadata for external application services and entity definitions.

The Administration object model can also be used to build custom tools to manage Business Data Catalog metadata. Unlike the Runtime object model, which supports only the reading of metadata, the Administration object model has full access to the data.

The Business Data Catalog in SharePoint is implemented as a shared service and is designed to be consumed by more than one portal site within a deployed server farm. The underlying metadata database is stored along with the Shared Services database with a MOSS deployment. The Shared Services database contains the Business Data Catalog tables that contain the metadata, but not the underlying business data defined by the metadata.

Page Model for Web Content Management

Master pages in SharePoint are a foundation of the page publishing model. As discussed earlier in the chapter, the master page is used to provide consistency for the look and feel of a site and can also be used to provide items that need to be shared by multiple pages on a site, including navigation elements, search features, logos, and other standard text or controls that apply to many or all pages on the site. Here we'll take a deeper look at how master pages function in SharePoint's Web Content Management functions.

Master pages make use of ASP.NET 2.0 user controls, web server controls, and Web Parts. A master page is combined with a page layout to provide the content for a rendered page.

> We referred to these rendered pages as "content pages" earlier in the chapter. In some SharePoint references, you'll see these referred to as simply "pages."

The page layout references a master page and includes field controls, which enable content to be edited and displayed on a site. Default pages for Microsoft Office SharePoint Server and Windows SharePoint Services do ship out of the box; however, you can easily modify those master pages to change the overall appearance of a site. Modifications for master pages can be made in text or script editors such as Visual Studio and Notepad or in the Office SharePoint Designer.

The architecture of the Web Content Management services in SharePoint 2007 is shown in Figure 2-3. Because SharePoint is based on ASP.NET 2.0, there is a mix of file-system–based content and SharePoint content stored in the content database for a site. IIS provides the primary interface for web content functions and works in conjunction with the ASP.NET 2.0 HTTP runtime to support page processing and rendering functions. As mentioned earlier, the key templates for page rendering are the master pages and page layouts containing content.

When an instance of SharePoint receives an HTTP request, it is handled by the ASP.NET engine. As an example, consider a page hosted in SharePoint. Upon request for that page, the ASP.NET engine retrieves the page layout based on the supporting `PageLayout` class, which in turn implements a file provider feature in SharePoint to render the page directly from the file system if it's a static page, render a dynamic page such as an `.aspx`, and provide the response to the user. This provides a consistent method of processing pages whether they are comprised of static HTML, are dynamic ASP.NET pages with Web Parts, or are out-of-the-box SharePoint pages based on a default master page.

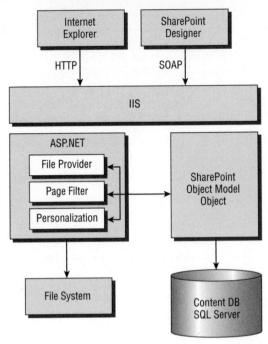

Figure 2-3

Document Converters for Content Management

In addition to providing a page model for web content management, SharePoint also provides support for converting documents to web-based content to make it easier for users to publish existing documents to their web sites or to make the process of building content for web-based publishing easier.

To support the conversion of documents, SharePoint supports an object model to implement document conversion utilities into the publishing process. This enables a document to be published in a Word format and automatically converted to a PowerPoint document in a SharePoint site. The following listing shows the document converter schema definition to enable conversion of a file type for converting Foo files to Bar format:

```xml
<?xml version="1.0" encoding="utf-8" ?>
<Elements xmlns="http://schemas.microsoft.com/sharepoint/">
  <DocumentConverter ID="{F9168C5E-CEB2-4faa-B6BF-329BF39FA1E4}"
    Name="Foo to Bar Conversion"
    App="fooToBar.exe"
    From="foo"
    To="far"
    ConverterUIPage="fooToBar.aspx"
    ConverterSpecificSettingsUI="fooToBarConfig.aspx"
    ConverterSettingsForContentType="fooToBarConfig.ascx"
  />
</Elements>
```

The document converter program is invoked using a command-line syntax that provides arguments specifying the input location, any special settings that are defined in an XML settings file, an output location, and a log to write the result of the conversion. From a deployment perspective, there are two ways to deploy the document conversion service: You can configure the service to process content synchronously or asynchronously and by a single instance of the conversion service or through a load-balanced configuration. In cases where documents need to be converted in real time and there is a high throughput of documents for conversion, the architecture can scale to meet that need.

The concept of document conversion is useful for ensuring that there is a consistent format for documents that are published to SharePoint. However, this capability can also be used to specify the conversion of traditional Office-style (and other formats) documents to web pages within a web content management environment as well. The process for converting documents to web pages using the "DocConversion" services is outlined in Figure 2-4.

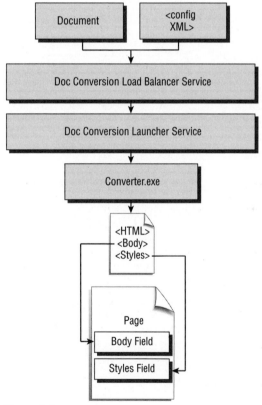

Figure 2-4

In this case, a Word document is converted to HTML. Out of the box, SharePoint supports the conversion of Word documents (.docx) to web pages, InfoPath files to web pages, and XML files to web pages using a user-defined XSLT. The document conversions can be completed individually by an end user publishing to a web page or can be queued for conversion by the DocConversionLauncherService.

With a range of content services for hosting web content, accessing business data, and converting document-based data into web content, it is very important to have a comprehensive set of tools for searching to help users locate the content that they need. In the next section, we cover the search capabilities of SharePoint and look at the key things that you need to know about SharePoint Search in order to make it work for your applications.

Search Services

While Windows SharePoint Services 3.0 provides search only for its own sites, and search that is limited in scope, Microsoft Office SharePoint Server 2007 provides a search capability that supports the indexing and retrieval of data from SharePoint sites, web-based applications and sites, Exchange folders, network file systems, and other business data sources. The Advanced Search user interface is shown in Figure 2-5.

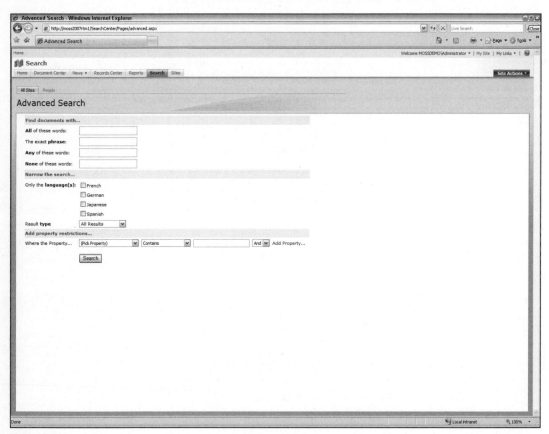

Figure 2-5

The search feature supports keyword and content-based searching, a "best bet" concept that enables a search administrator to identify the preferred result for a particular search term or phrase, and the capability to search data related to people within an organization. The people data search is based on organizational data from Active Directory and the content of individual My Sites that contain a user's profile information and other keywords related to their interests and organizational responsibilities.

An overview of the Search system's architecture is shown in Figure 2-6.

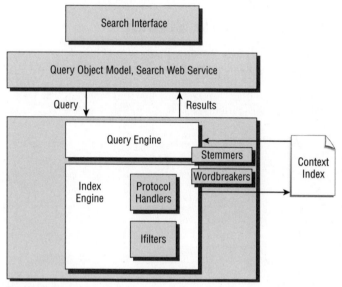

Figure 2-6

A number of core elements work together to comprise the SharePoint Enterprise Search architecture. The *Index Engine* of SharePoint performs the processing of text and property data from the various content sources, and stores data into a content index for later queries.

The *Query Engine* is used to process keyword and syntax-based queries to return the results of the content index based on search configuration.

Protocol Handlers and *IFilters* support gaining access to content sources using native protocols and documents in their native format for purposes of indexing content.

Of greatest importance to developers is the execution of search queries and techniques to execute queries from other applications. To understand this better, first we need to look at how a search query is executed. When a query is executed in SharePoint, the query engine processes the query and applies a wordbreaker. *Wordbreakers* are language specific and are used by both the query and index engine to break down words and phrases, making it easier to index and query against that content. After this step, the resulting words from the query are processed by a stemmer. A *stemmer* reduces words to their root (or stem) form. As an example, the word "dogs" can be reduced to "dog." Similarly, "dogged" and "doggy" could also be reduced. SharePoint search applies stemmers to both content indexing and query processing to support improved relevance for search results. The result of the query is then provided to end users in a list that is ordered based on the results' relevance to the query word or phrase. In SharePoint, if a user doesn't have access to a document or item returned in the results, then the query engine filters the document out of the returned list.

Shared Service Provider for Search

Search is configured as a Shared Service Provider in Microsoft Office SharePoint Server 2007. *Shared Server Providers* are groupings of services that are available to multiple sites within a farm. Unlike what you had to deal with in SharePoint Portal Server 2003, you don't need to manage crawling and indexing for each individual site or be concerned with redundant crawling and indexing. The Central Administration console for configuring Shared Service Providers is shown in Figure 2-7.

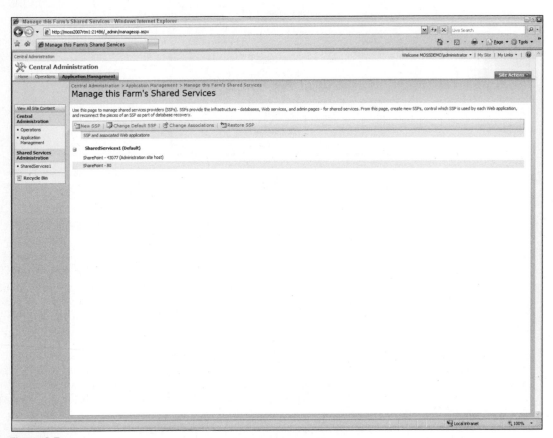

Figure 2-7

You should be aware of two components associated with the search shared service: *content sources* and *shared scopes*. Content sources provide a set of sources to crawl and index by default, including defining the behavior and schedule for the content to be crawled. Shared scopes are search scopes that group content items together based on content address, properties, or content source. This enables you to deliver search results that are specialized or targeted within a single content source.

Query Web Service

Microsoft Office SharePoint Server supports a Query web service that enables developers to perform queries from applications outside of SharePoint. The web reference to the Query web service is `http://Server_Name/[sites/][Site_Name/]_vti_bin/search.asmx`. A similar Query web service is supported by Windows SharePoint Services. Numerous methods are exposed by the web service, including support for search queries, search service status and methods to access search metadata, and search scope information. For building applications, most developers will focus on the query and service status methods, and deal with the other parameters through SharePoint administration tools. A few of the key methods to understand include the following:

❑ `Query` — Returns query results in an XML string

❑ `QueryEx` — Returns query results in an ADO.NET `DataSet` object

❑ `Status` — Returns the availability of the search service (ONLINE or OFFLINE)

The input for both the `Query` and `QueryEx` methods are an XML string specifying the query parameters. `QueryEx` supports some advanced scenarios, including the return of multiple result types. In addition, the default properties returned for the `QueryEx` method are much more extensive than the `Query` method, which returns only Title, Description, Date, and Relevance.

When specifying a search query, the XML string that specifies the query parameters can be built in a few different ways. This input format can be built using plaintext (supporting strings) for the query or you can use the SQL-based query language for SharePoint's full-text searching capabilities. The format of the query string is as follows:

```
<QueryText language=xml:lang type={STRING|MSSQLFT}>
searchQuery
</QueryText>
```

The `language` attribute is used to determine the language to be used for the keyword search. When this value isn't specified, the default language is the language of that site. The `type` attribute of `STRING` specifies a keyword query, and `MSSQLFT` specifies a SQL full-text syntax query. To compare the syntax for full-text queries versus keyword queries, take a look at the two versions side by side:

```
<QueryText language=xml:lang type='MSSQLFT'>
SELECT Title, Path, Description, Write, Rank, Size FROM Scope() WHERE
CONTAINS(Description, 'dogs')
</QueryText>


<QueryText language=xml:lang type='STRING'>
dogs
</QueryText>
```

In this listing, the SQL full-text query syntax is shown along with the plaintext query format. When using the SQL full-text syntax with the Query method (versus QueryEx), specific properties must be specified for the return results.

While the Search capability is a major feature for developers looking to leverage SharePoint, most developers want to focus on the APIs and the specific application services, such as Excel Services, the managing and publishing of APIs, and the features for mobile applications. In the next section, we will look into some of the SharePoint APIs and programming support features to help build and manage applications that are deployed on SharePoint.

SharePoint APIs and Application Services

The Microsoft Office SharePoint Server 2007 and Windows SharePoint Services 3.0 platform offers an extensive set of APIs for managing sites, publishing data such as pages, lists, and documents, and managing search and administration functions. To understand some of the potential uses of the SharePoint APIs and to set the stage for later chapters in this book, we'll take a look at the most commonly used top-level classes in the Microsoft.SharePoint namespace, the FrontPage Server Extensions remote procedure calls (RPCs), and the features of Excel Services.

Microsoft.SharePoint Namespace

The Microsoft.SharePoint and Microsoft.SharePoint.Administration namespaces contain many of the most commonly used classes for managing SharePoint configuration settings for a farm of load-balanced web applications and for managing site collections and individual sites, lists, files, and security settings. In this section, we'll take a closer look at the classes that support accessing a SharePoint site collection, drilling down to an individual site, and then accessing a specific list. While there are a many classes, these will give you an idea of what is required to programmatically access site data for a specific site, and expose you to some of the methods for manipulating list data.

To perform actions on a SharePoint site list, you need an instance of the SPWeb object to access a named list within that site. You can also use one of the SPSite constructors to return objects from a site collection to access a web site. The following listing includes code for creating an instance of SPSite and SPWeb for site and list data manipulation:

```
//Return a site collection using the SPSite constructor providing the site URL
SPSite siteCollection = new SPSite(siteUrl);

//Return the target web site based on site name
SPWeb site = siteCollection.AllWebs[siteName];
```

In this example, the SPSite constructor takes a string input parameter with the URL for the site collection. Then you can assign an instance of SPWeb to a named site within the site collection to enable access to specific lists and data elements within that site.

Accessing list data is an important function, if not the most important, of the SharePoint object model. The following example includes code for adding item data to a list:

```
//Return a collection of list items based on the Items property
//for the specific named list
```

```
SPListItemCollection listItems = site.Lists[listName].Items;

//Add a list item to the list items collection
SPListItem item = listItems.Add();
```

Lists or item data that is changed through the object model doesn't persist immediately and requires that you call the Update method for the changed item instance or collection of items.

In the Microsoft.SharePoint.Administration namespace, a commonly used class is SPWebApplication. A number of methods and properties are supported by SPWebApplication. A few examples of methods that are supported include the following:

❑ Delete — Removes the SharePoint web application and related resources owned by the application

❑ OnBackup — Called when a web application is being backed up

❑ Update — The web application will serialize state and promote changes within a farm configuration

❑ UpdateMailSettings — Updates e-mail settings, including outbound SMTP server name, From address and Reply to address

Additional information on the object model support for SharePoint lists items and libraries, along with detailed code samples, is included in Chapter 3.

FrontPage Server Extensions and WSS RPCs

Several remote procedure calls (RPCs) are supported within SharePoint. The history of RPCs in Share-Point dates back to the supported RPCs in FrontPage Server Extensions and in early versions of SharePoint that did not provide web services support. RPCs have been used as a point of integration from third-party web applications, from the Office client and from other custom applications accessing data in SharePoint. Because these RPCs do not include any functionality that is not supported through another more current API or web service, this section only looks at a few examples of each based on common usage, including those that have a corresponding replacement in the newer web services APIs.

Going forward, many of the Windows SharePoint RPCs have been deprecated and may not be supported in future versions. The following bulleted list describes a few examples of deprecated RPCs that are commonly used to create lists and update or delete list items. We're including a few that have a direct replacement in the newer APIs. For a full description of the RPCs, including those deprecated and still supported, you'll want to look at the Windows SharePoint Services 3.0 SDK. In addition to the RPCs, we have included their newer web services functional replacements:

❑ Delete — Deletes a specified item from a list or document library. Replaced by the UpdateListItems web service method.

❑ DeleteList — Deletes a specified list. Replaced by the DeleteList web service method.

❑ NewList — Creates a list such as Discussions, Contacts, or Surveys based on XML element, specifying list type. Replaced by the AddList web service method.

❑ UPDATEFIELD — Modifies the schema of an existing field in a SharePoint list. Replaced by the UpdateList web service method.

To support backward compatibility and legacy code from previous versions of SharePoint and other web-based collaboration tools, FrontPage RPCs are still supported in Windows SharePoint Services 3.0. The following list provides examples of FrontPage RPCs related to creating and deleting web sites and documents, some of the more commonly used functions:

- ❑ `create service`: Creates a web site
- ❑ `get document`: Retrieves a document
- ❑ `put document`: Writes a document to a web site directory
- ❑ `remove service`: Removes a specified site

The following listing shows an example of executing the `create service` procedure call to create a site named `createsample`:

```
POST /site_url/_vti_bin/_vti_adm/admin.dll HTTP/1.0
.
.
.
method=create+service:6.0.n.nnnn
&createsample=/site_url
```

The HTTP `POST` verb is used to specify the RPC method call for the specified resource. The server version is specified as `6.0.n.nnnn`, where `n.nnnn` specifies the phase version and incremental version number for the server.

While most functions supported by FrontPage RPCs can be implemented in another manner, either by using existing web services or the SharePoint object model, the RPCs do provide a method of access to SharePoint data that may be used by legacy applications that you'll run across, including legacy Microsoft tools designed to work with the original SharePoint Team Services infrastructure circa 2001. While you shouldn't look to these RPCs as the solution for writing new code, you do want to be aware of them in case you run across specific cases where they are implemented and may require some troubleshooting or maintenance.

Excel Services

Another mode of using SharePoint as an application server involves use of the new Excel Services for hosting Excel-based reports and providing data and calculation services. Excel Services is a feature of Microsoft Office SharePoint Server 2007 and includes several functional components, including Excel Web Access, Excel Web Services, and Excel Calculation Services. We'll look at each of these areas to understand how they support reporting and application services from an architectural perspective.

More detailed functional examples and scenarios are included in Chapter 7.

Excel Web Access

Excel Web Access provides front-end access to render the web content for Excel workbooks that are published in Excel Services. Excel Web Access is based on SharePoint Web Parts and supports the end-user functionality for these services. Using the `Microsoft.Office.Excel.Server.WebUI` namespace, you can also programmatically add an Excel Services Web Part to a SharePoint site, which enables you to place a dynamically rendered Excel workbook on a SharePoint site.

Excel Web Services

The Excel Web Services component of Excel Services provides access to the core Excel server web services. From an architectural perspective, these web services are the programmatic front end for the overall Excel Services technology stack. The WSDL definition for the Excel Web Services is shown in Figure 2-8.

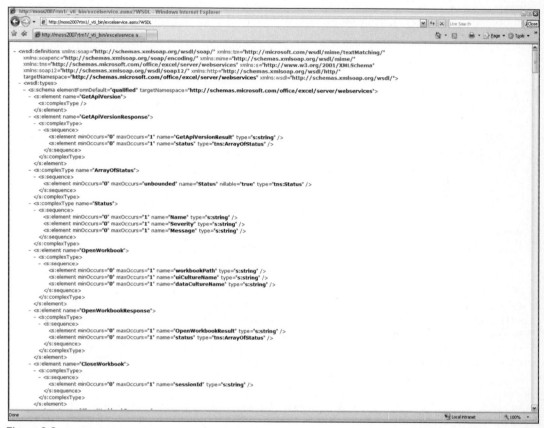

Figure 2-8

These web services can be used to extract values from a workbook, to refresh data connections to external sources, and to incorporate server logic contained in a workbook in an external application. The primary scenarios for implementing Excel Web Services involve the use of server-side Excel calculations, accessing those from external applications, and the automation of updates to server-based workbooks for extraction by a custom application. In this case, the server processes the workbook updates and then provides a data feed to the external application.

Excel Calculation Services

Excel Calculation Services are part of the application tier of the Excel Services architecture and are used to load workbooks, perform calculations in workbooks, access user-defined functions that represent custom code or logic, and refresh data from external sources. Excel Calculation Services also maintains a session for an individual user, or session state, subject to the user closing the session or experiencing a timeout. Excel Calculation Services also provides caching support for Excel workbooks, calculations, and queries to external data sources to improve performance in multi-user scenarios for workbooks.

Infrastructure for Excel Services

The infrastructure for Excel Services can be configured to run on a single server in a simple scenario or can support the division of components to run on web front-end servers and back-end application servers as needed. The basic infrastructure architecture for Excel Services is shown in Figure 2-9.

The Excel Web Access and Web Services functions can be run on a web front-end server, and the Calculation Services components reside on a back-end application server. Assemblies supporting user-defined functions (UDFs) would also reside in the application layer. The architecture supports load balancing to enable requests across multiple instances of Excel Calculation Services or the Excel Web Access and Excel Web Services components. The front-end and back-end services can be scaled out independently of the other infrastructure layer.

In addition to the application services and APIs supported by SharePoint, another new architectural component is the deeper integration of mobile site support and configuration support for mobile applications. In the next section we'll look at the mobile application services supported by SharePoint.

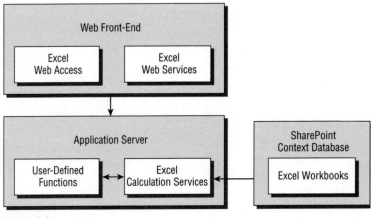

Figure 2-9

Mobile Application Services

The SharePoint architecture is designed to support mobile access to SharePoint sites and list data. Mobile users can be redirected to a mobile site when accessing a SharePoint site when "/m/" is appended to the standard URL. The redirected mobile site, mbllists.aspx, then displays the lists accessible for the mobile user. An example of a standard SharePoint site rendered as a mobile site is shown in Figures 2-10 and 2-11.

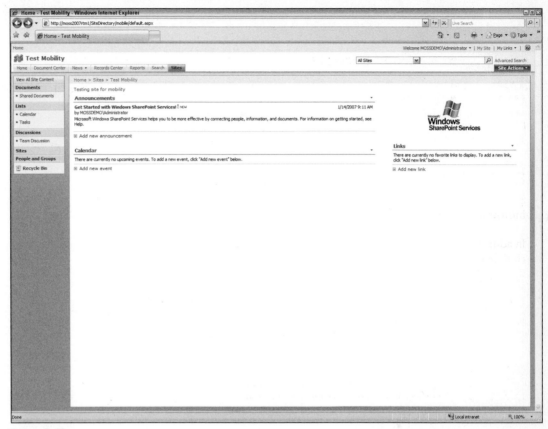

Figure 2-10

By default, sites of the STS or BLOG site definition will automatically redirect users to a mobile site version. In the case of blog sites, users are redirected to `bloghome.aspx`. The redirection support is based on SharePoint Features, a mechanism designed to make it easier to add custom features to sites and to ensure that those customizations are propagated in a consistent way and are persistent during upgrades and other infrastructure tasks. Features is covered in greater depth in Chapter 5.

The `Microsoft.SharePoint.MobileControls` namespace supports a rich set of mobile controls and navigation elements to enable the development of custom mobile sites. One of the commonly used classes for mobile development is `SPMobileListView`, which renders lists for consumers using mobile devices. Other mobile controls classes designed for greater user interaction and template-based customization options are available for mobile application developers.

Figure 2-11

Architecting Your Applications with SharePoint 2007 Services

With the significant number of services available to developers in SharePoint 2007, it is important to think about how you can leverage these services and build applications based on SharePoint features from the perspective of architecture. To some, SharePoint's specific infrastructure requirements and limitations related to deployment and configuration may make it unattractive to software architecture purists. On the other hand, there is a lot of functionality to be gained by using SharePoint in your applications.

While this book focuses more on solutions and functionality for developers rather than on architecture, it is nevertheless important to have a basic framework to think about the architectural components in this platform. To consider how we can fit SharePoint into an overall application architecture, we'll break things down into the following perspectives: the user presentation, business logic, and data tier services that can be delivered with SharePoint and the service-oriented and business process management applications that can be based on SharePoint.

SharePoint Tiers

As is the case with most applications, SharePoint can be thought of as a series of tiers that deliver the user experience, the business logic, and the data services for the application. In the context of an application architecture, SharePoint doesn't deliver a purist's approach to architecture but can be reconciled from an architectural perspective based on how the services are delivered. We'll take a look at SharePoint from the perspective of user presentation services, business logic services, and data services.

User Presentation Services in SharePoint

From the perspective of user presentation services, SharePoint can deliver the user experience for an application through the following mechanisms:

❑ **Web Parts:** SharePoint can provide out-of-the-box Web Parts or can host custom Web Parts that are used to deliver presentation services to end users. Examples of out-of-the-box Web Parts that can be used include a list view Web Part, a document library Web Part, or more advanced Web Parts such as business data Web Parts that can present data from the BDC.

More detail on building custom Web Parts is included in Chapter 5.

❑ **Web Content Management:** SharePoint provides a Web Content Management framework to deliver a full web site experience. In the case of a pure web application, SharePoint can provide the overall framework for the site, with all custom features and functions being integrated via ASP.NET 2.0 extensions or through links to other web applications.

A full Web Content Management scenario is demonstrated in Chapter 9.

❑ **Mobile Services:** While this is a variation on the web content features supported through SharePoint, there are specific features in place to support rendering of content for mobile devices that can be configured at the site level. This may make SharePoint an attractive alternative for some applications with mobile collaboration requirements.

❑ **Office Integration:** SharePoint supports integration with the Office client to help provide the user presentation experience. For example, documents associated with a content type in SharePoint that require users to enter metadata for newly created documents are presented with the Document Information Panel through Office. In SharePoint, the corresponding functionality is exposed through the document library itself. In addition, workflow assignments can be shown through a corresponding Office application or directly from SharePoint as part of a list of tasks associated with any given workflow.

In short, for applications that require user presentation (which is essentially all applications), SharePoint provides web content features, the use of custom or existing Web Parts, or integration points with the Office client. For special cases such as mobility, SharePoint may provide some distinct advantages in comparison to other existing web applications, which could differentiate its use within your organization. When building an application based on SharePoint or integrated with SharePoint, you'll want to consider these features when determining how to best deliver the end user interaction features for the application.

Business Logic Services in SharePoint

For the purists focused on software architecture, SharePoint poses a problem because it doesn't segregate the user interface and the "application" tier or business services from each other. With SharePoint, you essentially have a two-tier architecture with the SharePoint server and the database. In addition, even that it isn't a clear representation, as the SharePoint object model is a heavy abstraction of the underlying SharePoint database. Given those limitations, here are some business services that you can think about when considering SharePoint for an application:

❑ **BDC Runtime object model:** For services mapped within the Business Data Catalog, the Runtime object model can expose those services via SharePoint to other applications. Whether it is an existing SharePoint application or a custom Windows Forms application, the BDC services can extend the relevant business data from an Enterprise Resource Planning (ERP), CRM, or other custom application to the application of your choosing.

❑ **Excel Web Services:** While it is considered a "front-end" service in the Excel Services landscape, from an application architecture perspective you can consider employing Excel Web Services to access Excel Calculation Services from a remote application. This enables you to leverage formulas and logic embedded in an Excel workbook that is hosted in Excel Services from any application.

More examples related to Excel Web Services are included in Chapter 7.

❑ **SharePoint Collaboration Services:** When you are developing a SharePoint-centric application for collaboration, the SharePoint APIs and web services to support the creation of new sites, generation of new documents, and many other key site configuration, administration, and publishing activities can be automated. In this case, the business services themselves are related to the functions of SharePoint collaboration sites that support a larger series of integrated application components.

❑ **Search Services:** If you're developing an application that will make decisions based on the results of search data or will largely automate the search experience for a specific user-driven or automated process, the Search services in SharePoint provide support through the object model and existing web services to deliver this business logic to an application.

While the business services supported by SharePoint tend to be very specific to the types of services that SharePoint supports (for example, Excel Services, search, and collaboration), the support for the Business Data Catalog provides SharePoint with a general framework for providing business data and logic to both SharePoint and non-SharePoint applications.

Data Services in SharePoint

While SharePoint provides a high level of abstraction between the SharePoint object model and applications and its SQL Server data store, there are logical data services in SharePoint that you can consider as part of an architecture when building an application. Here are some logical data services to consider:

❑ **SharePoint Object Model:** The SharePoint object model's support for list and document data, specifically `SPList` and `SPDocumentLibrary`, enable developers to access existing lists and document libraries or to create new lists and libraries to store data.

A detailed example of how to use a SharePoint list as a data store for an application outside of SharePoint is shown in Chapter 3.

❑ **Search Services:** Again, here SharePoint will cause some purists concern when it comes to software architecture. Having said that, the search services index and query services that are supported to access data do provide a reasonable set of capabilities to read and consume existing index data within the SharePoint search infrastructure.

In addition to the breakdown of SharePoint services by user presentation, business, and data tier services, we can also look at SharePoint architecture from the perspective of functional services enabling service-orientation and business process management applications that can extend beyond the boundaries of SharePoint. The next section does just that.

Service Orientation and Business Process Management in SharePoint

For application architects today, two of the most important characteristics for new applications are as follows:

❑ Support for consuming existing services or producing services for consumption by other applications

❑ Support for business process management, or process automation, with little or no source code required for consuming services within those applications

In the SharePoint world, both of these characteristics are embedded in the platform.

Service Orientation in SharePoint

In SharePoint, there are a number of ways to consume or produce services that are based on XML or rely on the use of the HTTP protocol as a transport mechanism. Here are the ways that services are manifested within SharePoint:

❑ **Web Services APIs:** SharePoint supports a broad range of web services APIs. Examples shown earlier in this chapter include support for search queries and access to Excel Services. Additional examples are shown in Chapter 3. Any application that can consume web services should be a candidate for integration with SharePoint.

❑ **RSS Feeds:** SharePoint can produce RSS feeds for any list item data. RSS feeds provide an XML representation of the list item data that can be parsed by any third-party application outside of SharePoint.

An example of consuming RSS data from an application outside of SharePoint is included in Chapter 3.

❑ **XML/RSS Feeds Web Parts:** SharePoint can consume XML or RSS feeds using out-of-the-box Web Parts. This means that any external service that produces XML, including RSS, can be directly connected to SharePoint and integrated from a data display perspective.

❑ **Remote Procedure Calls:** The RPCs supported by SharePoint leverage HTTP to perform lightweight procedure calls that can be easily scripted or integrated into third-party applications, regardless of platform. While this mode of integration is not necessarily sound going forward and may be subject to removal in future product releases, it still exists today for legacy support.

In addition to SharePoint's support for services, SharePoint also supports a range of business process management functions.

Business Process Management in SharePoint

Business process management refers loosely to a category of applications, or in many cases application platforms, that can consume or produce services, enable workflow processes between humans and systems, and provide visibility to business processes in progress. SharePoint delivers capabilities for business process management in the following ways:

❑ **Integrated support for Windows Workflow Foundation:** SharePoint's out-of-the-box support for workflow based on the .NET Framework's workflow implementation is the key enabler for business process management in the platform. This workflow can be leveraged using out-of-the-box workflows, using workflows built with SharePoint Designer, or using custom workflows based on workflows developed using Visual Studio.

❑ **InfoPath Forms Services:** InfoPath Forms Services deliver the electronic forms capabilities that, along with integrated workflow, provide the most likely used human interface for business process management applications. These lightweight user input applications, combined with the support for consuming existing web services, make the InfoPath Forms client a strong development tool for SharePoint BPM applications.

❑ **Services support:** SharePoint's support for consuming and producing various web services and XML data formats as described in the previous section provide the basis for a rich set of functional scenarios that are required for a business process management tool.

From an architectural perspective, we are simply scratching the surface of possible scenarios and combinations of functions that can be performed in SharePoint. As you consider how SharePoint applications can be architected for your business and as you focus on how to reconcile SharePoint development with sound principles of software architecture, consider this framework as a starting point to help you determine what services within SharePoint you might leverage for an application and how the architecture of SharePoint can best be leveraged to meet your functional and technical goals.

Summary

This chapter on the architecture of Microsoft Office SharePoint Server 2007 and Windows SharePoint Services 3.0 has provided a broad survey of many significant services and architectural components of the SharePoint platform.

The SharePoint content services reviewed include components of the ASP.NET architecture integrated into the web content management infrastructure, including master pages. We also examined the document conversion features that support web and document publishing as part of the content management platform.

Business data services, including the Business Data Catalog that enables developers to leverage the functionality of SharePoint in custom applications and to extend the functionality as required for various business processes, also provide a significant base of functionality for developers. The combination of these services with SharePoint's Enterprise Search create some interesting new scenarios that support user interaction and better organization of data from Line of Business applications.

The SharePoint APIs available for site management, list and document access, and other administrative functions were also reviewed, along with the remote procedure calls supported through Windows Share-Point Services and FrontPage Server Extensions. In addition, SharePoint mobile services enable site administrators and developers to leverage SharePoint features in applications accessed by mobile devices, including Windows Mobile 5.0 phones and PDAs, among others.

From an architectural perspective, SharePoint delivers a set of services for user presentation, business services, and data-related functions. SharePoint can also be a key enabler for service-oriented applications, based on its ability to produce and consume web services and XML data. With the integration of a workflow engine and a forms management tool, along with all of the other platform services, including security, business data services, and more, the key components of a business process management platform are now in place with SharePoint.

In the next chapter, we'll take a deeper look at programming SharePoint to understand what level of customization is possible within the product and which APIs and development techniques can be harnessed to extend lists, items, and libraries. We'll also look at how this customization can be leveraged to build applications that integrate seamlessly with SharePoint as a data source, and how the supporting object model can be leveraged to build automated processes that are tied into SharePoint list and document data.

3

Programming SharePoint Lists and Libraries

SharePoint technologies are designed for users, applications, and systems to create, store, and track data related to various teams, projects, and other business processes or activities. Similarly, there are libraries in SharePoint that are a specialized type of list that enable users or applications to manage files. These are the foundational features of a SharePoint site.

The core functionality in Windows SharePoint Services 3.0 (WSS) and Microsoft Office SharePoint Server (MOSS) are based on these list and library functions and most out-of-the-box user features are based on or rely upon the use of lists and libraries. As you might expect, the programmability of lists and libraries in SharePoint is a key part of the functionality that developers need to learn in order to build applications that integrate effectively with SharePoint.

This chapter covers the programmability features of lists and libraries within SharePoint so that as a developer you better understand these features and what is supported for programmatic interaction with SharePoint data. In addition, we'll review some practical examples regarding how to build end-to-end solutions based on list and library data in SharePoint.

To understand SharePoint lists and libraries from a developer's perspective, we'll dive into the following topics:

- ❑   Understanding SharePoint lists and libraries
- ❑   Programming SharePoint lists
- ❑   Programming SharePoint libraries
- ❑   Examining web services for list and libraries
- ❑   Using list events
- ❑   Using RSS for list data retrieval

At the end of this chapter, developers should have a good understanding of both how lists and libraries work in SharePoint technologies and what developer feature areas and technology components related to lists and libraries can be leveraged in your applications. This chapter also serves as a technical foundation for later examples related to SharePoint UI development, including the creation of new Web Parts and methods to customize list rendering in various SharePoint sites.

Understanding Lists and Libraries

Lists are a key part of the SharePoint architecture. For a developer, a list is the main mechanism through which data stored in SharePoint is presented to a user for data input or retrieval. This means that in many cases customizing a SharePoint site requires some customization to the list data or programmatic access to the data from an application that resides outside of SharePoint.

Before getting started with coding against the SharePoint lists and libraries, it's important to understand a little bit more about how lists work, how they're created within the SharePoint web interface, and what properties they have that can impact how you write code for a list. To do this, we'll walk through an example involving the creation of a list.

Creating a SharePoint List

As a developer, you'll often need to write code against existing SharePoint lists to extend sites or add new functions to those sites already in place. You create a new list using the `Add` method of the `SPListCollection` class, and specify the type of list using the `SPListTemplateType` enumeration. As a developer using SharePoint lists, you will focus most of your effort on adding, updating, and deleting list items to synchronize your SharePoint content with another data store or to serve as a primary store for application data. Most lists that are used in SharePoint either will be default lists or will be deployed as part of a Feature for a site that is configured in advance, as opposed to creating lists on-the-fly for your specific application. For those reasons, this chapter focuses primarily on the manipulation of data for existing lists. A little later in the chapter, we'll revisit creating new lists through the APIs.

To better understand lists and how they're used in SharePoint we'll create a sample list using the SharePoint UI that includes part numbers. Once that list is created, we'll focus on the features and functions of that list as defined by the SharePoint web interface, and then we'll look at using the SharePoint object model and the web services that can be used to access the list.

Here are the steps to follow to create the parts list:

1. On a sample SharePoint server, use an existing site or create a new site (for example, standard Team site) on which to add the list. For the example in the book, we have created a new team site named "Parts List Sample."

2. Using the Site Actions menu, select the Create option to create a new list, library, or web page on the SharePoint site.

3. As shown in Figure 3-1, select the Custom List option to create a new custom list that doesn't correspond to the existing SharePoint list templates.

4. Specify the name and description for the new list. For this sample, the name of the list will be "Parts List."

5. Once the list is created, select Settings to add columns, create new views, or change other list settings.

To create a new column, select the List Settings item. On the next page, look for the Columns section and select the Create Column link. Once you're at the Create Column page, you can select the type of data the column supports, along with other parameters for configuring this data element. An example showing the creation of the Price column definition is shown in Figure 3-2.

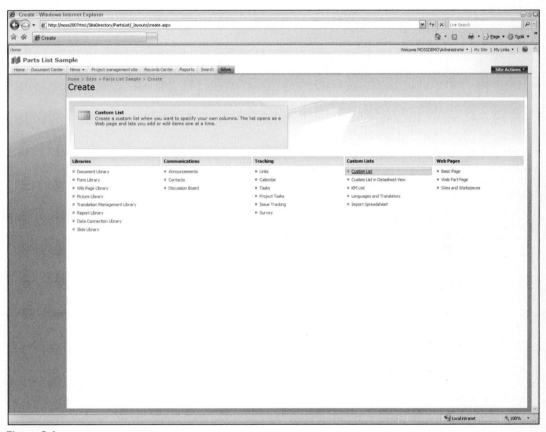

Figure 3-1

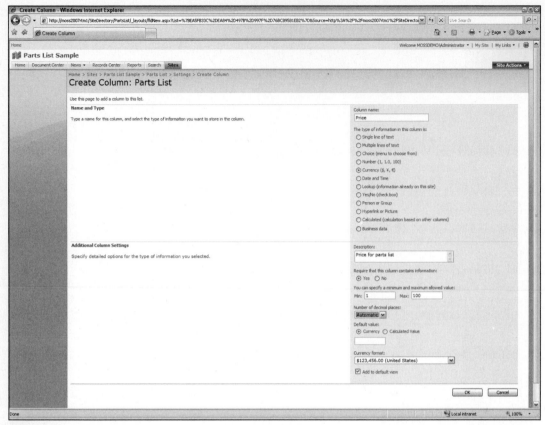

Figure 3-2

6. The sample parts list uses the Title field for the part name and includes the additional columns Part Number and Price. The Part Number field is restricted to a number type and the Price field reflects currency. For the example, configure the Price field with a constraint not to exceed $10.00 and set the field to be required as shown in Figure 3-3.

7. Enter some sample parts, as shown in the completed list in Figure 3-4.

With the completion of these steps you now have a custom list. You can either save this list as a template or you can use content types in SharePoint to define lists for your sites. *Content types* are used to help the users of SharePoint sites better organize their data and essentially capture the requirements for the types of data to be entered. Basically, a content type acts as a form of schema. While a list template enables you

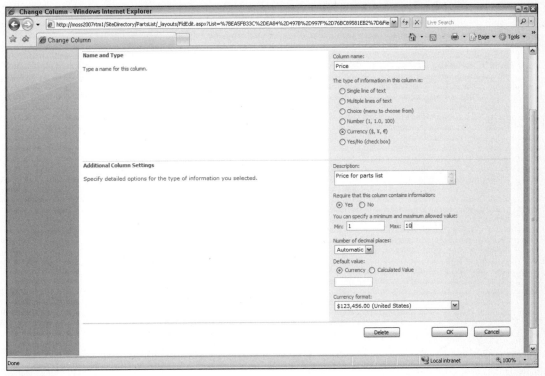

Figure 3-3

to export the definition of a list, a content type enables you to define the elements that are included in a list. There are custom content types and standard content types. In the case of this parts list, we'll add additional content types to the list to demonstrate how this works.

1. From the Parts List, select Settings and the List Settings option. On the Customize Parts List page, select Advanced Settings.

2. On the List Advanced Settings: Parts List page, select the Yes radio option for "Allow management of content types?" This will return you to the Customize Parts List page.

3. The current list definition will be defined under Content Types as Item. This denotes the standard list item that is created when a new item is added to this list. To add a new content type, select the Add from Existing Content Types link in the Content Types section.

4. On the Add Content Types: Parts List page, choose to select content types from List Content Types in the drop-down list and add the Announcement content type, as shown in Figure 3-5.

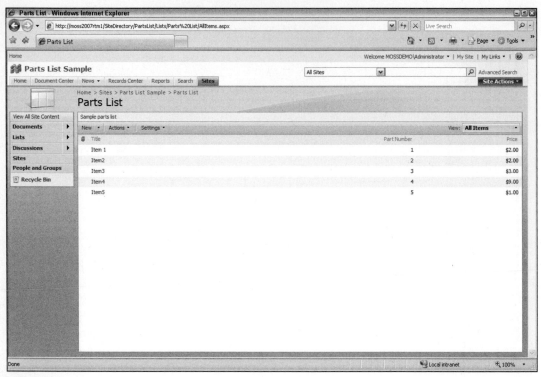

Figure 3-4

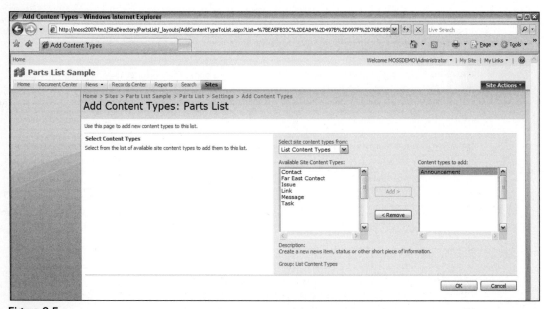

Figure 3-5

5. Once the Announcement content type is added, you can return to the main parts list and select the option to add a new item to the list. An example of an Announcement content type added to the list in addition to the other list items supporting the Item content type is shown in Figure 3-6.

Figure 3-6

In this case, the list now supports the custom defined items for the parts list along with the standard Announcements content type. This provides you with the flexibility to maintain a single list with multiple types of list items supported. Content types are designed to support settings for the following types of information:

❑   Metadata or properties that you want to assign to the list or document library

❑   Custom forms for use within the SharePoint interface for the content

❑   Workflows available for this type of content, including support to define automated workflow based on event, content, or other conditions

❑   For specific document content types, the template associated with this type of document

❑   Custom data for applications that can be stored as XML as part of the content type

Content types can be created in several ways. A user can create and deploy a content type through the SharePoint user interface. Content types can also be created through the SharePoint object model, or they can be created and deployed as Features in an XML definition file format.

For the purposes of this chapter, we're going to proceed with writing code against our manually created lists. In Chapter 5, we'll dig a little deeper into content types to understand how they fit into a SharePoint infrastructure with defined custom templates for sites, custom lists, and custom Web Parts. We'll also look into Features to understand how they can be used to deploy content types, Web Parts, and site templates throughout your SharePoint infrastructure.

Now that you have a sample list to work with, we can look into the object model for dealing with lists in SharePoint and create some custom applications to modify those lists.

Programming SharePoint Lists

With a new custom list in place for the SharePoint site, you can now dig into the object model and think about what you'd like to do in code with that list. Before getting started, you need to understand a bit more about the features of lists in Windows SharePoint Services 3.0.

As shown in the earlier sample list example, you can create lists to support specific data types with field constraints to ensure data integrity for new data entered into the list. Similar functionality for lists has existed in previous SharePoint versions, including support for out-of-the-box lists and the capability to create custom lists. Windows SharePoint Services 3.0 offers several new list features, including the following:

❑   Cross-list queries

❑   List indexing

❑   Folders in lists

❑   List item versions

❑   Item-level security

Some of these features speak to performance in Windows SharePoint Services 3.0. For example, the SPSiteDataQuery class enables you to use a SQL-like syntax to query lists across multiple SharePoint sites in a very efficient manner. In addition, list indexing enables you to index specific list columns, making the data store more scalable for querying by third-party applications. Item-level security is a great feature that has long been wished for in SharePoint, enabling a site administrator to now provision access to a library or list to a group of individuals but to still maintain security of each item in the list. Moreover, with the security trimming features supported within SharePoint, now sites can be configured to allow users to see only content to which they have permissions.

In this chapter we'll take a closer look at folders in lists and the versioning features for lists that enable some new scenarios not previously supported in Windows SharePoint Services 2.0 from a programmer's perspective. In the next section, we'll walk through the basics for creating a program to programmatically access a SharePoint list.

Adding a List Item

To get started with using the SharePoint object model, we use the example of adding an item to a list. To do this, we can create a simple Windows forms application on a test SharePoint server. We can use the list that we created to track parts as the list to program against in this example.

This example creates a base Windows application with a single form and a tab control. This enables us to maintain one project with a single form and compiled binary to support different list access scenarios. As you follow the examples in the next few sections, note that the screenshots reflect the use of this tab control and tabs that are in place for upcoming samples in the chapter. As you follow along and write your own samples, feel free to ignore the tab control or to use it as part of your solution. The code shown in this chapter is also available as a download at www.wrox.com.

Here are the steps to get started with programmatically adding an item to a list:

1. Create a Windows application in Visual Studio. For the samples in this book, we used Visual Studio 2005 Professional Edition.

2. Add a reference to `Microsoft.SharePoint` to enable access to this namespace.

3. Add controls to accept text input for the fields and a button to initiate uploading of the data to the list.

4. For the button click member function associated with the form, we'll add code to connect to the SharePoint site, retrieve a collection of the list items associated with the parts list, and then update that item in the list. This includes the following steps:

   a. Return a site collection using the `SPSite` constructor providing the site URL.

   b. Return the target web site based on the `SPWeb` class based on the site name.

   c. Return a collection of list items (`SPListItemCollection`) based on the `Items` property for the specific named list.

   d. Add a list item to the list items collection using the following syntax: `SPListItem item = listItems.Add()`.

   e. Populate the item data using the input values from the form.

   f. Call the `Update` method of the `SPListItem` class to update the SharePoint database.

5. Check the Parts List to see the updated list item.

The sample application to drive the list automation is shown in Figure 3-7. The results of the list update and the Visual Studio project code are shown in Figures 3-8 and 3-9, respectively.

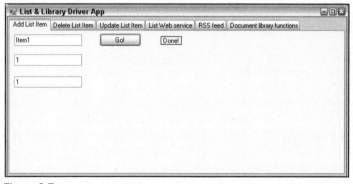

Figure 3-7

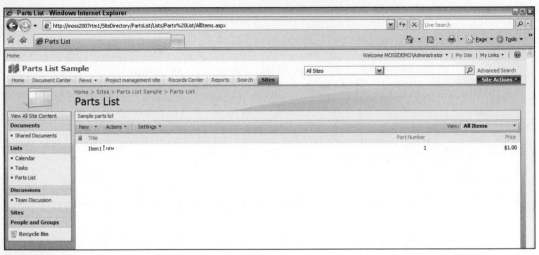

Figure 3-8

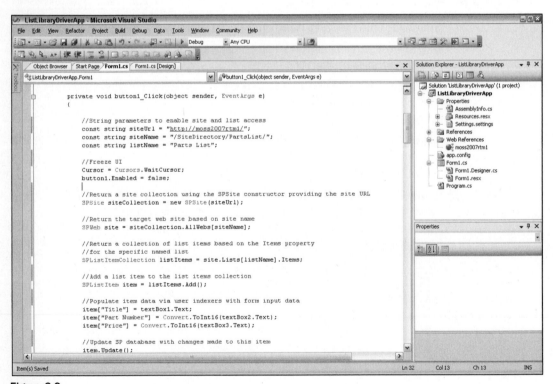

Figure 3-9

```
                //String parameters to enable site and list access
                const string siteUrl = "http://moss2007rtm1/";
                const string siteName = "/SiteDirectory/PartsList/";
                const string listName = "Parts List";

                //Freeze UI
                Cursor = Cursors.WaitCursor;
                buttonAddItem.Enabled = false;

                //Return a site collection using the SPSite constructor providing the
     site URL
                SPSite siteCollection = new SPSite(siteUrl);

                //Return the target web site based on site name
                SPWeb site = siteCollection.AllWebs[siteName];

                //Return a collection of list items based on the Items property
                //for the specific named list
                SPListItemCollection listItems = site.Lists[listName].Items;

                //Add a list item to the list items collection
                SPListItem item = listItems.Add();

                //Populate item data via user indexers with form input data
                item["Title"] = textBoxPartName.Text;
                item["Part Number"] = Convert.ToInt16(textBoxPartNumber.Text);
                item["Price"] = Convert.ToInt16(textBoxPrice.Text);

                //Update SP database with changes made to this item
                item.Update();

                //UI clean-up
                labelResult.Text = "Done!";
                Cursor = Cursors.Default;
                buttonAddItem.Enabled = true;
```

In this sample, we've used a Windows application that runs on the SharePoint server to test these functions via the object model. This same functionality can be leveraged from within ASP.NET code running on the SharePoint server or from a client application that references SharePoint binaries on another machine. However, for applications that are not resident on the server, you can look to web services for updating the list data or to other mechanisms such as RSS feeds for data retrieval. (Note that error-handling code has not been added to this function.)

Now that we've accomplished the addition of a list item into an existing SharePoint list, let's take a look at how you can update and delete list items and examine the structure of the list data within exposed by the object model.

Deleting and Updating List Items

Now that you've added an item to a SharePoint list programmatically, it is also important that you understand what is involved in updating and deleting list items. This, along with adding folders and moving data within lists, provides you with much of the functionality that you need to control list functions through external programs or custom extensions to SharePoint web applications, and to synchronize SharePoint list data with third-party data sources.

Similar to the list item add feature in the earlier example in this chapter, we'll use a Windows application to drive the updates and deletes. Remember that in the sample application for this chapter we have used a series of tabs from the standard Visual Studio tab control to demonstrate different functions on the main application form.

Here are the steps to perform a list update from your custom application:

1. Add the UI elements to the new form tab for deleting a list item. Include a text box for the part number and a button to initiate a delete based on an item match with that part number.

2. Upon clicking the form button, enter the code to delete an item using the following steps:

    a. Use the `SPSite` class constructor with the URL for the main site to return a site collection.

    b. Return the target web site based on site name.

    c. Return a collection of list items based on the `Items` property for the specific named list (for example, `"Parts List"`).

    d. Assign the value of the `Count` property of the list items collection to `counter variable`.

    e. Loop through the list items collection and compare the part number entered on the form to the part number associated with the item from the list. Upon a match, delete that item from the list.

3. Clean up the form and check the corresponding list for a delete.

The form for the delete scenario is shown in Figure 3-10. The corresponding code follows.

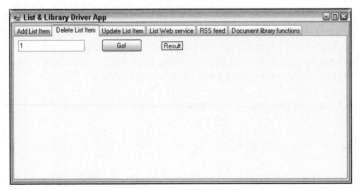

Figure 3-10

```
            //String parameters to enable site and list access
            const string siteUrl = "http://moss2007rtm1/";
            const string siteName = "/SiteDirectory/PartsList/";
            const string listName = "Parts List";

            //"Freeze" UI
            Cursor = Cursors.WaitCursor;
            buttonDeleteItem.Enabled = false;

            //Use SPSite constructor with URL for main site to return a site
collection
            SPSite siteCollection = new SPSite(siteUrl);

            //Return the target web site based on site name
            SPWeb site = siteCollection.AllWebs[siteName];

            //Return a collection of list items based on the Items property
            //for the specific named list
            SPListItemCollection listItems = site.Lists[listName].Items;

            //Assign the value of the Count property of the list items collection
to counter variable
            int itemCount = listItems.Count;

            //Loop through list items collection and compare the part number
entered on form
            //to the part number associated with the item from the list. Upon
match, delete.
            for (int i = 0; i < itemCount; i++)
            {
                SPListItem item = listItems[i];

                if (textBoxPartNumber1.Text == item["Part Number"].ToString())
                {

                    //Delete list item and decrement counter variable
                    listItems.Delete(i);
                    itemCount -- ;
                }
            }

            //UI clean-up
            labelResult1.Text = "Done!";
            Cursor = Cursors.Default;
            buttonDeleteItem.Enabled = true;
```

The code for updating an item within a SharePoint list is similar. Here are the steps to update a list item from your application:

1. Add UI elements for part number matching for the item, a price field for new price input, and a button to initiate the update.

2. For the form button click function, implement the following steps in your code:

    a. Use the `SPSite` class constructor with the URL for the main site to return a site collection.

    b. Return the target web site based on site name.

    c. Return a collection of list items based on the `Items` property for the specific named list (for example, `"Parts List"`).

    d. Assign the value of the `Count` property of the list items collection to `counter variable`.

    e. Loop through the list items collection and compare the part number entered on the form to the part number associated with the item from the list. Upon a match, update the price input data for the matching item.

3. Clean up the form and check the corresponding SharePoint list for the updated data.

The Windows form application for the list update is shown in Figure 3-11, and the corresponding code follows.

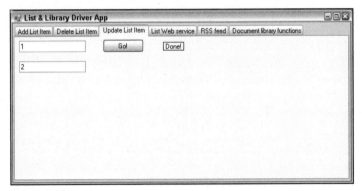

Figure 3-11

```
        //String parameters to assign site and list
        const string siteUrl = "http://moss2007rtml/";
        const string siteName = "/SiteDirectory/PartsList/";
        const string listName = "Parts List";

        //Freeze UI
        Cursor = Cursors.WaitCursor;
        buttonUpdateItem.Enabled = false;

        //Use SPSite constructor with URL for SP site to access site collection
        SPSite siteCollection = new SPSite(siteUrl);

        //Return the target web site based on site name
        SPWeb site = siteCollection.AllWebs[siteName];

        //Return collection of list items based on Items property for the
specified list
        SPListItemCollection listItems = site.Lists[listName].Items;
```

```
            int itemCount = listItems.Count;

            //Loop through list items collection and compare the part number
entered on form
            //to the part number associated with the item from the list. Upon
match, update price.
            for (int i = 0; i < itemCount; i++)
            {
                SPListItem item = listItems[i];

                if (textBoxPartNumber2.Text == item["Part Number"].ToString())
                {
                //Update price for part number match
                    item["Price"] = Convert.ToInt16(textBoxPrice2.Text);
                    item.Update();
                }
            }

            //UI clean-up
            labelResult2.Text = "Done!";
            Cursor = Cursors.Default;
            buttonUpdateItem.Enabled = true;
```

At this point, we have examined the code and supporting object model to deal with SharePoint lists. We have added data to a list, updated list data, and deleted items from a list.

Another type of list data that is supported in SharePoint is a document library. A *document library* is another form of a list, and while it has many similar properties to a standard list, it also includes additional functions to enable document uploads, retrieval, and other functions to support document management and collaboration. In the next section, we'll look at document libraries and some of the corresponding functionality for developers.

Programming SharePoint Document Libraries

This chapter had an earlier section focused on programming lists in SharePoint, and as we just indicated, the document library itself is just another form of a list. There is a default content type associated each type of list available in SharePoint, and as a document library is just a form of list, it comes with an associated content type to define the properties of list items in the document library.

Focusing on document libraries separately from lists make sense for a couple of reasons:

❑   The way that users and programs will access SharePoint to deal with list data versus document data (or document libraries) is different.

❑   The document libraries expose features and events that are different from the standard list templates, so they are worth considering as separate functions within the SharePoint architecture.

Uploading a Document to a Library

To upload a document to a library, you need to build a test application that can take the location of a file for upload and use the APIs to add that file to the existing site. For this example, the code to upload the file is designed to run on a file system accessible to the local SharePoint server. Remember that this sample code is available for download at www.wrox.com. Other mechanisms to upload remote files or to copy files from another SharePoint URL are also available.

Take a look at the steps to upload a document to an existing library in code:

1. Create a Windows application with a text box to input the file URL and a button to initiate the upload to the server. The sample form is shown in Figure 3-12.

2. For the example, create a sample document to upload on the file system. You'll then need to remember the location and name of this file later when running the sample application.

3. For the button click, add code for the following implementation:

   a. Use the `SPSite` constructor to create a site collection based on the URL.

   b. Assign a target web site to an instance of `SPWeb` based on the target site name (for example `"/SiteDirectory/Parts List/"`).

   c. Create an instance of `FileStream` based on the file location entered in the form's text box.

   d. Create a byte array to correspond to the `FileStream` instance and write the contents of the file stream into the array.

4. Add the file to the document library using the `SPFileCollection Add` method.

5. Check the Parts List document library to ensure file upload.

The code for this sample routine is included in the listing that follows:

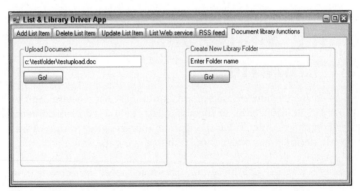

Figure 3-12

```
//String parameters to assign site and document library destination URL
const string siteUrl = "http://moss2007rtm1/";
const string siteName = "/SiteDirectory/PartsList/";
const string destination = "http://moss2007rtm1/SiteDirectory/
PartsList/Shared%20Documents/TestUpload.doc";
```

```
                    //Freeze UI
                    Cursor = Cursors.WaitCursor;
                    buttonNewFolder.Enabled = false;
                    buttonUploadDocument.Enabled = false;

                    //Use SPSite class constructor to create site collection based on URL
                    SPSite siteCollection = new SPSite(siteUrl);

                    //Assign target web site based on site name
                    SPWeb site = siteCollection.AllWebs[siteName];

                    //Assign file stream to file opened based on file location form input
                    FileStream fStream = File.OpenRead(textBoxFilePath.Text);

                    //Instantiate byte array to correspond to the file stream object
                    byte[] contents = new byte[fStream.Length];

                    //Read data from file stream and write into byte array
                    fStream.Read(contents, 0, (int)fStream.Length);
                    fStream.Close();

                    //Add file to document library based on destination URL and content of
        byte array
                    site.Files.Add(destination, contents);

                    //UI clean-up
                    Cursor = Cursors.Default;
                    buttonNewFolder.Enabled = true;
                    buttonUploadDocument.Enabled = true;
```

The `SPFileCollection Add` method is overloaded to take various forms of file input (either `Stream` or `Byte` arrays) along with metadata for the file and other settings. The date and time for the upload can be specified, and a flag can be set to determine whether or not the file upload can overwrite a file of the same name.

Creating a Document Library Folder

One of the important features of a document library is the folder feature. This enables a user or program to create and organize folders for storing documents. Because list security can be controlled at the item level, a folder can provide a clean way of separating content from a security perspective. In addition, a folder can enable an existing site with a large number of documents to have a simple view of a subset of the documents that corresponds to a static URL based on folder location.

Follow these steps to extend the sample code for creating a folder within a document library:

1. Create the UI elements on a Windows form to take the folder name to create in the document library and a button to initiate the folder creation.

2. On the button click, implement the following steps in code:

    a. Use the `SPSite` constructor to assign a site collection based on the top-level URL for SharePoint.

b. Assign an instance of `SPWeb` based on the target site name (for example, `"/SiteDirectory/PartsList/"`).

c. Create an instance of `SPDocumentLibrary` based on the named instance of the document library (for example, `"Shared Documents"`).

d. Create an instance of `SPFolderCollection` based on the folders for this site.

e. Use the `Add` method for the folders collection to add the named folder. Concatenate the input folder name value along with the relative path to create the named folder at the document library location.

f. Call the `Update` method of the instance of `SPDocumentLibrary` to update the library with the added folder.

3. Check the Shared Documents site for the added folder. See Figure 3-13 for an example.

Figure 3-13

```
//String parameters for site URL and site name
const string siteUrl = "http://moss2007rtm1/";
const string siteName = "/SiteDirectory/PartsList/";

//Freeze UI
Cursor = Cursors.WaitCursor;
buttonNewFolder.Enabled = false;
```

```
                buttonUploadDocument.Enabled = false;

                //Use SPSite constructor to assign site collection based on top-level URL
                SPSite siteCollection = new SPSite(siteUrl);

                //Assign target site (SPWeb instance) based on site name
                SPWeb site = siteCollection.AllWebs[siteName];

                //Create an instance of SPDocumentLibrary based on the named doc
       library list
                SPDocumentLibrary docLibrary = (SPDocumentLibrary)site.Lists["Shared
       Documents"];

                //Create instance of SPFolderCollection and add a named folder based on
       forms input,
                //then update library to reflect added folder
                SPFolderCollection myFolders = site.Folders;
                myFolders.Add("http://moss2007rtm1/SiteDirectory/PartsList/
       Shared%20Documents/" + textBoxFolderName.Text + "/");
                docLibrary.Update();

                //UI clean-up
                Cursor = Cursors.Default;
                buttonNewFolder.Enabled = true;
                buttonUploadDocument.Enabled = true;
```

Based on this code you can then upload documents to the documents library based on folder locations, or move existing content into the library and folder structure. The SPDocumentLibrary class also enables you to set administrative options on this library, including allowing moderation for posted documents, enabling or disabling crawling for Search, and many other options.

One limitation of the code samples shown so far for managing lists and libraries is that the code is designed to run locally to the SharePoint server. In some cases, this solution will work quite well. In others, it is extremely limiting and not useful for clients that are remote to the SharePoint infrastructure. In the next section, we'll examine the web services support within SharePoint for accessing and managing lists and libraries. We'll also walk through an example related to accessing list data via web services.

Web Services for Lists and Libraries

While the SharePoint object model provides a rich set of functionality for accessing lists and document library data from within SharePoint, it is often not feasible to use this object model. The object model works very well for SharePoint site extensions and custom ASP.NET applications deployed within the SharePoint infrastructure, but it is not a great solution for third-party systems and client applications that need to abstract the features of SharePoint from the end user while creating or updating lists or document data within the SharePoint infrastructure.

SharePoint exposes a variety of web services supporting functions for list access and many other application and administrative functionality. This section briefly reviews all of the SharePoint web services provided in the platform and then walks through some specific examples related to SharePoint lists.

WSS Web Services

A number of web services are supported within Windows SharePoint Services for basic system functions. In addition, Microsoft Office SharePoint Server exposes other web services for specific applications, including Excel Services and Search. The following table lists the core WSS web services:

WSS Web Services	Web Reference URL
Administration	`http:// MyServer:Port_Number/_vti_adm/Admin.asmx`
Alerts	`http://MyServer/[sites/][MySite/][MySubsite/]_vti_bin/alerts.asmx`
Copy	`http://MyServer/[sites/][MySite/][MySubsite/]_vti_bin/copy.asmx`
Document Workspace	`http://MyServer/[sites/][MySite/][MySubsite/]_vti_bin/dws.asmx`
Forms	`http://MyServer/[sites/][MySite/][MySubsite/]_vti_bin/forms.asmx`
Imaging	`http://MyServer/[sites/][MySite/][MySubsite/]_vti_bin/imaging.asmx`
List Data Retrieval	`http://MyServer/[sites/][MySite/][MySubsite/]_vti_bin/dspsts.asmx`
Lists	`http://MyServer/[sites/][MySite/][MySubsite/]_vti_bin/lists.asmx`
Meetings	`http://MyServer/[sites/][MySite/][MySubsite/]_vti_bin/meetings.asmx`
People	`http://MyServer/[sites/][MySite/][MySubsite/]_vti_bin/people.asmx`
Permissions	`http://MyServer/[sites/][MySite/][MySubsite/]_vti_bin/permissions.asmx`
Search	`http://MyServer/[sites/][MySite/][MySubsite/]_vti_bin/search.asmx`
Site Data	`http://MyServer/[sites/][MySite/][MySubsite/]_vti_bin/sitedata.asmx`
Sites	`http://MyServer/[sites/][MySite/][MySubsite/]_vti_bin/sites.asmx`
Users and Groups	`http://MyServer/[sites/][MySite/][MySubsite/]_vti_bin/usergroup.asmx`

WSS Web Services	Web Reference URL
Versions	`http://MyServer/[sites/][MySite/][MySubsite/]_vti_bin/versions.asmx`
Views	`http://MyServer/[sites/][MySite/][MySubsite/]_vti_bin/views.asmx`
Web Part Pages	`http://MyServer/[sites/][MySite/][MySubsite/]_vti_bin/webpartpages.asmx`
Webs	`http://MyServer/[sites/][MySite/][MySubsite/]_vti_bin/webs.asmx`

Given the large number of web services endpoints exposed by SharePoint, it is important to understand what function each of these performs. The following list summarizes the functions of various SharePoint web services:

- ❑ **Administration:** This service provides methods for managing a deployment of WSS, including creating or deleting sites.

- ❑ **Alerts:** This service enables you to work with alerts for list items in a SharePoint site.

- ❑ **Copy:** This service supports copying files from or within a given SharePoint site.

- ❑ **Document Workspace:** This service exposes methods for managing Document Workspace sites and data.

- ❑ **Forms:** The Forms service methods return forms used in the SharePoint UI when working with list contents.

- ❑ **Imaging:** This service supports methods for creating and managing picture libraries in SharePoint.

- ❑ **List Data Retrieval:** This service enables you to perform queries against WSS lists.

- ❑ **List:** Use the methods of this service to work with lists and list data, including retrieving all list items in a collection.

- ❑ **Meetings:** This supports methods to create and manage Meeting Workspace sites.

- ❑ **People:** This service supports managing information about people.

- ❑ **Permissions:** Use this service for support when working with permissions for a site or list.

- ❑ **Search:** This service supports search queries. It was introduced earlier in Chapter 2.

- ❑ **Site Data:** This service provides methods that return list data and metadata from sites or lists in WSS.

- ❑ **Sites:** This service supports returning information about site templates for a given site collection.

- ❑ **Users and Groups:** This service enables you to work with users, site groups, and cross-site groups.

- ❑ **Versions:** This service provides methods for working with file versions.

- ❑ **Views:** This service supports methods for working with views of lists.

❏ **Web Part Pages:** This service provides methods to send and retrieve information from web services.

❏ **Webs:** This service enables you to work with sites and subsites.

Out of this laundry list of services, we'll look specifically at Lists and List Data Retrieval services.

Retrieving List Data via Web Services

To look at how you can leverage web services to work with list data, we'll take a look at an example for retrieving list schema definitions through the Lists web service. In order to retrieve SharePoint data using the web services, you must first create a web reference either statically or dynamically within your code to access the service. To perform a list function using web services in this example, follow these steps:

1. Create a Windows form with a button to call the web services, and a text box to provide the web services output.

2. Create a web reference for the Lists service in SharePoint. Right-click Web References in the Solution Explorer within Visual Studio and follow the prompts. When prompted for a web service reference URL, enter the URL in this format: `http://MyServer/[sites/]` `[MySite/][MySubsite/]_vti_bin/lists.asmx`. The result should be similar to that shown in Figure 3-14.

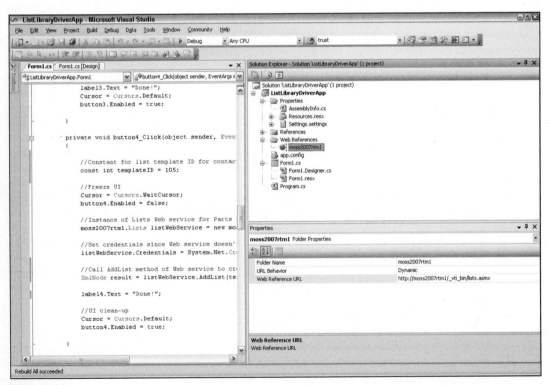

Figure 3-14

3. Implement the following for the form's button click function:

   a. Create an instance of the Lists web service based on the previously created web reference.

   b. Establish credentials for the web service to use. For the sample, you can use the default credentials for the security context of this application.

   c. Call the `GetList` method of the web service to return list items based on a named list.

   d. Write the output of the `GetList` method to the form's text box.

The output from the `GetList` method of the web service is shown in Figure 3-15. In this case, we have retrieved the list schema for the Tasks list associated with the site's web service. The code is as follows:

Figure 3-15

```
                    //String for list name
                    const string listName = "Tasks";

                    //Freeze UI
                    Cursor = Cursors.WaitCursor;
                    buttonRetrieveList.Enabled = false;

                    //Instance of Lists Web service for Parts List site
                    moss2007rtm1.Lists listWebService = new moss2007rtm1.Lists();

                    //Set credentials since Web service doesn't allow anonymous access
                    listWebService.Credentials =
System.Net.CredentialCache.DefaultCredentials;

                    //Call GetList method of Web service to return list items
                    System.Xml.XmlNode listItems = listWebService.GetList(listName);

                    //Write output of GetList method to form
                    textBoxResults.Text = listItems.OuterXml;

                    //UI clean-up
                    Cursor = Cursors.Default;
                    buttonRetrieveList.Enabled = true;
```

In this last example, you can see how to retrieve the schema of the Tasks list. This kind of list operation is useful for retrieving schema definitions for a list to render an alternative user interface for list input in an external application. This kind of integration could help you to render a SharePoint list-based application in a PHP web site. A real-world example might include synchronizing SharePoint list data with an open-source issue tracking tool.

Creating a New List via Web Services

At the beginning of this chapter, we started by looking into how lists work and what is involved with creating a new list through the SharePoint user interface. Now we'll walk through the use of the Lists web service to create a new list programmatically. To do so, we'll follow these steps:

1. Create a Windows form with a button to call the web service, and a text box to capture the name and description for a new list.

2. As in the previous example, create a web reference to the Lists web service in your project.

3. Implement the following for the form's button click function:

    a. Create an instance of the Lists web service based on the previously created web reference.

    b. Establish credentials for the web service to use. For the sample, you can use the default credentials for the security context of this application.

    c. Call the AddList method of the web service to create a new list. In this sample, we'll create a new Contacts list based on the list template for Contacts.

The code for this list creation example using web services is as follows:

```
//Constant for list template ID for contacts list
const int templateID = 105;

//Freeze UI
Cursor = Cursors.WaitCursor;
buttonCreateContactList.Enabled = false;

//Instance of Lists Web service for Parts List site
moss2007rtm1.Lists listWebService = new moss2007rtm1.Lists();

//Set credentials since Web service doesn't allow anonymous access
listWebService.Credentials =
System.Net.CredentialCache.DefaultCredentials;

//Call AddList method of Web service to create new list
XmlNode result = listWebService.AddList(textBoxListName.Text,
textBoxListDescription.Text, templateID);

labelResult3.Text = "Done!";

//UI clean-up
Cursor = Cursors.Default;
buttonCreateContactList.Enabled = true;
```

With this code for creating a new list, you can use SharePoint as a repository for creating new lists automatically based on a trigger from another business process or based on a rule invoked from an external application. As in the previous example, the use of an external issue tracking or project management tool outside of SharePoint could present the need for integration or synchronization of a list of project issues or tasks.

Because the web services within SharePoint can be targeted at the site level and are fairly granular, you need to investigate the various methods supported by different services to determine the best strategy for creating web references for your applications. In general, you'll look to web services as the preferred integration technique for heterogeneous scenarios and distributed applications that need to support integration across network boundaries. In the next section, we'll look at the use of a more data-centric integration technique that's new within SharePoint 2007.

Using RSS for List Data Retrieval

While a number of different methods have been reviewed in this chapter for retrieving list data, one of the more intriguing new features in SharePoint is the capability to generate RSS data for any list. RSS stands for Really Simple Syndication (formerly Rich Site Summary) and is an XML data format designed to enable you to syndicate content and retrieve updates for lists of content and various types of articles and data sources. While generally thought of as a data format to be used to read blogs, RSS can be applied in a number of different application development scenarios and is a good, lightweight protocol for data publishing and syndication.

The sample list that we created earlier in the chapter can be turned into an RSS feed through a single click in the browser, as shown in Figure 3-16. The RSS feed output is shown in Figure 3-17.

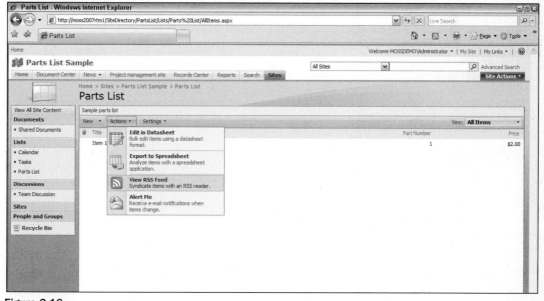

Figure 3-16

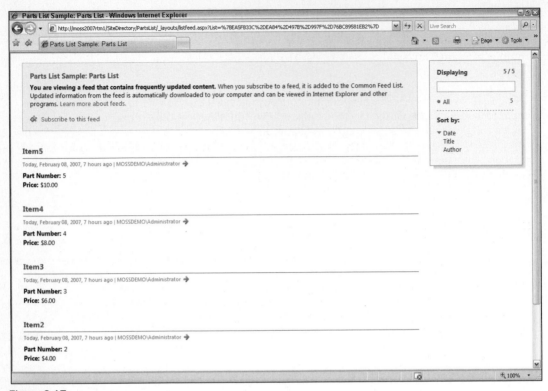

Figure 3-17

SharePoint enables administrators to determine whether or not to publish RSS data for lists, but RSS will be published by default. For applications that simply need to retrieve list data and reformat it or perform some logic against it, RSS is a good alternative.

To better understand how you can use RSS data in applications, the following example walks through a sample to retrieve RSS data. To implement code to grab RSS data, do the following:

1. Create a Windows form application with a button to retrieve the RSS data feed and a text box to contain the RSS output.

2. On button click, implement the following logic (shown in the code listing that follows shortly):

   a. Use the `SPSite` constructor to retrieve the site collection for a specified URL.

   b. Create an instance of the specific site (for example, `"PartsList"`) based on the site name.

   c. Create an instance of the list to retrieve as RSS based on list name.

   d. Create an instance of `XmlDocument` and a new stream object.

   e. Call the `WriteRssFeed` method of the list to populate the stream object.

f. Load the stream data into the instance of `XmlDocument`.

g. Use the `OuterXml` property of `XmlDocument` to load text into the form's text box.

3. View form output to verify results.

```
//String parameters to access site and list data
const string siteUrl = "http://moss2007rtm1/";
const string siteName = "/SiteDirectory/PartsList/";
const string listName = "Parts List";

//Freeze UI
Cursor = Cursors.WaitCursor;
buttonGetRSS.Enabled = false;

//Use SPSite constructor to get site collection
SPSite siteCollection = new SPSite(siteUrl);

//Create instance of site (SPWeb) based on site name
SPWeb site = siteCollection.AllWebs[siteName];

//Create instance of list based on list name
SPList listRSS = site.Lists[listName];

//Create instance of XmlDocument
XmlDocument xmlOutput = new XmlDocument();

//Create new stream
Stream rssStream = new MemoryStream();

//Call WriteRssFeed method of list to populate stream
listRSS.WriteRssFeed(rssStream);
rssStream.Seek(0, SeekOrigin.Begin);

//Load stream data into Xml document instance
xmlOutput.Load(rssStream);

//Use form's text box to render xml document data
textBoxResults2.Text = xmlOutput.OuterXml;

//Clean-up UI
Cursor = Cursors.Default;
buttonGetRSS.Enabled = true;
```

The RSS output is shown in Figure 3-18.

RSS data from SharePoint can be consumed through custom applications, rendered through the web browser, or synchronized with the Windows RSS platform components. Because the data is just an XML format, it can be parsed as required or simply transformed and rendered to fit the needs of your particular tool.

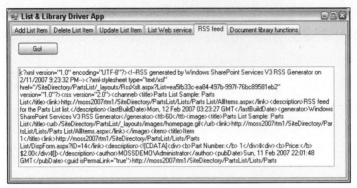

Figure 3-18

For scenarios in which the client application is remote to the SharePoint server, both the web services options and RSS feeds can be useful tools.

❑   In the case of web services, the developer has a comparable set of functions to the various classes under the Microsoft.SharePoint namespace and can perform a considerable number of functions within SharePoint from both a site administrative and data management perspective.

❑   With RSS data, the main advantage for developers is the capability to access list data using a very lightweight protocol that can be easily parsed or rendered as the consuming application requires. In addition, developers can leverage out-of-the-box features in Internet Explorer 7.0, the Windows RSS platform, and applications such as Office 2007 in order to leverage the RSS format. Moreover, there are a significant number of third-party RSS aggregator applications and APIs for integration with various web content management and document publishing tools.

Using List Events

In the previous version of SharePoint, Windows SharePoint Services 2.0, support for events was limited to document libraries and form libraries. Those events were tied to completed actions, and in some cases were too limiting to support the kinds of behaviors developers would like to implement. Just as databases support triggers for inserts, updates, and deleted data, SharePoint data needed to provide similar events for developers to access when SharePoint list data was updated. In the SharePoint 2007 technologies, a richer infrastructure for events is supported, enabling developers to build in data validation and business logic to support events associated with lists and sites.

As a SharePoint developer, you can override methods of the SPItemEventReceiver class to handle specific events. SharePoint supports events that fire synchronously during actions such as updates and deletes, and events that occur after actions occur. Here are some of the methods of SPItemEventReceiver that you can override to handle specific events:

❑   ItemAdding

❑   ItemAdded

❑   ItemUpdating

❑   ItemUpdated

❑   `ItemDeleting`

❑   `ItemDeleted`

❑   `ItemCheckingIn`

❑   `ItemCheckedIn`

❑   `ItemCheckingOut`

❑   `ItemCheckedOut`

Other methods supporting undoing checkouts, renaming files, and adding or deleting attachments are also supported. Note that the method names denote whether or not the method is executed during an action or after an action. For example, `ItemAdding` is accessible during the addition of an item to a list, and `ItemAdded` fires after the item is added to the list. With these methods you have access to the data associated with `SPListItem` through an object that is provided as an input parameter when the method is called. This object is of type `SPItemEventProperties` and much like a database trigger, exposes the data associated with the list and the affected list item.

To build an event handler for a list, do the following:

1. Create a new Visual Studio project for a class library. In this example, we'll name this project (and ultimately the namespace and class), `ListEventHandler`.

2. Add a reference to `Microsoft.SharePoint` to access the SharePoint namespace.

3. Add the `Microsoft.SharePoint` directive.

4. Change the name of the class to `ListEventHandler` and inherit from `SPItemEventReceiver`.

5. Override the `ItemAdding` method to trap the event that fires while an item is being added to the list but prior to the addition action being completed.

6. Use the `Cancel` property of `SPItemEventProperties` to cancel the event and list item action. This will prevent an item from being added to the list.

The code for this example is shown in the following listing:

```
using System;
using System.Collections.Generic;
using System.Text;
using Microsoft.SharePoint;

namespace ListEventHandler
{
    //Class inherits from SPItemEventReceiver
    public class ListEventHandler : SPItemEventReceiver
    {
        //Override ItemAdding method, which will wire up the event that fires when
adding an item to the list
        public override void ItemAdding(SPItemEventProperties properties)
        {
            //Access SPListItem data exposed through SPItemEventProperties
            //Cancel the list item action
            properties.Cancel = true;
```

```
            //Write back an error for adding this item to the list- append the list
name,
            //list item name and URL
            properties.ErrorMessage = "Adding data disabled for this list: "
                + properties.ListTitle + "; " + properties.ListItem.DisplayName +
"; "
                + properties.WebUrl;


        }
    }
}
```

In this example, we're handling an event that occurs when an item is added to the list, and canceling the event and the list item action. While you could use security features to prevent users from adding items to lists, this event-based technique can be used to drive the user of the site to another behavior, such as going to a different area of the site to add the content. The event handler can also invoke some logic external to SharePoint, such as kicking off a workflow that is resident on another platform or synchronously looking up data in an external system to perform a data validation action.

To deploy an event handler such as this one, you need to use Features within SharePoint to register the event handler and ensure that it is deployed to the appropriate sites with all of the dependencies that are required to support the desired end user experience. We cover Features in subsequent chapters such as Chapter 5, in the Web Parts examples; Chapter 8, on workflow; and in the content management scenario featured in Chapter 10.

Summary

This chapter has provided an overview of the programmability features for SharePoint lists and document libraries. We looked at the features for document libraries and lists, including requirements to create these lists and the APIs that are available to facilitate access to the data from third-party applications.

In looking at SharePoint lists and libraries for documents, you learned the characteristics of a list and how lists are used throughout the SharePoint architecture. With respect to programming SharePoint lists, you examined the object model and looked at examples for adding, updating, and deleting data from these lists. This involved using the object model to establish a connection to a site collection, an individual site, and a named list or document library within a site. In addition, you looked at the methods used to identify specific list items within a collection, and the techniques used to create new objects, such as folders, in a list. For SharePoint document libraries, you created a sample application to upload a new document to the server and looked at the object model's support for creating new folders within a library.

You also examined the mechanisms for programming against SharePoint lists for clients that do not have direct access to the SharePoint object model and applications running on non-Windows platforms. This included the various SharePoint web services available for list functions and out-of-the-box support for RSS feed generation on a per-list basis. You also learned about the events associated with SharePoint lists.

In the next chapter, you'll take a deeper look at the XML developer features in Microsoft Office 2007 and the document format improvements, including Open XML, along with other programmability features based on XML data and the extensible user interface of Office 2007.

4

XML, Web Services, and Extensibility in Office 2007

Modern information workers need the ability to work across systems seamlessly with tools that are familiar in order to make them more productive. Using Office as a front end to custom and line-of-business applications enables users to work with tools they already know, and it gives users the ability to customize and reuse information to suit their needs.

Office is also the client for most document management activities that take place in Windows SharePoint Services (WSS) and Microsoft Office SharePoint Server (MOSS). Since the time of the first version of SharePoint, Office has been extended to seamlessly support working with online repositories; and with the Office 2007 release, the Office client applications even expose workflow capabilities to provide a seamless experience.

In this chapter, we'll look at how Office 2007 can be used as part of a custom solution:

❑ First we'll look at Content Controls and Custom XML Parts, new features of Word 2007 that make it easier to separate document design from data.

❑ Then we'll look at the new OpenXML file formats, first by examining their structure and then by generating a Word document in an ASP.NET application and populating it from data in a SQL Server database.

Most of the examples in this chapter focus on Word and Excel because they are the most popular applications in the Office suite. In addition, they are the most common targets for developers building custom solutions. We will also translate several concepts to PowerPoint along the way. Because Word, Excel, and PowerPoint all share the same rules regarding file structure, the concepts used to manipulate Word and Excel files will help when you are doing the same things with PowerPoint files.

Word 2007 Customization and XML Capabilities

As XML has become the standard for exchanging structured data, Word has embraced XML in different ways to enable users to create more structured documents that can be used by other applications. In this section, we'll examine two new capabilities of Word 2007, Content Controls and Custom XML Parts, and how they create new ways for building data-driven applications.

Content Controls

Microsoft Word is probably the host of more electronic forms than any other application. The reasons for this are simple: Just about everyone has it installed, most computer users are proficient in it, and Word makes it easy to build a simple form. The problem with most of these forms is that they are pure unstructured data. You may put a blank on a form where someone should enter a date, but there's no easy way to confirm that what was entered was really a date.

In addition, most documents are unprotected, so users are free to make changes to the form structure at will, making it difficult to achieve consistency on a widely used form. Word has supported some basic form field and protection capabilities in prior versions, but these fields still didn't offer the kind of flexibility and control that most form authors need.

Word 2007's new Content Controls provide the capability to create more structured documents by defining the parts of a document that are editable, and locking those parts of the document that are fixed (e.g., formatting, structure). These controls are available from the developer ribbon (which is enabled in the Word Options dialog) and are shown in Figure 4-1.

Figure 4-1

There are seven types of Content Controls:

- ❑ **Rich Text:** Many formatting and layout options, including fonts, styles, tables, and paragraphs
- ❑ **Text:** Plain text
- ❑ **Image:** An embedded image
- ❑ **Combo Box:** Selection from a list of options
- ❑ **Drop-Down List:** Selection from a list of options
- ❑ **Date Picker:** Validated date entry with a pop-up calendar
- ❑ **Building Block Gallery:** Selection of a prebuilt, reusable document section

Once inserted into a document, these content types can be configured through the Properties button in the Content Controls ribbon area, which opens the Content Control Properties dialog box, shown in Figure 4-2. The Title property puts a small title over the field in the form, and the tag property is used to reference that property in code or in the OpenXML file itself.

Once a form has Content Controls embedded, the form author can lock down the rest of the form outside of the Content Controls by using the Protect Document feature of Word. The author can also lock down specific Content Control fields from user editing.

Content Controls frames are visible on a form so that users know where they can enter data, but they are not obtrusive and do not affect form printing. Figure 4-3 shows a good example of how a Content Control (a date picker, in this case) can be used to support rich form entry in a document without changing the flow or printed format of a document. The title and selection control appear over the document only when they are being used.

Content Controls are a powerful tool for simple forms, but they are limited to flat, nonrepeating data sets. Content Controls have no provision for repeating sections, so tables and repeating lists need to be generated in the document itself by generating OpenXML and inserting it directly into the document.xml within the document package. We'll do exactly that later in this chapter in the "Example: MS Word Report Generation with OpenXML" section.

Figure 4-2

Figure 4-3

For complex forms with expanding sections, advanced validation, and data-driven lists, InfoPath forms are usually a better choice.

Custom XML Parts

Custom XML Parts are actually XML files embedded in a Word document file and made available from within Word itself when the document is open. This is a key extensibility feature of the OpenXML document format. Although not exposed anywhere in the Word 2007 user interface, the Word object model provides access to creating, manipulating, and removing Custom XML Parts.

Word also has the built-in capability to support binding between Content Controls and data within Custom XML Parts. This support enables a scenario in which a data set from a line-of-business application is embedded into a Word document and then exposed in a form using binding. When the document needs to be updated with the latest version of this data, the Custom XML Part is simply refreshed and the document reflects the new data in its Content Controls. This clean separation also makes it easy to support customizing the appearance of the document and the placement of Content Controls without breaking the capability to refresh the document as needed.

In this next section, you'll see this capability in action using Word's object model to load Custom XML Parts and declare bindings to Content Controls, all within Word and Visual Basic for Applications.

Example: Creating a Word Document with a Custom XML Part and Content Controls

In this example scenario, we will create a status report document that uses XML data from a project management system.

Creating the Blank Status Report Document and Adding Content Controls

We'll start by using Word to create a new status report document containing places for key project statistics that we'll later populate from an XML document. In this case, this is a brand-new document, not a built-in Word template, so we just structure it so that it has places to insert the data we want to display. (This is available for download with the rest of the book's source code at www.wrox.com.) To create the Content Controls, we'll switch to the developer ribbon (enabled in the Word Options dialog). Most of the fields on this form are plaintext except for the Activity Report section that needs to hold layout and styles, so the Rich Text Content Control is a better choice there. For the date fields, we'll use the Date Content Control. Figure 4-4 shows the status report document after the Content Controls have been added.

The properties ribbon button opens the dialog in which control tiles and tags can be defined. If they are entered, then the title is displayed over the field to control when it has focus, and the tag is used to define the XML tag used when this document is saved. (Entering titles and tag names isn't necessary for the user-entered controls, but it can become useful when you want to work with the content in the raw WordprocessingML format.) However, before you can bind these controls to a data source, you need to load that XML data source into your Word document.

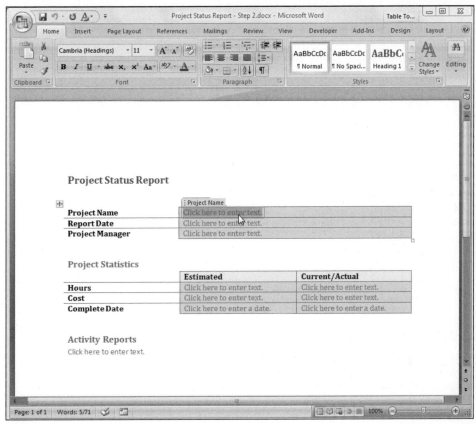

Figure 4-4

Loading the XML Data Source with VBA

Word has no user interface for adding or manipulating Custom XML Parts, so you launch Visual Basic for Applications by opening Word, going to the Developer tab, and clicking the Visual Basic button. From there, the Immediate window enables you to work with the document interactively (see Figure 4-5).

We'll be using the following component in the Word object model:

Component Reference	Description
`ActiveDocument.CustomXmlParts`	A collection of Custom XML Parts in the current document. Depending on the template used to create the document, this collection will contain one or more Custom XML Parts when working with a blank, new document.

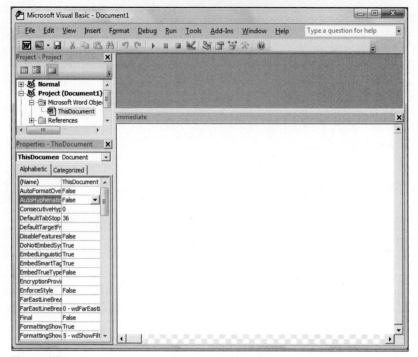

Figure 4-5

We start by loading a custom XML document from the local file system into the active Word document:

```
ActiveDocument.CustomXMLParts.Add.Load("C:\projectStatus.xml")
```

We take a shortcut here by calling the `Add` method, which returns a `CustomXMLPart` object, which we call the `Load` method on in the same line:

```
? ActiveDocument.CustomXMLParts.Count
2
```

By querying the count of the `CustomXMLParts` collection, we find the index of the part we just added. Why 2 instead of 1? Word creates a default Custom XML Part for every document, so our new part took the second position.

Now we'll read the Custom XML Part back out to make sure it's there (the question mark is a shortcut for the Print command):

```
? ActiveDocument.CustomXMLParts(2).XML
<projectStatus>
    <baselineCost>329200</baselineCost>

    <baselineHours>2730</baselineHours>
```

```
        <baselineCompleteDate>2007-6-1</baselineCompleteDate>
        <actualCost>310150</actualCost>
        <actualHours>2610</actualHours>

        <estimatedCompleteDate>2007-6-12</estimatedCompleteDate>
</projectStatus>
```

Although we are creating Custom XML Parts in Microsoft Word here, Excel and PowerPoint also contain the same capability for storing custom XML data within their respective file formats. However, this does not mean they have support for Content Controls that bind to this data like Word does. These parts are accessed in Excel and PowerPoint using VBA.

Binding the Content Controls

Now that you have this XML data in your Word document, you can bind the Content Controls you inserted into the document. Again, we do this through VBA and the Word object model. The collection is indexed using the ordinal index of these controls relative to their placement in the document. (Writing a function to find Content Controls by title or tag wouldn't be difficult.) The control is mapped to the XML using an XPath expression. For this example, we're using simple path expressions, but it is possible to use attribute filters and other XPath constructs to work with more complicated XML documents:

```
ActiveDocument.ContentControls(4).XMLMapping _
  .SetMapping("/projectStatus/baselineCost")

ActiveDocument.ContentControls(5).XMLMapping _
  .SetMapping("/projectStatus/actualCost")

ActiveDocument.ContentControls(6).XMLMapping _
  .SetMapping("/projectStatus/baselineHours")

ActiveDocument.ContentControls(7).XMLMapping _
  .SetMapping("/projectStatus/actualHours")

ActiveDocument.ContentControls(8).XMLMapping _
  .SetMapping("/projectStatus/baselineCompleteDate")

ActiveDocument.ContentControls(9).XMLMapping _
  .SetMapping("/projectStatus/estimatedCompleteDate")
```

As soon as these binds are added, the values in the document should reflect the actual data from the embedded XML document. Word's XML mapping is a two-way connection, so if the Content Controls are edited in the document, the Custom XML Part's values reflect these changes immediately.

If this is not desired behavior, the Content Controls can be locked so that editing is not allowed either through the property page for the control or programmatically through the LockContents *and* LockContentControl *properties on the* ContentControl *object.*

In an example later in the chapter, we'll use Content Controls again, this time inserting WordprocessingML text into a Content Control with a server-based application. Although it doesn't use Custom XML Parts, it does enable us to load repeating data into a Word document through code.

Automating the Process

The real benefit of creating these data-driven documents is the capability to now take this simple template and replace or refresh the data within it without losing the user-entered content. There are two basic ways to do this, depending on how the template will be used: server-generated documents or a refresh macro built into the document template:

❏ In a server-generation scenario, a simple .NET application can leverage the OpenXML file structure and the .NET Framework to simply replace or extract the Custom XML Part in a document. This is powerful in a server scenario because it doesn't have a dependency on Word itself, so the manipulation and reading of these documents on the server scales well.

❏ If the deployment scenario calls for the capability for users to refresh the data in existing documents, then the easiest place to do that might be within Word itself. Using the same object model we used to load the Custom XML Part in the first place, we can replace that XML data or even read that data and load it into another system.

Office 2007's OpenXML File Formats

Office 2007 introduced a new file format across the three core applications, Word, Excel, and PowerPoint, all standardized into the OpenXML specification. The OpenXML formats replace the legacy binary formats as the default format for saving files. Microsoft took the additional step of submitting OpenXML for standardization, opening many new opportunities for using Office documents outside of the Office suite.

Previous versions of Office supported several levels of text-based formats, with varying degrees of capabilities. For example, Excel could open and convert delimited text files or HTML tables. However, consistently supporting basic features such as styles and formulas was difficult, and that's not considering the absence of advanced features such as graphs and refreshable data connections.

Office 2003 added XML file formats for Word and Excel with WordML and ExcelML, but these came with a few limitations. Because they were uncompressed monolithic XML documents, they were complicated and larger than their native binary equivalents. Even though these files were in XML format, every bit of the document, including binary data, was stored in the single XML document. Images embedded into the document were base-64 encoded and stored as text, which make files even larger and more difficult to modify. This design made the format complicated, making it difficult to work with just a single piece of a document. Another challenge was that this XML format support was not extended to PowerPoint. Due to these limitations, the Office 2003 XML file formats provided flexibility for developers but were not adopted as the default file format over the existing binary document file format.

With a full-fidelity XML file format in Office 2007, parsing, creating, and modifying Office documents is greatly simplified. With the old binary formats, developers either used third-party components to work with the formats or controlled the Office applications through automation interfaces. While this automation sometimes works, these programs are activated as out-of-process executables with large memory and processor overhead, even for simple operations. Many developers have experienced the pain of automating Word or Excel in a server application with mixed results; and while some versions of Office supported loading documents from XML, HTML, CSV, or some other text-based format, this support varied between versions and didn't support the full range of capabilities of the binary documents.

Other benefits of the new OpenXML file formats are decreased file sizes (due to compression), more robust file storage (because the files are more modular), and easier extensibility (with manageable XML-based files and a structure that enables customization).

Before looking at how you can use the Office 2007 OpenXML file formats, let's take a closer look at how these files are structured. While documenting the complete structure and syntax of the OpenXML file formats is out of the scope of this book, we'll look at a few key aspects that developers can leverage for custom solutions.

Open Packaging Convention

OpenXML files (`docx`, `xlsx`, `pptx`) are actually ZIP archive files that hold XML documents defining the structure and content of the document and any additional non-XML content such as images and embedded audio. Because these are industry-standard ZIP files, there are already numerous tools and libraries that work with the document archive format.

OpenXML documents use the an existing data packaging structure called Open Packaging Convention (OPC), which is an ECMA standard and is also used in the XML Paper Specification (XPS) format introduced with Windows Vista. (However, don't confuse XPS and OpenXML — even though they share the same packaging standard, they are separate formats with different purposes.) OPC defines some common terminology that carries over into the OpenXML document formats. The self-contained document archive file is called a *package*. Packages contain *parts*, which are files within the archive that define specific aspects of the document.

OPC defines rules regarding the following:

- ❑ A method for describing relationships between items within the package and to external resources
- ❑ Rules on naming parts within a package
- ❑ Address formats for addressing parts within a package
- ❑ Digital signatures

OPC supports a broad range of applications, so OpenXML defines a few rules about which aspects of OPC are used. For example, whereas OPC supports various archive methods, all OpenXML documents are stored in ZIP archives.

You can explore and manipulate OpenXML documents manually by adding a .ZIP extension to an OpenXML document file and extracting its contents to analyze it (see Figure 4-6).

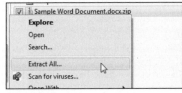

Figure 4-6

To automate this process, Microsoft added to .NET Framework 3.0 the System.IO.Packaging name-space, which supports working with OPC archives. We'll look at that later when we look at manipulating OpenXML documents programmatically.

Exploring the Word OpenXML format (WordprocessingML)

Figure 4-7 provides an overview of a simplified Word OpenXML document. Note the relationships between the parts in the document archive.

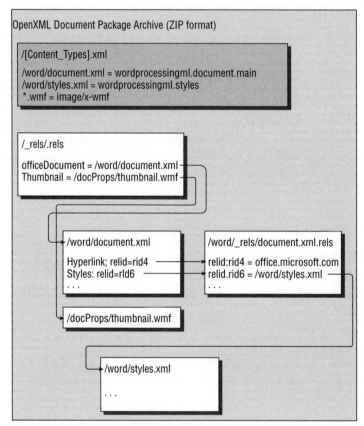

Figure 4-7

The [Content_Types].xml maps all of the part types in the package. The root.rels file (/_rels/.rels) identifies the root resources in the package. The main document part (/word/document.xml) defines content and links (via its own rels file) to other document parts (/word/styles.xml). In this next section, we'll take a closer look at the structure and contents of OpenXML documents, with the goal of building a program that generates OpenXML documents.

A little history: As one would expect, the OpenXML file format for Office 2007 changed in several ways from early preview releases. One example is that `/word/document.xml` was previously named `/word/worddocument.xml`. Some online documentation still references this deprecated filename.

Word organizes OpenXML document archives in a consistent folder and file structure. The following table shows the complete contents of a Word document package:

Folder or File Name	Description
`_rels (folder)`	Root relationship folder. A `_rels` folder maps relationships only for documents in its parent directory. If a document in a subdirectory has relationships, then it will have its own `_rels` directory.
`.rels`	The package relationship item. Other `.rels` files will have a filename before the `.rels` extension, but the root `.rels` file is a special case.
`customXml (folder)`	The `customXml` folder stores Custom XML Parts associated with the Word document. These are the same Custom XML Parts we embedded through VBA code in the previous section.
`item1.xml`	The first Custom XML Part in this document. This first Custom XML Part is a generic one generated by Word.
`item2.xml`	This is a user-added Custom XML Part.
`itemProps1.xml`	Properties associated with the first Custom XML Part
`itemProps2.xml`	Properties associated with the second Custom XML Part
`_rels (folder)`	Relationship folder for the `customXML` folder
`item1.xml.rels`	Relationship item for the `item1.xml` part
`item2.xml.rels`	Relationship item for the `item2.xml` part
`docProps (folder)`	Document properties folder. Stores metadata on the document.
`core.xml`	Common OpenXML file metadata (consistent across the different OpenXML file formats)
`App.xml`	Application-specific metadata
`Thumbnail.wmf`	Document thumbnail; used by the Windows shell for live document icons
`word (folder)`	Folder containing the word processing content in this package

Continued

Folder or File Name	Description
document.xml	The core document
settings.xml	Various document preferences and state information
fontTable.xml	Information about the fonts used in this document
styles.xml	The styles used and defined in this document
webSettings.xml	Information related to web publishing
_rels (folder)	Relationship folder for /word/
Document.xml.rels	Relationship item for document.xml, including hyperlinks, images, styles, comments, and settings
media (folder)	Folder for media embedded in the document
Image1.jpeg	An image embedded in the document
theme (folder)	Folder for any themes used in the document
theme1.xml	A theme used in this document

Note that some aspects of the folder structure are strictly organizational — it is possible to create a working OpenXML document with a flatter structure, but Office uses the folders to organize content. The only required folder is the _rels folder, which we'll look at shortly.

Content Types

Every OpenXML document is organized into granular components instead of being stored in one large file. This makes it easy to modify or replace a single piece of a document without affecting the rest of the document. To help track these various files, OpenXML uses a [Content_Types].xml file that defines the types for the files contained in the document archive. Here's an example of a [Content_Types].xml file from a Microsoft Word document:

```xml
<?xml version="1.0" encoding="UTF-8" standalone="yes"?>
<Types xmlns="http://schemas.openxmlformats.org/package/2006/content-types">
  <Default Extension="jpeg" ContentType="image/jpeg"/>
  <Override PartName="/word/comments.xml" ContentType="application/
vnd.openxmlformats-officedocument.wordprocessingml.comments+xml"/>
  <Default Extension="wmf" ContentType="image/x-wmf"/>
  <Default Extension="rels" ContentType="application/vnd.openxmlformats-
package.relationships+xml"/>
  <Default Extension="xml" ContentType="application/xml"/>
  <Override PartName="/word/document.xml" ContentType="application/
vnd.openxmlformats-officedocument.wordprocessingml.document.main+xml"/>
  <Override PartName="/word/styles.xml" ContentType="application/
vnd.openxmlformats-officedocument.wordprocessingml.styles+xml"/>
  <Override PartName="/docProps/app.xml" ContentType="application/
vnd.openxmlformats-officedocument.extended-properties+xml"/>
```

```
    <Override PartName="/word/settings.xml" ContentType="application/
vnd.openxmlformats-officedocument.wordprocessingml.settings+xml"/>
    <Override PartName="/word/theme/theme1.xml" ContentType="application/
vnd.openxmlformats-officedocument.theme+xml"/>
    <Override PartName="/word/fontTable.xml" ContentType="application/
vnd.openxmlformats-officedocument.wordprocessingml.fontTable+xml"/>
    <Override PartName="/word/webSettings.xml" ContentType="application/
vnd.openxmlformats-officedocument.wordprocessingml.webSettings+xml"/>
    <Override PartName="/docProps/core.xml" ContentType="application/
vnd.openxmlformats-package.core-properties+xml"/>
    </Types>
```

Having this type mapping is important because many of the files in an OpenXML document archive have the same XML extension, so this file maps filenames or extensions to a specific file type. In addition, because file extension associations vary between machines, this eliminates any problems with variances between file extension registrations across different machines. An OpenXML document will have a single `[Content_Types].xml` in the archive root.

Relationship Files

From the file structure shown earlier in Figure 4-7, you can see a _rels folder in several of the archive folders. The _rels folder contains one or more .rels files, which are XML documents that define the relationships connecting the different files in the OpenXML document. Each OpenXML document has one root rels file, named simply .rels, stored in a _rels folder off the root. Here's an example of a root rels file in a Word document:

```
<?xml version="1.0" encoding="UTF-8" standalone="yes"?>
<Relationships xmlns="http://schemas.openxmlformats.org/package/2006/
relationships">
    <Relationship Id="rId3"
        Type="http://schemas.openxmlformats.org/package/2006/relationships/metadata/
core-properties"
        Target="docProps/core.xml"/>
    <Relationship Id="rId2"
        Type="http://schemas.openxmlformats.org/package/2006/relationships/
metadata/thumbnail"
        Target="docProps/thumbnail.wmf"/>
    <Relationship Id="rId1"
        Type="http://schemas.openxmlformats.org/officeDocument/2006/relationships/
officeDocument"
        Target="word/document.xml"/>
    <Relationship Id="rId4"
        Type="http://schemas.openxmlformats.org/officeDocument/2006/relationships/
extended-properties"
        Target="docProps/app.xml"/>
</Relationships>
```

This root rels file defines the parts related to the entire document. Each `<relationship>` tag in the .rels document defines a relationship ID, a relationship type, and a target:

❑   The relationship ID is unique to the .rels file and provides a unique identifier for that relationship. Document parts with references then refer to this relationship ID instead of linking directly.

❑   Relationships are also assigned a type to provide an indication of how the relationship is used.

❑   Finally, the target attribute defines the actual location of the related resource.

In the root `rels` file in an OpenXML document, the relationship of type `"http://schemas .openxmlformats.org/officeDocument/2006/relationships/officeDocument"` always points to the actual document part. When an application opens this document, this is how it determines where to start reading content. This relationship is common across all Office OpenXML documents but may point to a different part name — for example, depending on whether it is a Word or an Excel document. The next step in opening this file would be to look up the target in `[Content_Types].xml` to determine what kind of document this actually is and then read that part appropriately depending on its type. This root `rels` file also has links to metadata parts within this document in the `docProps` folder. We'll look at these later in this section.

A document can have additional `rels` files that correspond to parts in the document. Here's a `rels` file from the document part in a Word document:

```xml
<?xml version="1.0" encoding="UTF-8" standalone="yes"?>
<Relationships xmlns="http://schemas.openxmlformats.org/package/
2006/relationships">
  <Relationship Id="rId8"
    Type="http://schemas.openxmlformats.org/officeDocument/2006/relationships/
theme"
    Target="theme/theme1.xml"/>
  <Relationship Id="rId3"
    Type="http://schemas.openxmlformats.org/officeDocument/2006/relationships/
webSettings"
     Target="webSettings.xml"/>
  <Relationship Id="rId7"
    Type="http://schemas.openxmlformats.org/officeDocument/2006/relationships/
fontTable"
    Target="fontTable.xml"/>
  <Relationship Id="rId2"
    Type="http://schemas.openxmlformats.org/officeDocument/2006/relationships/
settings"
    Target="settings.xml"/>
  <Relationship Id="rId1"
    Type="http://schemas.openxmlformats.org/officeDocument/2006/relationships/
styles"
    Target="styles.xml"/>
  <Relationship Id="rId6"
    Type="http://schemas.openxmlformats.org/officeDocument/2006/relationships/
comments"
    Target="comments.xml"/>
  <Relationship Id="rId5"
    Type="http://schemas.openxmlformats.org/officeDocument/2006/relationships/
image"
    Target="media/image1.jpeg"/>
  <Relationship Id="rId4"
    Type="http://schemas.openxmlformats.org/officeDocument/2006/relationships/
hyperlink"
    Target="http://office.microsoft.com"
```

```
            TargetMode="External"/>
</Relationships>
```

In the last relationship tag in the example, this Word document has a hyperlink that connects to an external resource, so the `TargetMode` attribute is specified as `External` (the default being internal).

By looking at this file, you may notice how the `.rels` file serves as a centralized roadmap or outline of the contents within a document. Using separate relationship files makes it easy for an application to map out all of the related elements of an OpenXML document without reading and parsing every part in the document. Additionally, changes to the location of a resource can be handled by simply changing the `rels` file instead of scanning for every reference.

Document Metadata

Another common aspect across all OpenXML document formats is the `docProps` folder, which contains document metadata. The `core.xml` part stores general document attributes such as create date, author, and revision. Then a separate `app.xml` defines application-specific metadata. For Word, this includes details such as word count, page count, editing time, and the version of Word used to create the document. For Excel, this data includes sheet names and information about data tables in the sheet. If a document has custom metadata, this data would be stored in the third document property part, named `custom.xml`.

The following shows `/docprops/app.xml`, a document properties example from a Word document:

```
<?xml version="1.0" encoding="UTF-8" standalone="yes"?>
<Properties xmlns="http://schemas.openxmlformats.org/officeDocument/2006/
extended-properties"
            xmlns:vt="http://schemas.openxmlformats.org/officeDocument/2006/
docPropsVTypes">
  <Template>Normal.dotm</Template>
  <TotalTime>5</TotalTime>
  <Pages>1</Pages>
  <Words>32</Words>
  <Characters>193</Characters>
  <Application>Microsoft Office Word</Application>
  <DocSecurity>0</DocSecurity>
  <Lines>17</Lines>
  <Paragraphs>14</Paragraphs>
  <ScaleCrop>false</ScaleCrop>
  <Company>Microsoft</Company>
  <LinksUpToDate>false</LinksUpToDate>
  <CharactersWithSpaces>211</CharactersWithSpaces>
  <SharedDoc>false</SharedDoc>
  <HyperlinksChanged>false</HyperlinksChanged>
  <AppVersion>12.0000</AppVersion>
</Properties>
```

Now that we have looked at the common and peripheral pieces of an OpenXML document, let's look at how the document content itself is stored. Since this is different for each document type, we'll start with Word and then look at Excel. (PowerPoint follows many conventions of these two so we won't go into details on that format here.)

Exploring Word's Main Document Part (/word/document.xml)

A word processing OpenXML document uses the following part types to represent the content of the document:

- ❑ Main document (`document.xml`)
- ❑ Subdocument (these can be stored within the parent document archive or stored in their own document packages)
- ❑ Header
- ❑ Footer
- ❑ Comment
- ❑ Frame
- ❑ Text box
- ❑ Footnote
- ❑ Endnote

The main document part is the only required one and it contains the core text of the document. For the purposes of this book, we'll focus on this part type and not cover the others because the main document part is what would be manipulated when generating, parsing, or modifying documents through code.

Although Word uses `/word/document.xml` as the name of the main document part, `[Content_Types]` `.xml` actually defines where the main document part is in this line:

```
<Override PartName="/word/document.xml"
    ContentType="application/vnd.openxmlformats-officedocument
.wordprocessingml.document.main+xml"/>
```

The rest of the document parts are also declared in `[Content_Types].xml` and linked into the main document part in the `/word/_rels/document.xml.rels` file.

The formatting in the main document resembles HTML with some tags, but these similarities don't go too far beyond a few core tags. WordprocessingML is the main schema used to define the content of a Word OpenXML document. Here's a short example of a simple main document part:

```
<?xml version="1.0" encoding="UTF-8" standalone="yes"?>
<w:document xmlns:w="http://schemas.openxmlformats.org/wordprocessingml/2006/main">
  <w:body>
    <w:p>
      <w:r>
        <w:t>Hello World!</w:t>
      </w:r>
    </w:p>
  </w:body>
</w:document>
```

A single `<document>` tag is required. In this case, it only references the main WordprocessingML namespace, but a typical Word document will include references to several other namespaces to support tags for vector art, equations, and other advanced document elements. Within the `<document>` tag is a single

`<body>` tag that serves as the main container for the text of the document. In this document, we have a single paragraph identified with the `<p>` tag. The Run `<r>` tag defines a range with common properties (such as formatting). Within the run tag, a text tag `<t>` represents a line of text.

A complete reference of the tags supported in WordprocessingML is available from the EMCA site (www .ecma-international.org/publications/standards/Ecma-376.htm), but we'll take a closer look at a few of the most useful ones for program-generated documents.

Formatted Text

Formatting can be applied for an entire paragraph or for a span of characters (a run) in the document. In this sample, we look at a single paragraph with several sections for formatting. It also includes a hyperlink that we'll save for the next section.

```
<w:p w:rsidR="003A4A8D" w:rsidRDefault="003A4A8D" w:rsidP="000F2C2C">
  <w:pPr>
    <w:pStyle w:val="IntenseQuote"/>
    <w:outlineLvl w:val="0"/>
  </w:pPr>
```

The block starts with a paragraph tag (`<w:p>`) tag that, just like the previous example, works as a paragraph container. This one also includes a set of `rsidR` attributes that track revisions within the document. Blocks with the same `rsidR` were created or modified in the same editing session.

The `<w:pPr>` is the paragraph properties tag that specifies settings for the entire parent paragraph. In this example, it references the `IntenseQuote` style with the `<w:pStyle>` element:

```
    <w:r>
      <w:t xml:space="preserve">It can hold </w:t>
    </w:r>
    <w:r w:rsidRPr="003A4A8D">
      <w:rPr>
        <w:color w:val="FF0000"/>
      </w:rPr>
      <w:t>character</w:t>
    </w:r>
    <w:r>
      <w:t xml:space="preserve"> and paragraph formatting as well as attachments
and </w:t>
    </w:r>
    <w:hyperlink r:id="rId5" w:history="1">
      <w:r w:rsidRPr="003A4A8D">
        <w:rPr>
          <w:rStyle w:val="Hyperlink"/>
        </w:rPr>
        <w:t>links</w:t>
      </w:r>
    </w:hyperlink>
    <w:r>
      <w:t>.</w:t>
    </w:r>
  </w:p>
```

The rest of the paragraph contents are runs (`<w:r>`) and the hyperlink reference. The second run element contains a Run Properties element (`<w:rPr>`) that, just like the paragraph properties element, contains properties for its parent. In this case, it is character-level formatting applied to the entire run. Here, a color element (`<w:color>`) is applied using the hex value for red.

If predefined styles are used (like `IntenseQuote` in this example), these are defined in the `/word/styles.xml` part, which is referenced through the `/word/_rels/document.xml.rels` relationship file.

Hyperlinks

Here is a main document part with a hyperlink tag:

```
<?xml version="1.0" encoding="UTF-8" standalone="yes"?>
<w:document xmlns:w="http://schemas.openxmlformats.org/wordprocessingml/2006/main"
            xmlns:r="http://schemas.openxmlformats.org/officeDocument/2006/
relationships">
  <w:body>
    <w:p>
      <w:r>
        <w:t>You can click on this: </w:t>
        <w:hyperlink r:id="rId4">
          <w:r>
            <w:rPr>
              <w:rStyle w:val="Hyperlink" />
            </w:rPr>
            <w:t>This is a hyperlink</w:t>
          </w:r>
        </w:hyperlink>
      </w:r>
    </w:p>
  </w:body>
</w:document>
```

This listing builds on the previous one in several ways. First, we add another namespace reference because the `<hyperlink>` tag uses a `r:id` attribute to map the actual link URL. Remember that relationships are stored in `rels` files, not in document parts; and a hyperlink is just another type of relationship. In this case it's an external reference, so it links to an area outside of the OpenXML document structure. In this case, the URL is stored in the `rels` file for `document.xml`. Here's the corresponding element in `/word/_rels/document.xml.rels` that this element references:

```
<Relationship Id="rId4"
              Type="http://schemas.openxmlformats.org/officeDocument/2006/
relationships/hyperlink"
              Target="http://office.microsoft.com" TargetMode="External"/>
```

Here you can see the actual URL of the link referenced by the resource ID `"rId4"`.

Inside of the hyperlink tag in the Run `<r>` tag is the Run Properties tag (`<rPr>`), which defines the Run (`<r>`) tag. This section applies the `Hyperlink` style to the run so the link looks like a hyperlink. Without this formatting, the link would work but a user wouldn't recognize this is a link. (When you create a link in Word, this is done automatically.)

Tables

Tables are the most common way to present data in Word, and WordprocessingML includes a rich set of tags for creating tables within documents. Because WordprocessingML is a full-fidelity format supporting every table structure and construct in Word, we'll focus on the core features of defining a table, its borders, and its content.

Consider the following WordprocessingML fragment for a simple 4 × 4 table:

```
<w:tbl>
  <w:tblW w:w="2000" w:type="fixed"/>
  <w:tblBorders>
    <w:top w:val="single" w:sz="4" w:space="0" w:color="auto" />
    <w:left w:val="single" w:sz="4" w:space="0" w:color="auto" />
    <w:right w:val="single" w:sz="4" w:space="0" w:color="auto" />
    <w:bottom w:val="single" w:sz="4" w:space="0" w:color="auto" />
  </w:tblBorders>
  <w:tableGrid>
    <w:gridCol w:w="1000" />
    <w:gridCol w:w="1000" />
  </w:tableGrid>

  <w:tr>
    <w:tc>
      <w:p><w:r><w:t>A1</w:t></w:r></w:p>
    </w:tc>

    <w:tc>
      <w:p><w:r><w:t>B1</w:t> </w:r></w:p>
    </w:tc>
  </w:tr>

  <w:tr>
    <w:tc>
      <w:p><w:r><w:t>B1</w:t></w:r></w:p>
    </w:tc>

    <w:tc>
      <w:p><w:r><w:t>B2</w:t></w:r></w:p>
    </w:tc>
  </w:tr>

</w:tbl>
```

The table starts with the `<tbl>` tag, which serves as the overall table container. The Table Width (`<tblW>`) tag then defines the desired width of the table with the `w` attribute. The `type` attribute selects which table layout algorithm Word should use for this table. In this case, we're using the fixed layout, which uses the `w` attribute to size the table in twentieths of a point, or 1,440 units per inch.

The Table Borders (`<tblBorders>`) tag defines the outside borders of the table. Here we're using a single 4-point border on all sides with no spacing and `"auto"` as the color, so Word uses the default based on the current style.

The Table Grid tag (`<tableGrid>`) section defines the horizontal Grid Columns (`<gridCol>` elements) used to lay out the table's cells across the page. In this example, we have a one-to-one mapping of grid columns and horizontal cells, but cells can span grid columns for more advanced table layouts.

Finally, the Table Row elements (`<tr>`) start to actually define the contents of the table. The `<tr>` elements define rows in the table, which contain Table Cell (`<tc>`) elements. This is similar to how HTML tables are built using HTML `<tr>` and `<td>` elements. The `<tc>` elements themselves each contain paragraphs, runs, and text areas, just like in our first simple document example.

These simple table tags are a great way to lay out data in Word documents, as we'll do in next in the Report Generation example.

Example: MS Word Report Generation with OpenXML

Developers have many reasons for generating output in Microsoft Office formats. If end users need to manipulate a generated document, there's a good chance they have Microsoft Office installed and that they know how to use it. Microsoft Office file formats provide richer formatting options than simple text or HTML provide, along with the ability to embed rich capabilities such as logic, markup, comments, document properties, and other useful metadata.

This next example addresses a common scenario for developers. Whether it's a web-based line-of-business solution that needs to create price quotes or a rich client application that generates letters, the need to generate Microsoft Word documents programmatically is easy to understand. While Word itself, as well as most enterprise reporting tools, can output reports to MS Word format, developers sometimes need complete management of the document-generation process and fine-grained control over the output.

In this example we'll generate a simple product list report in Microsoft Word 2007 OpenXML format in an ASP.NET application. Whereas in the Content Control/Custom XML Part example we were limited to a single record of data, in this scenario we have full flexibility to work with any numbers of rows of data and multiple data sets. With this new flexibility, however, we are now writing more code, as we are generating WordprocessingML ourselves. We also lose the clean separation between formatting and data we had in the prior example using Custom XML Parts. Nonetheless, for situations in which the data structures are too complicated for Content Control XML mappings, this approach provides a solution.

In this example we'll use Microsoft Word to create a document as a starting point. This way, we have a nicely formatted document without worrying about generating all of this formatting through code. With this approach, we have the best capabilities of Word at our disposal to design the document.

Creating the Template Document

Our report will have a static header with a company logo and a title. In this example, we won't be plugging in any other data into the report other than the data table itself, but the same exact techniques that we'll use here can be extended to populate any number of data elements in a document.

The most important step in designing the report is defining where the actual data should be inserted. Ideally, the approach we use will be robust enough not to be affected by formatting and layout changes in the document. This is another great use of Content Controls, this time the Rich Text Content Control that can hold formatted text and tables. Using the developer ribbon (enabled in Word Options), click the

Rich Text Content Control button (see Figure 4-8) to insert the control; and while the control is selected, click the Properties button in the Controls toolbar area to bring up the property dialog for the new control.

Figure 4-8

We'll use the Title property on the Content Control to locate this control in the main document part (see Figure 4-9). This time, instead of Word's object model, which we used to set XML mappings to Content Controls, we'll use these to modify the XML of the document itself.

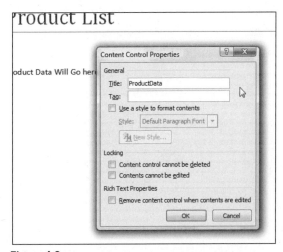

Figure 4-9

Now, at this point, we have a Word document that has headers, layout, and formatting that we want, and a Rich Text Content Control that holds the place where we are going to add our report.

Populating the Template Document

In this solution, the ASP.NET-based server application will actually contain most of our application logic. Our code here is structured in a procedural layout to make it easier to understand, but it can be reorganized into an object-oriented structure to support cleaner reuse. The heavy lifting is done in a function named CreateWordDocument, declared with the following signature:

```
public static void CreateWordDocument(string templateFilePath, string
replaceFieldName, IDataReader sourceReader, Stream outputStream)
```

The parameters are as follows:

Parameter	Description
templateFilePath	A file system path referencing the template document
replaceFieldName	The title of the Content Control into which we are inserting data
sourceReader	A prepared ADO.NET Data Reader that we'll use as a data source for our report
outputStream	A stream that will serve as a destination for our completed document

The sequence of activities in this method is as follows:

1. Read the template document from disk into memory for manipulation.
2. Navigate the OpenXML document structure and load the `document.xml` part for editing.
3. Find the XML element representing the Content Control placeholder we defined in our template document.
4. Create a table with header and data rows to insert in the content placeholder by looping through each row in our supplied result set and output a Word table row for each one.
5. Replace the `document.xml` inside of the OpenXML document package with the newly modified version.
6. Route the resulting document archive to a defined destination stream.

The next sections look at the code one step at a time.

Step 1: Reading the Template Document

This first step involves reading the template document from disk into memory for manipulation. Consider the following:

```
// create a copy of the template in memory
MemoryStream docxCopy = new MemoryStream();
byte[] docxBytes = File.ReadAllBytes(templateFilePath);
docxCopy.Write(docxBytes, 0, docxBytes.Length);
```

We're using a `System.IO.MemoryStream` to hold the document in memory. While working with the document in memory adds some overhead to the web application, the alternative of using the disk as a workspace would introduce additional I/O overhead in the web application. The static `File.ReadAllBytes` method is used to load the file from disk, which is then written to the stream.

Step 2: Loading the document.xml Part for Editing

The second step involves navigating the OpenXML document structure and loading the `document.xml` part for editing, like so:

```
// Open the template document archive and the document.xml part
```

```
Package docxPackage =
    Package.Open(docxCopy,FileMode.Open,FileAccess.ReadWrite);
Uri documentUri = new Uri("/word/document.xml",UriKind.Relative);
PackagePart documentPart = docxPackage.GetPart(documentUri);
Stream documentPartStream = documentPart.GetStream();

// create an XmlDocument to manipulate document.xml
XmlDocument documentPartXml = new XmlDocument();
documentPartXml.Load(documentPartStream);
```

Using the `System.IO.Packaging` namespace (added with .NET Framework 3.0) for working directly with Open Packaging Convection (OPC) document archives, we start by creating a `Package` object, opening the in-memory working document from our `docxCopy` memory stream. The `FileMode` and `FileAccess` enumerations work the same way as they do in other `System.IO` components that work with files. In this case, we are opening an existing file (`FileMode.Open`) and getting read/write access (`FileAccess.ReadWrite`).

Next, we declare a URI object with the path of the `document.xml` in a WordprocessingML document and use that URI to open a `PackagePart` object from the document archive and then a `Stream` with the actual contents of the `document.xml` part. Lastly, an `XmlDocument` is initialized and loaded from this stream.

Step 3: Finding the XML Element Representing the Content Control Placeholder

To find the XML element representing the Content Control placeholder we defined in our template document, we use the following code:

```
// set up the namespace manager (needed for XPath queries on a document using XML
namespaces)
string namespaceUri_w = "http://schemas.openxmlformats.org/wordprocessingml/
2006/main";
XmlNamespaceManager xmlnsManager = new XmlNamespaceManager(documentPartXml
.NameTable);
xmlnsManager.AddNamespace("w", namespaceUri_w);

// This xpath query will find the Content Control
XmlElement targetContentControl = (XmlElement)documentPartXml.SelectSingleNode
("//w:sdt[w:sdtPr/w:alias/@w:val='" + replaceFieldName + "']", xmlnsManager);
```

Because we are now working with an `XmlDocument` object, our manipulation is done using standard .NET System.Xml tools. We'll use an XPath query to find our Content Control placeholder, but first we need to declare an `XmlNamespaceManager`, as WordprocessingML is composed of multiple namespaces. We're only working with the WordprocessingML namespace, so we'll add just this namespace and alias it to `"w"`.

The actual XML we're looking for in the document looks like this:

```
<w:sdt>
  <w:sdtPr>
    <w:alias w:val="ProductData" />
    <w:tag w:val="ProductData" />
    <w:id w:val="1062475713" />
    <w:placeholder>
```

```
        <w:docPart w:val="DefaultPlaceholder_22675703" />
      </w:placeholder>
    </w:sdtPr>
    <w:sdtContent>
      <w:p w:rsidR="00BD658E" w:rsidRDefault="00BD658E">
        <w:r>
          <w:t>Product Data Will Go here</w:t>
        </w:r>
      </w:p>
    </w:sdtContent>
  </w:sdt>
```

The `<sdt>` tag defines a Structured Document Tag, which is basically a cell that contains content related to structured data of some kind. In Microsoft Word 2007, the editor exposes these as Content Controls. A `<sdt>` tag contains two child elements: a properties definition `<sdtPr>` and a content definition `<sdtContent>`. Because the document may have multiple Content Control placeholders, we'll search by the title, which is contained in the `<alias>` element in the properties section. Our XPath query is built as follows:

`//`	Search the subtree of the document starting at the root
`w:sdt`	Look for and return the `sdt` element (in the "w" namespace)

`[w:sdtPr/w:alias/@w:val='ProductData']`	Restrict results to `w:sdt` elements that contain the following structure:
	 `<w:sdtPr>`
	    `<w:alias w:val="ProductData" />`

As long as our document has a Content Control with the targeted title, we now have a reference to the element in which we need to insert our new content:

```
if (targetContentControl != null)
{
    XmlElement targetContainer = (XmlElement)targetContentControl.SelectSingleNode
("w:sdtContent", xmlnsManager); // had: [w:p/w:r/w:t]

    targetContainer.InnerText = "";

    // build the Word table from the data in the reader and put it into the field
    XmlElement wordDataTable = DataReaderToWordTable(sourceReader,documentPartXml);
    if (wordDataTable != null)
    {
        targetContainer.AppendChild(wordDataTable);
    }
}
```

The next part of this step checks ensures that we did find the target Content Control and selects the content tag so we can replace its contents. The actual table is created in a separate method named `DataReaderToWordTable` defined in Step 4. The container is emptied and the generated Word-processingML XML code is inserted in its place.

Step 4: Creating a Table

This fourth step involves creating a table with header and data rows to insert in the content place-holder. This is done by looping through each row in our supplied result set and outputting a Word table row for each one.

Consider the following:

```
static XmlElement DataReaderToWordTable(IDataReader sourceReader, XmlDocument
workingDoc)
```

The code for this step is contained in the DataReaderToWordTable method, declared to accept the supplied Data Reader and a reference to the XmlDocument that will eventually hold the XmlElement this method returns:

```
string namespaceUri_w = "http://schemas.openxmlformats.org/wordprocessingml/
2006/main";

if (sourceReader == null || sourceReader.IsClosed)
{
    return null;
}

XmlElement tbl_el = workingDoc.CreateElement("w", "tbl", namespaceUri_w);

// define the table borders
XmlElement tblBorders_el = (XmlElement)tbl_el.AppendChild(workingDoc.CreateElement
("w", "tblBorders", namespaceUri_w));

// create border elements for each edge and inside
tblBorders_el.AppendChild(workingDoc.CreateElement("w", "left", namespaceUri_w));
tblBorders_el.AppendChild(workingDoc.CreateElement("w", "top", namespaceUri_w));
tblBorders_el.AppendChild(workingDoc.CreateElement("w", "right", namespaceUri_w));
tblBorders_el.AppendChild(workingDoc.CreateElement("w", "bottom", namespaceUri_w));
tblBorders_el.AppendChild(workingDoc.CreateElement("w", "insideH", namespaceUri_w));
tblBorders_el.AppendChild(workingDoc.CreateElement("w", "insideV", namespaceUri_w));

// set all border attributes the same (add same attributes to each)
foreach (XmlElement el in tblBorders_el.ChildNodes)
{
    el.SetAttribute("val", namespaceUri_w, "single");
    el.SetAttribute("sz", namespaceUri_w, "1");
    el.SetAttribute("space", namespaceUri_w, "0");
    el.SetAttribute("color", namespaceUri_w, "auto");
}
```

We repeat the namespaceUri_w variable definition (this would be declared as a constant in production code, but it is repeated here to keep the code samples as self-contained as possible). After some simple error checking on the source Data Reader, the <tbl> element itself is created, as is a <tblBorders> element that defines the borders of the generated table. After that, things get a bit repetitive, declaring a separate element for every edge, adding each of these elements to the <tblBorders> element. Finally, we take a shortcut by looping through all of the elements in <tblBorders> and setting all of their attributes to the same values — in this case, single border, 1 point with no spacing, and the default (auto) color assignment.

With the formatting out of the way, we can move on to actually putting content in the table. The header is generated first:

```
// generate the field titles
XmlElement tr_title_el = (XmlElement)tbl_el.AppendChild(workingDoc.CreateElement
("w", "tr", namespaceUri_w));
for (int fieldIndex = 0; fieldIndex < sourceReader.FieldCount; fieldIndex++)
{
    // create the <w:tc><w:p><w:r><w:t> element hierarchy
    XmlElement tc_el = (XmlElement)tr_title_el.AppendChild(workingDoc.CreateElement
("w", "tc", namespaceUri_w));
    XmlElement p_el = (XmlElement)tc_el.AppendChild(workingDoc.CreateElement
("w", "p", namespaceUri_w));
    XmlElement r_el = (XmlElement)p_el.AppendChild(workingDoc.CreateElement
("w", "r", namespaceUri_w));
    XmlElement t_el = (XmlElement)r_el.AppendChild(workingDoc.CreateElement
("w", "t", namespaceUri_w));
    // assign the field value to the <t> element
    t_el.InnerText = sourceReader.GetName(fieldIndex);
}
```

The row is defined in a `<tr>` element, and we loop for each cell generating a Table Cell element `<tc>`, Paragraph `<p>`, Run `<r>`, and Text `<t>` tags, finally containing the text we want in the cell:

```
while (sourceReader.Read())
{
    // create the row element (<w:tr>) and add it to the table element (<w:tbl>)
    XmlElement tr_el = (XmlElement)tbl_el.AppendChild(workingDoc.CreateElement
("w", "tr", namespaceUri_w));

    for (int fieldIndex = 0; fieldIndex < sourceReader.FieldCount; fieldIndex++)
    {
        // create the <w:tc><w:p><w:r><w:t> element hierarchy
        XmlElement tc_el = (XmlElement)tr_el.AppendChild(workingDoc.CreateElement
("w", "tc", namespaceUri_w));
        XmlElement p_el = (XmlElement)tc_el.AppendChild(workingDoc.CreateElement
("w", "p", namespaceUri_w));
        XmlElement r_el = (XmlElement)p_el.AppendChild(workingDoc.CreateElement
("w", "r", namespaceUri_w));
        XmlElement t_el = (XmlElement)r_el.AppendChild(workingDoc.CreateElement
("w", "t", namespaceUri_w));
        // assign the field value to the <t> element
        t_el.InnerText = sourceReader.GetValue(fieldIndex).ToString();
    }
}
return tbl_el;
```

The code to generate the table rows themselves is similar to the header, just wrapped with a `while` loop that runs through the entire result set.

Step 5: Replacing the document.xml

Next comes replacing the `document.xml` inside of the OpenXML document package with the newly modified version:

```
// Update document.xml with our new version
docxPackage.DeletePart(documentUri);
PackagePart newDocumentPart = docxPackage.CreatePart
(documentUri, "application/xml");


Stream documentPartOutStream = newDocumentPart.GetStream
(FileMode.Create,FileAccess.Write);
StreamWriter documentPartOutStreamWriter = new StreamWriter(documentPartOutStream);

documentPartXml.Save(documentPartOutStreamWriter);  // save the XML content back out
documentPartOutStream.Flush();
documentPartOutStream.Close();
```

Here we delete the old `document.xml` part (using the `documentUri` reference we used to retrieve it earlier). Then, creating a new `PackagePart` takes the same URI reference and a content type identifier (`"application/xml"`). Now, using the stream from the document part, we populate it by saving using the `Save` method of the `XmlDocument`. Then we clean up by flushing buffers and closing streams.

Step 6: Routing the Resulting Document Archive

The sixth and final step involves routing the resulting document archive to a defined destination stream, like so:

```
// Save the archive back out to the supplied stream
docxCopy.WriteTo(outputStream);

// clean up
outputStream.Flush();
docxCopy.Close();
```

All that's left to do is to save the newly populated OpenXML document to the supplied output stream and clean up streams and buffers.

All of the work of loading and modifying the document was handled within this `CreateWordDocument` method, but there are a few things that still need to be done in the code-behind of our ASP.NET page.

Outputting the OpenXML Document in ASP.NET

The ASP.NET page itself is responsible for setting up the required data connection and result set, as well as setting some HTTP headers. This is done so the browser knows that the response is coming and should be opened in Microsoft Word. This is all done in the `Page_Load` event:

```
protected void Page_Load(object sender, EventArgs e)
{
```

The `ClearContent` and `ClearHeaders` method calls cancel any content that may have been buffered to go back to the client. Because this request will not use the `.aspx` page itself, we want to make sure that pieces of that content aren't mixed in with the binary data of the OpenXML document archive.

The next step is to set two HTTP headers, `ContentType` and `Content-Disposition`, both of which are required for proper handling:

```
Response.ClearContent();
Response.ClearHeaders();
Response.ContentType = "application/vnd.ms-word.document.12";
Response.AddHeader("Content-Disposition", "attachment;filename=test.docx");
```

The rest is just a set up of the data source, a call to the `CreateWordDocument` method, and a cleanup of the Data Reader to make sure the connection is released:

```
IDataReader dr = GetSqlData();

WordReport.CreateWordDocument(Server.MapPath("Product List Template.docx"),
"ProductData", dr, Response.OutputStream);

dr.Close();
}
```

Now when a browser hits this page, the resulting document is populated from the database (see Figure 4-10).

Product List

Product Data			
ProductID	Name	ProductNumber	Size
680	HL Road Frame - Black, 58	FR-R92B-58	58
706	HL Road Frame - Red, 58	FR-R92R-58	58
717	HL Road Frame - Red, 62	FR-R92R-62	62
718	HL Road Frame - Red, 44	FR-R92R-44	44
719	HL Road Frame - Red, 48	FR-R92R-48	48
720	HL Road Frame - Red, 52	FR-R92R-52	52
721	HL Road Frame - Red, 56	FR-R92R-56	56
739	HL Mountain Frame - Silver, 42	FR-M94S-42	42
740	HL Mountain Frame - Silver, 44	FR-M94S-44	44

Figure 4-10

What we did here is just the beginning of what can be done with this technique of generating rich documents from applications. In addition to text and tables like this example shows, other pieces of the .NET Framework can be used to enhance this capability. For example, the `System.Drawing` namespace can be used to generate graphics that can be embedded in the document (with the image stored in the `/word/`

`media` folder in the document archive), or data can be embedded into a document from remote systems using web services. In addition, because this technique is not limited to being hosted in ASP.NET, this same capability can be built into a Windows Forms application or into a web service that returns fully formed Office documents.

Next, we'll take a look at Excel's file format, which provides similar capabilities and flexibility for integration into applications and services.

Exploring the Excel OpenXML format (SpreadsheetML)

Just like Word, Excel 2007 implements a new OpenXML file format. Using the same Open Packaging Conventions structure, the overall process for working with OpenXML spreadsheets is similar to what we did earlier in this chapter with Word files. Although the Excel OpenXML format is fairly easy to understand, a few details of its implementation make it more complicated to generate without a library to encapsulate some of its complexities.

To understand this format, let's create a simple spreadsheet in Excel (shown in Figure 4-11) and examine the corresponding document archive.

	A	B	C	D
1	Hello World			
2	Jan	1	2	$ 3.00
3	Feb	2	4	$ 5.00
4	Jan	3	3	$ 2.00
5	Feb	6	9	$ 10.00

Figure 4-11

We'll extract this document (just as before, add a .ZIP extension to the file and extract) and get following files. (We have omitted the details on the folders that mirror the Word format.)

Folder or File Name	Description
`_rels` (folder)	Relationship mapping (same as Word)
`docProps` (folder)	Document metadata (same as Word, but `app.xml` contains Excel-type information)
`xl` (folder)	Excel document contents
`_rels` (folder)	Relationship mapping for the spreadsheet
`theme` (folder)	Contains any themes used in this document
`workbook.xml`	Workbook-related information; index of worksheets
`worksheets` (folder)	Contains worksheets
`sheet1.xml`	The first worksheet in the document

Starting at `xl/workbook.xml`, you can see the list of worksheets in this workbook:

```
<?xml version="1.0" encoding="UTF-8" standalone="yes"?>
<workbook xmlns="http://schemas.openxmlformats.org/spreadsheetml/2006/main"
xmlns:r="http://schemas.openxmlformats.org/officeDocument/2006/relationships">
  <fileVersion appName="xl" lastEdited="4" lowestEdited="4" rupBuild="4505"/>
  <workbookPr defaultThemeVersion="124226"/>
  <bookViews>
    <workbookView xWindow="120" yWindow="105" windowWidth="11415" windowHeight="4845"/>
  </bookViews>
  <sheets>
    <sheet name="Sheet1" sheetId="1" r:id="rId1"/>
    <sheet name="Sheet2" sheetId="2" r:id="rId2"/>
    <sheet name="Sheet3" sheetId="3" r:id="rId3"/>
  </sheets>
  <calcPr calcId="124519"/>
</workbook>
```

As with any other reference, the `<sheet>` tags use a resource ID (`r:id`) to link to the actual worksheet contents. A quick examination of `/xl/_rels/workbook.xml.rels` resolves these links as well as some other related parts. There isn't much to this item other than these sheet references and some housekeeping details on window positions and last edit locations. The `calcPr` element tracks calculation details for the workbook (the `calcId` attribute is the version number of the calculation engine last used to calculate the worksheet).

/xl/workbook.xml.rels

The `workbook.xml.rels` file (stored in the `/xl/_rels/` folder) is what ties the `workbook.xml` we just looked at to the actual worksheet parts that contain the cells and other contents of the worksheets. Because this is a `.rels` file, it follows the same conventions you saw in our Word document example, with the same benefits of abstracting the actual file/part names and letting an application quickly analyze document the structure.

This `.rels` file maps the worksheet files, named `/xl/worksheets/sheet1.xml` through `sheet3.xml`, as well as a few other related files. We'll look at `sharedStrings.xml` in more detail when we start working with the sheet parts next.

The `calcChain.xml` is the calculation chain, which describes the order in which the spreadsheet cells can be calculated. This saves Excel the work of rebuilding this on reloads. If it isn't available or something causes it to be out of date, Excel will analyze the dependencies of cells in the workbook and regenerate this.

The `styles.xml` and `themes.xml` files are shared styles and themes that are applied to the document. They are embedded so that the document is portable across different machines, which may have different themes installed locally:

```
<?xml version="1.0" encoding="UTF-8" standalone="yes"?>
<Relationships xmlns="http://schemas.openxmlformats.org/package/2006/
relationships">
  <Relationship Id="rId3"
    Type="http://schemas.openxmlformats.org/officeDocument/2006/relationships/
worksheet"
    Target="worksheets/sheet3.xml"/>
  <Relationship Id="rId7"
```

```
        Type="http://schemas.openxmlformats.org/officeDocument/2006/relationships/
calcChain"
        Target="calcChain.xml"/>
  <Relationship Id="rId2"
        Type="http://schemas.openxmlformats.org/officeDocument/2006/relationships/
worksheet"
        Target="worksheets/sheet2.xml"/>
  <Relationship Id="rId1"
        Type="http://schemas.openxmlformats.org/officeDocument/2006/relationships/
worksheet"
        Target="worksheets/sheet1.xml"/>
  <Relationship Id="rId6"
        Type="http://schemas.openxmlformats.org/officeDocument/2006/relationships/
sharedStrings"
        Target="sharedStrings.xml"/>
  <Relationship Id="rId5"
        Type="http://schemas.openxmlformats.org/officeDocument/2006/relationships/
styles"
        Target="styles.xml"/>
  <Relationship Id="rId4"
        Type="http://schemas.openxmlformats.org/officeDocument/2006/relationships/
theme"
        Target="theme/theme1.xml"/>
</Relationships>
```

Now, we'll start with a closer look at one of the worksheets linked here (/xl/worksheets/sheet1.xml).

/xl/worksheets/sheet1.xml

The worksheet parts contain most of the data in an Excel document, so this is the richest document type in Excel and the place where most developers will spend their time when building or modifying documents through code.

The document starts with a header <worksheet> element that mainly maps namespaces. This is followed by some general properties for the worksheet such as dimensions (<dimension>), which is a range that contains all content in the worksheet; worksheet views, such as position of split panes, zoom scale, and so on; and column/row header information. The <cols> element in this example contains one column definition that has the customWidth flag set, indicating that this column was sized manually:

```
<?xml version="1.0" encoding="UTF-8" standalone="yes"?>
<worksheet xmlns="http://schemas.openxmlformats.org/spreadsheetml/2006/main"
xmlns:r="http://schemas.openxmlformats.org/officeDocument/2006/relationships">
  <dimension ref="A1:D5"/>
  <sheetViews>
    <sheetView tabSelected="1" zoomScale="110" zoomScaleNormal="110"
workbookViewId="0">
      <selection activeCell="B1" sqref="B1"/>
    </sheetView>
  </sheetViews>
  <sheetFormatPr defaultRowHeight="15"/>
  <cols>
    <col min="1" max="1" width="18" customWidth="1"/>
  </cols>
```

This next section of the document contains the actual worksheet cell data, a set of <row> elements with cell (<c>) children:

```
<sheetData>
  <row r="1" spans="1:4">
    <c r="A1" t="s">
      <v>0</v>
    </c>
  </row>
  <row r="2" spans="1:4">
    <c r="A2" t="s">
      <v>1</v>
    </c>
    <c r="B2">
      <v>1</v>
    </c>
    <c r="C2">
      <v>2</v>
    </c>
    <c r="D2" s="1">
      <v>3</v>
    </c>
  </row>
  <row r="3" spans="1:4">
    <c r="A3" t="s">
      <v>2</v>
    </c>
    <c r="B3">
      <v>2</v>
    </c>
    <c r="C3">
      <v>4</v>
    </c>
    <c r="D3" s="1">
      <v>5</v>
    </c>
  </row>
  <row r="4" spans="1:4">
    <c r="A4" t="s">
      <v>1</v>
    </c>
    <c r="B4">
      <f>B2+B3</f>
      <v>3</v>
    </c>
    <c r="C4">
      <v>3</v>
    </c>
    <c r="D4" s="1">
      <v>2</v>
    </c>
  </row>
  <row r="5" spans="1:4">
    <c r="A5" t="s">
```

```
            <v>2</v>
         </c>
         <c r="B5">
            <f>SUM(B2:B4)</f>
            <v>6</v>
         </c>
         <c r="C5">
            <f>SUM(C2:C4)</f>
            <v>9</v>
         </c>
         <c r="D5" s="1">
            <f>SUM(D2:D4)</f>
            <v>10</v>
         </c>
      </row>
   </sheetData>
   <pageMargins left="0.7" right="0.7" top="0.75" bottom="0.75" header="0.3"
footer="0.3"/>
</worksheet>
```

Excel does a few things here that make it easy to work with the sheet data. First, each Row (`<row>`) is numbered in the r attribute. Every Cell (`<c>`) has an attribute (r) identifying its coordinates. The value elements (`<v>`) contain the actual cell values. Formula cells have a formula (`<f>`) element containing the formula in addition to the value element. Most of the contents in the sheet part are self-explanatory.

One interesting aspect, though, is what happens to our string value in the first cell (A1). The actual value in the cell was "a", but we don't see that text here. Because string values tend to repeat in a spreadsheet, Excel uses a string lookup table that stores each unique string value in the document. In OpenXML, this string table is stored in /xl/sharedStrings.xml in a simple XML list.

/xl/sharedStrings.xml

You can't insert or read string data from the actual worksheet document parts, so you usually have to work with the sharedStrings.xml part as well. Here is that part for this same document:

```
<?xml version="1.0" encoding="UTF-8" standalone="yes"?>
<sst xmlns="http://schemas.openxmlformats.org/spreadsheetml/2006/main" count="3"
uniqueCount="3">
   <si>
      <t>a</t>
   </si>
   <si>
      <t>b</t>
   </si>
   <si>
      <t>c</t>
   </si>
</sst>
```

This is one of the simplest document parts we have worked with so far. The document element, `<sst>`, contains the usual namespace reference and a count of the items in this string table. The rest of the document is a series of string item (`<si>`) elements. In our simple example there are just text (`<t>`) elements, but the format also supports rich text run elements (`<r>`), which can contain formatting properties (`<rPr>`).

Although this string storage mechanism provides efficiencies in storage and memory use, it complicates our job when generating Excel spreadsheets in custom applications. This is a major reason why it's easier to use a library to generate your spreadsheet documents, because a library can easy encapsulate the lookup and management of this string lookup table. Several such libraries are available, both from commercial vendors and as community projects. The OpenXML developer portal (www.openxmldeveloper.org) is a good place to find these resources.

PowerPoint's OpenXML File Format (PresentationML)

As with Word and Excel, PowerPoint has received the XML file format overhaul in the 2007 release as well. PresentationML is the new XML-based presentation file format and it uses the same packaging structure, relationship mapping files, and metadata formats as Word and Excel 2007. Also like Word and Excel, this is a full-fidelity file format, so it handles embedded media, animation, slide notes, and other advanced presentation content.

We won't go into as much detail here on PowerPoint (the complete format documentation is in the OpenXML ECMA specification), but we will look at the sections of XML that define the content of the presentation. The first of these is /ppt/presentation.xml, which contains the slide list (<p:sldIdLst>) element shown here:

```
<p:sldIdLst>
  <p:sldId id="323" r:id="rId2"/>
  <p:sldId id="353" r:id="rId3"/>
  <p:sldId id="351" r:id="rId4"/>
  <p:sldId id="348" r:id="rId5"/>
  <p:sldId id="349" r:id="rId6"/>
  <p:sldId id="350" r:id="rId7"/>
  <p:sldId id="352" r:id="rId8"/>
  <p:sldId id="354" r:id="rId9"/>
  <p:sldId id="312" r:id="rId10"/>
  <p:sldId id="332" r:id="rId11"/>
  <p:sldId id="343" r:id="rId12"/>
  <p:sldId id="344" r:id="rId13"/>
  <p:sldId id="342" r:id="rId14"/>
  <p:sldId id="346" r:id="rId15"/>
  <p:sldId id="315" r:id="rId16"/>
  <p:sldId id="331" r:id="rId17"/>
</p:sldIdLst>
```

Each Slide element (<p:sldId>) represents a single slide in the deck and points to its own XML definition file with a Resource ID, which is mapped through the /ppt/_rels/presentation.xml.rels file. Each slide definition file (by default named /ppt/slides/slideX.xml, where X is the ordinal position of the slide in the deck) contains the actual slide contents, and references embedded data such as styles, media, and themes.

Excel and Word are usually the first choices when you are integrating custom applications with Office client applications, but PowerPoint's new OpenXML keeps that product on a par with those applications and makes it easier for developers to read, create, and modify PowerPoint slide decks.

Customizing and Integrating Office Client Applications with Web Services

Up to this point in the chapter, most of our work has been at the file format level. Here we will build add-ins for the client applications themselves to integrate into other systems. In most new development, that kind of integration is best done with web services. Now we'll walk through extending the Office client applications (Excel in this case) to integrate into a back-end system and demonstrate several ways to customize the Office user interface.

The steps for building this project are as follows:

1. Build the core project and add the ribbon icons.

2. Connect to a web service to retrieve data and insert that data into the active Excel document.

3. Build a custom task pane and link that to the ribbon; this task pane can then be integrated into other systems and the active document.

Building the Core Project with Ribbon Icons

We'll start with Visual Studio Tools for Office, Second Edition, and create an Excel Add-in project, which is under the `Office/2007 Add-ins` project type folder. (This could just as easily be done with Word or PowerPoint, but we would choose a different project template to start with.) Visual Studio will create a solution with two projects. The first is the Excel Add-in itself, which compiles into a COM DLL; the second project is a setup project, which deploys the add-in.

The add-in project starts with one class with the default name `ThisAddIn`. At its starting point, it includes code to manage initialization and shutdown, and wires up a few events that we will use to register the ribbon. The default code for the `ThisAddIn.cs` file is as follows (region contents are omitted):

```
using System;
using System.Windows.Forms;
using Microsoft.VisualStudio.Tools.Applications.Runtime;
using Excel = Microsoft.Office.Interop.Excel;
using Office = Microsoft.Office.Core;

namespace ExcelAddIn
{
    public partial class ThisAddIn
    {
        private void ThisAddIn_Startup(object sender, System.EventArgs e)
        {
            #region VSTO generated code [hidden]
        }

        private void ThisAddIn_Shutdown(object sender, System.EventArgs e)
        {
        }
```

```
        #region VSTO generated code   [hidden]
      }
   }
```

The Excel Add-in project template we are using also has built-in support for custom ribbons. Start defining the ribbon by adding a Ribbon Support item to the project. (For this example, we'll name it `ErpRibbon` because we are going to integrate this into web services for an Enterprise Resource Planning (ERP) system. This item type is specific to the Office 2007 Add-in projects.

Once you add this item, you actually get two new items in the project. The first is an XML document (`ErpRibbon.xml`) that defines the ribbon UI elements we will add. The second is a class file (`ErpRibbon.cs`) that supports loading the ribbon definition and handles events from the ribbon. You want to add three elements to the Ribbon UI:

❑ A text box to allow a user to enter an order number (`orderIDBox`)

❑ A button to import order details for the supplied order ID (`importOrder`)

❑ A button to toggle the appearance of a custom task pane that supports more integration scenarios (`toggleOrderManagement`)

The final product of this ribbon customization is shown in Figure 4-12.

Figure 4-12

You modify `ErpRibbon.xml` as follows to add these controls to the Data tab:

```xml
<customUI xmlns="http://schemas.microsoft.com/office/2006/01/customui"
          onLoad="OnLoad">
  <ribbon>
    <tabs>
      <tab idMso="TabData">
        <group id="ERPData" label="ERP Data">

          <editBox id="orderIDBox" label="Order ID"
              screentip="Enter an order ID"
              onChange ="orderIDBox_OnChange"/>

          <button id="importOrder" label="Import Order Data"
              onAction="importOrder_OnAction" imageMso="Order" />

          <toggleButton id="toggleOrderManagement" label="Manage Orders"
              onAction="OrderManagement_OnToggle" />

        </group>
```

```
        </tab>
       </tabs>
      </ribbon>
    </customUI>
```

The `<customUI>`, `<ribbon>`, and `<tabs>` elements are left as is. The `<tab>` element is modified such that its idMso attribute, which reference a predefined ribbon element ID, is set to TabData, which is the standard Excel 2007 Data tab. This puts our new content into this existing tab. We could just as easily have defined our own id and label attributes and created a new tab.

The `<group>` element corresponds to a ribbon group, which clusters together controls in a ribbon. We're creating a new one, defining its id and label.

You define the three controls on the form using the appropriate tags on those controls. The following table shows the common ribbon control elements:

Control	Available event callbacks
Button	onAction
comboBox	onChange
Dropdown	onAction
editBox	onChange

For now, we're just combining the control name with the event name to name the callback. We'll come back later and connect our code to these event callbacks.

We added the ribbon class and XML to our project but it has no way of being loaded when our add-in is initialized. You set this up by adding this method to the ThisAddIn.cs file in the ThisAddIn class:

```
        protected override object RequestService(Guid serviceGuid)
        {
            // register the ribbon when we are requested to do so
            if (serviceGuid == typeof(Office.IRibbonExtensibility).GUID)
            {
                if (ribbon == null)
                    ribbon = new ErpRibbon(this);
                return ribbon;
            }

            return base.RequestService(serviceGuid);
        }
```

This method override enables us to respond to Excel's request for ribbon extensions. This code is actually boilerplate code that Visual Studio Tools for Office puts into a comment at the top of the ribbon class so that it can be cut and pasted from there. The only modification we have made is changing the constructor of the ribbon (the ErpRibbon class) to take a reference to the add-in and hold on to it so that ribbon event handlers can reference the add-in class. Because we are calling that constructor from the add-in, we pass in the "this" reference.

At this point, we have completed the steps required to display our custom ribbon extension. Visual Studio Tools for Office makes it easy to test the add-in by pressing the Run toolbar button or the F5 key. Excel appears with the new ribbon elements as shown earlier in Figure 4-12.

Connecting to a Web Service to Load Retrieve Data to Load into Excel

We have our ribbon buttons working in Excel but so far they don't actually do anything. We'll take the first step to change that here by adding a web service call to retrieve data from our ERP system. Before we look at the web service call, we need to wire the ribbon button event callbacks to methods in our ribbon class. We update our Ribbon Callbacks region in ErpRibbon.cs so the class looks as follows (note that this source file spans the next four listing blocks):

```
(standard namespace references omitted)

using System.Data;

namespace Excel_ERP_Task_Pane
{
    [ComVisible(true)]
    public class ErpRibbon : Office.IRibbonExtensibility
    {
        private Office.IRibbonUI ribbon;
        private ThisAddIn addIn;
        private int orderId = 0;
```

Here we have added private members to store a reference to the add-in class and the order ID the user has entered in the edit box in the ribbon.

```
        public ErpRibbon(ThisAddIn addIn)
        {
            this.addIn = addIn;
        }
```

This is the modified constructor to store the add-in reference.

The next region (IRibbonExtensibility Members) was autogenerated and not modified:

```
        [IRibbonExtensibility Members] (region hidden)

        public string GetCustomUI(string ribbonID)
        {
            return GetResourceText("Excel_ERP_Task_Pane.ErpRibbon.xml");
        }

        #endregion
```

This upcoming region holds our callbacks. The first one (OnLoad) is declared in the project template and wasn't modified:

```
#region Ribbon Callbacks

public void OnLoad(Office.IRibbonUI ribbonUI)
{
    this.ribbon = ribbonUI;
}

public void importOrder_OnAction(Office.IRibbonControl control)
{
    addIn.InsertOrderDetails(orderId);
}

public void orderIDBox_OnChange(Office.IRibbonControl control,
    string value)
{
    orderId = int.Parse(value);
}

public void OrderManagement_OnToggle(Office.IRibbonControl control,
    bool isPressed)
{
    addIn.SetCustomTaskPaneVisible(isPressed);
}

#endregion

[Helpers] (region hidden)
    }
}
```

The ribbon UI callback methods (importOrder_OnAction, orderIDBox_OnChange, and OrderManagement_OnToggle) are declared with their predefined parameters. They all accept IRibbonControl as the first parameter. editBox also passes in the string value of its contents, while the toggleButton passes in its toggle state.

In the orderIDBox_OnChange method, we capture the latest value of the Order ID so that we have its latest value when the importOrder button is clicked. The event handler for this click (importOrder_OnAction) uses that orderID value to call the InsertOrderDetails helper method that we need to create on our add-in. This is the method that actually calls the web service.

For the purposes of this example, our web service is a simple method call with the following signature:

```
public class ErpData : System.Web.Services.WebService
{

    [WebMethod]
    public DataSet GetOrderDetails(int orderID)
    { …. }
}
```

The details of this web service implementation are not important here, so we can assume that if we pass in a valid `orderID`, we will get an ADO.NET data set containing a single `DataTable` listing the line items on that order.

Before we can call this web service from our add-in project, we need to add a web reference to this web service through the Project/Add Web Reference menu and wizard in Visual Studio.

We will add two methods to our Excel add-in class to support the web service call and insert data into the active spreadsheet:

```
public void InsertOrderDetails(int orderID)
{
    localhost.ErpData service = new localhost.ErpData();
    DataSet ds = service.GetOrderDetails(orderID);
    DataTable dt = ds.Tables[0];

    InsertDataTableToExcel(this.Application, dt);
}

// utility class to insert the contents of an ADO.NET DataTable
// into the current worksheet
public static void InsertDataTableToExcel(Excel.Application app,
    DataTable dt)
{
    Excel.Range sel = (Excel.Range)app.Selection;
    int startRow = sel.Row;
    int startCol = sel.Column;

    int curRow = startRow;
    // add headers
    int curCol = startCol;
    foreach (DataColumn col in dt.Columns)
    {
        Excel.Range range = (Excel.Range)app.Cells[curRow, curCol];
        range.Value2 = col.ColumnName;
        curCol++;
    }

    // add data; for each row, loop through each column
    curRow++;
    foreach (DataRow row in dt.Rows)
    {
        curCol = startCol;
        foreach (DataColumn col in dt.Columns)
        {
            Excel.Range range = (Excel.Range)app.Cells[curRow, curCol];
            range.Value2 = row[col].ToString();
            curCol++;
        }
        curRow++;
    }
}
```

`InsertOrderDetails` calls the web service, extracts the `DataTable`, and calls `InsertDataTableToExcel`, a utility method that inserts a `DataTable` into the active spreadsheet by looping through the column headers and then looping through every row/column.

With these updates, we now have a working web service call. By entering a valid value into the Order ID textbox and clicking the Import Order Data button, our add-in calls a web service, retrieves that order data. Then that data is inserted into the active spreadsheet, as shown in Figure 4-13.

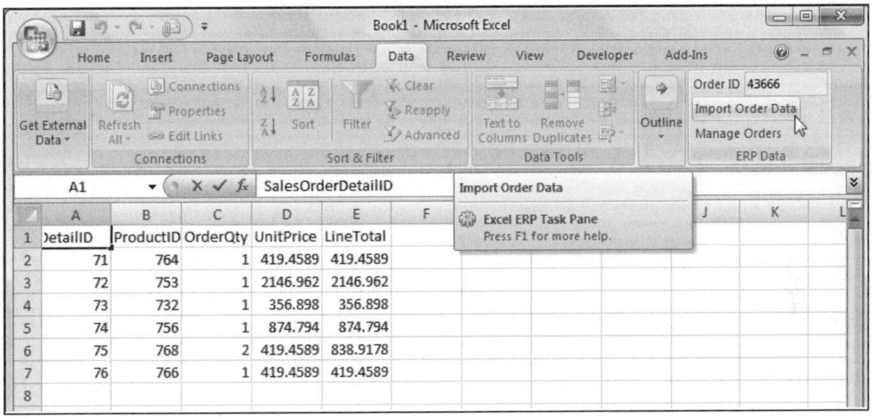

Figure 4-13

Adding a Custom Task Pane

As you have seen so far in this example, the Office ribbon provides rich capabilities for UI enhancement, well beyond what toolbars allowed in the past. However, even with the power of the ribbon, you run into situations where you need another way to embed a more customized user interface into Office. Custom task panes are one solution, providing you with a way to embed almost any kind of control right into the Office application, retaining access to the active document and the context of the Office application.

In this section we will work through the process of adding a custom task pane to our add-in example. We will not spend a lot of time adding functionality to the example because once we get to the point where we are dragging controls on a workspace, setting properties, and handling events, the work is basically like any other .NET development project.

Custom task panes are actually built using Windows Forms UserControls. UserControls are container controls, so they provide a space in which other controls can exist, very much like a normal Windows Form serves as a control container. The power of this approach is that it gives the developer a GUI designer and a broad choice of controls to be used.

In our example, we'll create a new UserControl in our add-in project, open it up in design mode, and drag a DataGridView and Button onto its surface. You can size these controls or even use Visual Studio 2005's capabilities in positioning and sizing the child controls based on the size of the parent UserControl.

Once we have designed our UserControl and added any code needed to make it interactive, we need to add some code to the `ThisAddIn` main class. We are only making small changes to the class, so instead of repeating the entire listing, we'll take a look at just the beginning of the class. Remember that regions are not expanded in this listing:

```
public partial class ThisAddIn
{
    ErpRibbon ribbon;
    private Microsoft.Office.Tools.CustomTaskPane erpTaskPane = null;

    private void ThisAddIn_Startup(object sender, System.EventArgs e)
    {
        #region VSTO generated code [Hidden]

        erpTaskPane = this.CustomTaskPanes.Add(new ErpPendingOrders(),
            "Pending Orders");
    }
```

The lines we added are marked in bold and highlighted in this listing. The only changes are a class-level private declaration for a `CustomTaskPane` field and a line in the `Startup` callback method that registers this custom task pane.

With those changes, the task pane is registered, but this doesn't actually make it appear. To do that, you need to add this method to the `ThisAddIn` class so the last custom ribbon toggle button can make this task pane appear and disappear:

```
public void SetCustomTaskPaneVisible(bool visible)
{
    erpTaskPane.Visible = visible;
}
```

This method just uses the `erpTaskPane` class variable we just declared when we registered the task pane. The ribbon's last remaining event handler will now call this method, passing in a `bool` depending on the value of the button's toggle state:

```
public void OrderManagement_OnToggle(Office.IRibbonControl control,
    bool isPressed)
{
    addIn.SetCustomTaskPaneVisible(isPressed);
}
```

Now with this event callback wired up to the custom task pane, you can test this last feature of the Excel add-in (see Figure 4-14).

Through this scenario we used a powerful combination of extensibility features that make Office much easier to customize for user requirements. Users get the productivity and familiarity of the industry-standard Office user interface with their proprietary applications exposed through the ribbon and custom task panes. Through managed wrappers on the Office object libraries, these add-ins can manipulate and read documents. Through the extensive web service library in SharePoint, developers can customize the behavior of the entire solution, better integrating the client applications with line-of-business applications and SharePoint.

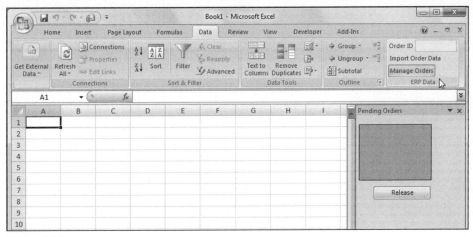

Figure 4-14

Summary

Content Controls provide a great way to create forms with Word, giving a form author the ability to define the type of data that should be entered into the form and a way to prevent changes on the fixed parts of a form.

Custom XML Parts are XML documents embedded in a Word document. These parts are available through the Word object model for macros and scripts. A developer can also create mappings between data elements in these parts and Content Controls in a form. Because this is a two-way binding, the user can even update content in the Custom XML Part, which can then be read back out from the document.

This integration of Custom XML Parts and Content Controls can be used to send users forms pre-populated with data from line-of-business systems. A server application would simply replace the Custom XML Part in the document before sending it to the user.

The OpenXML document formats create new options for developers. Because these documents are stored in ZIP file containers, a developer can easily read and manipulate these files using existing .NET APIs and XML. The published Open Packaging Convention (OPC) sets rules regarding how the different parts of an OpenXML document are linked together. OpenXML defines the schemas for the different document types. We looked at Word and Excel specifically.

When generating a Word OpenXML document, Content Controls are a good way to define placeholders in a document where generated data can go. This supports situations in which data structures are too complicated for the Content Control/Custom XML Part mapping approach.

Finally, this chapter demonstrated how rich and powerful the Excel OpenXML format is, supporting all features of Excel, including pivot tables and database connections. Generating Excel OpenXML documents from a database query is more complicated than our Word scenario because Excel streamlines some data storage that makes generating cells more complicated.

For scenarios beyond just generating and consuming Office documents, the Office client applications support a rich extension model that supports customization of the ribbon user interface as well as embedding custom task panes for more complicated user interface needs.

Building custom solutions with Office applications gives end users more flexibility and reduces the amount of custom development of building an entirely custom user interface. For data and document-driven applications, building on Office has many advantages.

5

User Interface Development in SharePoint and Office 2007

While there is extensive support for programming data sources, lists, and the file formats of Microsoft Office SharePoint Server (MOSS) 2007, Windows SharePoint Services (WSS) v3, and Office 2007, the user interface (UI) programming features are significant as well. It is important for developers to understand some of the key UI programming elements in Office and SharePoint, because this is how users will ultimately interact with the software and perform or initiate the functions supported by the backend services such as SharePoint or other business systems.

In SharePoint, the underlying technology is based on ASP.NET. This enables a developer to build Web Parts and to publish ASP.NET pages into the SharePoint environment to perform new functions. From a SharePoint perspective, the *Web Part* provides a small unit of user interface and functionality that can be published to a Web Part gallery and managed to some degree by the end user. With the ASP.NET page model being integrated into SharePoint, master pages can be customized in a site to provide some new functionality, to provide a consistent look and feel, or to ensure that specific content elements are provided in a consistent manner as required on a site.

To gain an understanding of the key user interface development features, we'll examine the following scenarios in Office and SharePoint 2007:

❑ Developing Web Parts in SharePoint 2007 to display content

❑ Using master pages in SharePoint for presentation, navigation, and content elements

❑ Creating new elements in the Office Ribbon

❑ Using the Document Information Panel to extend metadata to Office documents and SharePoint libraries and lists

For each of these areas we'll look at the functionality for developing and configuring the user interface to support application development. We'll also examine the underlying object models and APIs that enable this functionality.

At the end of this chapter, you will have an understanding of how to build custom solutions that expose new functionality to end users in Office and SharePoint 2007. In addition, the concepts related to content types and the document information panel in Office 2007 will apply in later chapters related to Forms Services and workflow (Chapters 6 and 10).

Web Parts in SharePoint 2007

The Web Parts supported in Microsoft Office SharePoint Server 2007 and Windows SharePoint Services v3 are designed to support customization, personalization, and the delivery of content to the end users of a web site. Users with the appropriate permissions can select Web Parts that are published to a gallery or as part of a site template and customize how those Web Parts are displayed on a page. For more complex Web Parts, configuration functions can be used to define how the Web Part displays data and how the user can interact with the content in the Web Part.

Web Parts can provide a number of advantages from a development perspective. Unlike other web-based development paradigms, Web Parts are compiled code and can provide some advantages from a rendering performance perspective. Web Parts also run in context with SharePoint and can easily render data based on the underlying SharePoint object model. In addition, SharePoint Web Parts can be connected with other Web Parts and integrated with other web content within SharePoint to provide an end user with some dynamic features related to content display and updates. Even techniques such as AJAX (Asynchronous JavaScript and XML) have been integrated into SharePoint to provide dynamic data updates and interaction without changing the browser state.

To understand how SharePoint Web Parts work, we'll look at what it takes to create a basic Web Part and deploy it within SharePoint. We'll also look at a more complex example of a Web Part that involves Web controls and how Web Parts can be deployed as Features within SharePoint.

Building a Web Part

To get started with SharePoint Web Parts, we'll build a Web Part that renders some basic content when viewed in SharePoint. A SharePoint Web Part is based on the `System.Web.UI.WebParts` namespace and consists of a class derived from `WebPart`. Building a Web Part can be as simple as providing basic code to render HTML output once Web Parts is instantiated.

The following listing shows the code for a basic Web Part.

```
using System;
using System.Text;
using System.Web.UI.WebControls.WebParts;

namespace SampleWebParts
{
    public class BasicWebPart : WebPart
```

```
        {
              protected override void Render(System.Web.UI.HtmlTextWriter writer)
              {
                  writer.Write("<a href='http://sharepoint.microsoft.com/'>SharePoint
    Community Site</a>");
              }
        }
    }
```

To build and deploy this Web Part, we go through the following steps:

1. Create a class in Visual Studio named Basic Web Part as shown in Figure 5-1. You can use the Class Library project type.

2. In the `Render` method, implement the `Write` method of `HtmlTextWriter` to render content for the Web Part. In the sample, we'll render some HTML content (in this case, a link to the SharePoint Community Site).

3. Compile the class library. In an upcoming step, you're going to copy the assembly from the bin directory under the Visual Studio project to a directory accessible by the SharePoint application on the server.

4. Look up your SharePoint application in the IIS console and under Properties find the Home Directory tab and find the value of the local path field. (See Figure 5-2.)

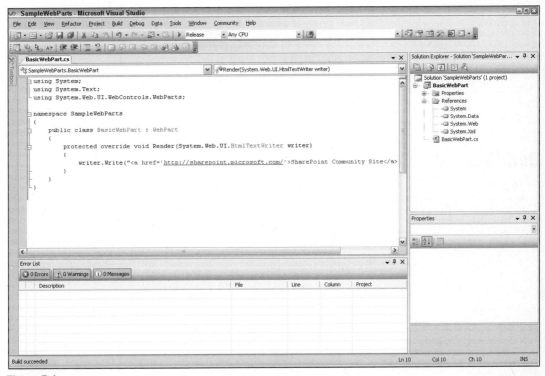

Figure 5-1

Figure 5-2

5. Now, copy the assembly that you compiled in Step 3 to the local path location you looked up in Step 4.

6. Raise the trust level of the bin directory. To do this, open the web.config file as part of the web application root and find the element that reads <trustLevel name="WSS_Minimal">. Change the trust level to WSS_Medium or add a new element that includes the trust level attribute being set to WSS_Medium. Both options work.

7. Add your Web Part data to the safe controls list under the web.config file as shown in Figure 5-3. The syntax should read as follows.

```
<SafeControl Assembly="BasicWebPart, Version=1.0.0.0, Culture=neutral,
PublicKeyToken=null" Namespace="SampleWebParts" TypeName="*" Safe="True"
AllowRemoteDesigner="True"/>
```

In this element of the web.config file, you're adding your Web Part to a list of other controls that are designated to be accessible by SharePoint. Attributes such as Assembly, Version, and Namespace enable you to designate your Web Part and specific version or namespace supported by the class for SharePoint accessibility. This SafeControl element can also be used at a solution level for larger groups of Web Parts or other SharePoint elements being deployed as a set of features to a SharePoint server. Once your Web Part is added to this list, it is then accessible to SharePoint. If you add this Web Part as part of a Feature, that will control where the Web Part is available and what other functions, such as custom lists or navigation elements, that Web Part is dependent on.

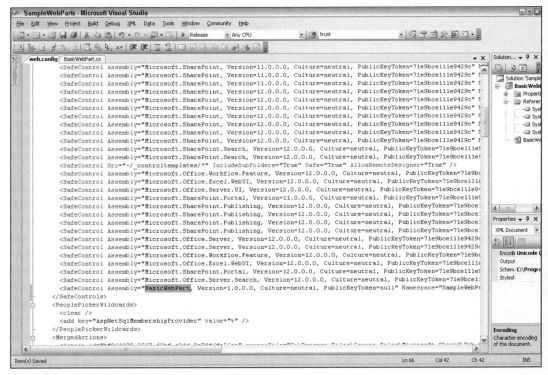

Figure 5-3

You can find more detail on how SharePoint Features work a little later on in this chapter and also in Chapters 9 and 10 of this book.

8. Go the new Web Parts listing under your SharePoint site, located here: `http://yoursitename/_layouts/newdwp.aspx`. Select your Web Part by checking the box to the left of the Web Part name.

9. Once you've selected your Web Part, populate your Web Part to the Web Part Gallery by clicking the Populate Gallery button. This will direct you to the Web Part Gallery page.

10. Once your Web Part is added to the gallery, it will appear in the list, as shown in Figure 5-4.

11. Now that the Web Part is published to the gallery, find a page to edit to add your Web Part. Select Add Web Part for one of the Web Part Zones on the page and select your Web Part as shown in Figure 5-5.

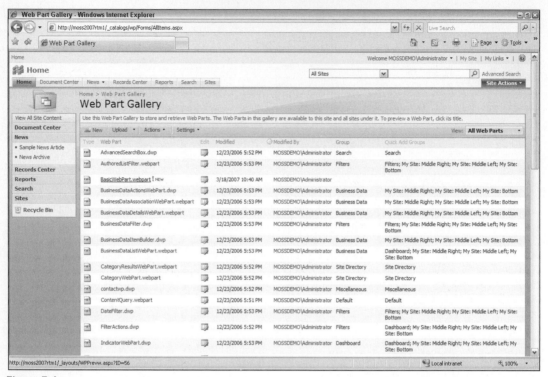

Figure 5-4

12. Once you've selected your Web Part, it will appear in the edit mode shown in Figure 5-6. Once you exit the edit mode, it should appear on the page as shown in Figure 5-7.

This demonstrates the basic steps required to create and deploy a new Web Part. For most organizations building applications on SharePoint, there are many factors you'll want to consider before building Web Parts. First, this Web Part sample includes only some basic text and a link. For those that are familiar with SharePoint, you'll know that there are other Web Parts that can provide the same functionality for delivering a static link. There are announcement Web Parts and other Web Parts that enable you to edit and add content, including hyperlinks, which would be suitable here. There is also the RSS viewer Web Part that consumes data from RSS feeds (which we looked at in Chapter 3) and can then display that content on a page. Essentially, the basic Web Part sample can be accomplished in a number of different ways without writing any code in SharePoint.

When you're thinking about building a new Web Part that isn't contained out of the box, think about the data sources that SharePoint can't easily connect to out of the box. If you have a business application that stores its data in a set of data files that can't be accessed through a web service or through the Business Data Catalog database connectivity but can be accessed through a .NET application, this might be a candidate for exposure through a SharePoint Web Part. In addition, if there is specific logic in the presentation of the content that can benefit from a custom Web Part or some specific look and feel that needs custom code to be accomplished, these are good areas to consider when thinking about building a Web Part. SharePoint has a lot of functionality that can be leveraged without custom Web Part development, so investigate your solution carefully to make sure you're not building something that's already easily accomplished another way in the product.

Figure 5-5

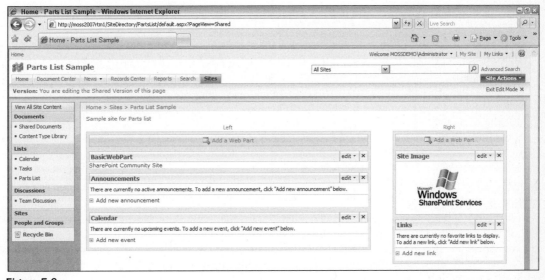

Figure 5-6

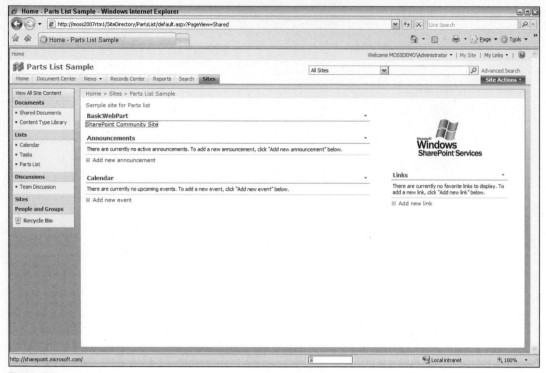

Figure 5-7

Now that we've created a basic functioning Web Part and added it to a page in SharePoint, the next section looks at a Web Part that includes web controls and additional functionality to better understand how these web parts can be extended.

Web Parts with Web Controls

To expand on the Web Parts concept, we'll build a part that supports controls and can be used for further interaction by the end user. Web controls are an important concept in ASP.NET since they deliver most of the user interface functionality that is ultimately presented to the end users of an ASP.NET application. Since SharePoint is based on ASP.NET, SharePoint takes advantage of these controls as well. Some of the commonly used web controls in ASP.NET include buttons, text boxes, images, labels, check boxes, and radio buttons. Other controls include panels to contain a set of controls; a check box list for multiselection check boxes; tables; datagrids, which display data in a grid format and can be bound to a data source; and hyperlinks. The main functions of controls are that they offer functionality that can be used to build your user interface in the web application and also offer events that make it easy to write code on the server to perform specific tasks for control action, such as clicking a button, without writing all of the HTML and script code required to execute these functions within a browser. By leveraging a control, the server then does the work for you to emit the appropriate HTML and script to your browser when the code is executed.

To get started, you'll follow the previous Web Part example and build out a class library derived from `WebPart`. The code for this Web Part is as follows:

```
using System;
using System.Text;
using System.Web.UI;
using System.Web.UI.WebControls;
using System.Web.UI.WebControls.WebParts;
using Microsoft.SharePoint;

namespace SampleWebParts
{

    public class CustomListWebPart : WebPart
    {

        protected TextBox txtListItem;
        protected Button btnSubmit;

        //Create child controls for Web Part
        protected override void CreateChildControls()
        {
            txtListItem = new TextBox();
            //Add control to Web Part
            this.Controls.Add(txtListItem);

            btnSubmit = new Button();
            btnSubmit.Text = "Submit List Item";

            //Add control to Web Part
            this.Controls.Add(btnSubmit);

        }

        // render HTML for Web Part
        protected override void Render(HtmlTextWriter writer)
        {
            //Render text box, button and HTML tag <p> for separation
            txtListItem.RenderControl(writer);
            writer.RenderBeginTag(HtmlTextWriterTag.P);
            btnSubmit.RenderControl(writer);

        }

    }

}
```

To build and test this Web Part on a SharePoint site, go through the following steps:

1. Create a class library named `CustomListWebPart`, as shown in Figure 5-8.

2. Use the `CreateChildControls` method to add web controls from the `System.Web.UI`
   `.WebControls` namespace to the Web Part. The `Render` method will be used to write the
   HTML contents represented by the controls to the Web Part user interface.

3. Compile the class library and copy the assembly from the bin directory under the Visual
   Studio project.

4. Look up your SharePoint application in the IIS console, and under Properties, find the Home
   Directory tab and find the value of the local path field as shown in the previous example.

5. Copy the assembly to the bin location.

6. Raise the trust level of the bin directory. To do this, open the `web.config` file as part of the
   web application root and find the element that reads `<trustLevel name="WSS_Minimal">`.
   Change the trust level to `WSS_Medium`.

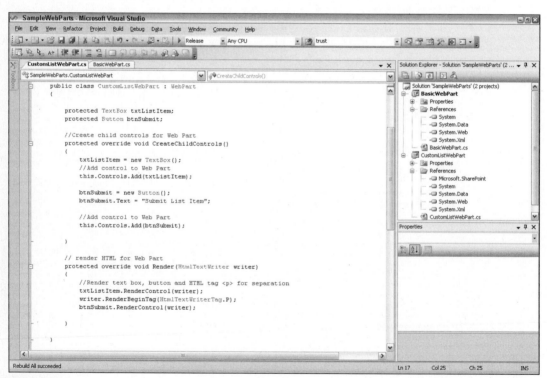

Figure 5-8

7. Add your Web Part data to the safe controls list under the `web.config` file as shown in the previous example. The syntax should read as follows:

```
<SafeControl Assembly="CustomListWebPart, Version=1.0.0.0, Culture=neutral,
PublicKeyToken=null" Namespace="SampleWebParts" TypeName="*" Safe="True"
AllowRemoteDesigner="True"/>
```

As with the previous example, you're specifying the assembly containing this Web Part so it will be accessible by SharePoint.

8. Go the new Web Parts listing under your SharePoint site, located here: `http://yoursitename/_layouts/newdwp.aspx`. Select your Web Part.

9. Once you've selected your Web Part, populate your Web Part to the Web Part Gallery by clicking the Populate Gallery button. This will direct you to the Web Part Gallery page.

10. Once your Web Part is added to the gallery, it will appear in the list as shown in Figure 5-9.

11. Now that the Web Part is published to the gallery, click on the link to view a preview of the Web Part. The output is shown in Figure 5-10.

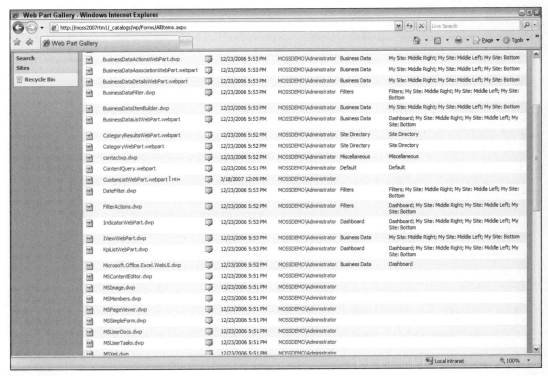

Figure 5-9

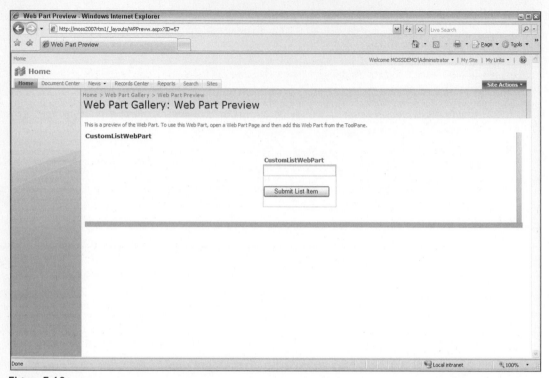

Figure 5-10

Using the Web Controls as part of the Web Parts, you can now take advantage of their supported functionality, including properties, methods, and events to further develop functionality as part of these Web Parts.

> *Remember that even though there is a broad range of web controls that you can use to build Web Parts, there are many out-of-the-box Web Parts that may provide functionality that you're looking for, so do some investigating before spending too much time writing code to build Web Parts.*

Now that you've looked at what it takes to create these Web Parts, it's time to take a look at Features in SharePoint to understand how you can bundle a Web Part as part of a Feature and manage its deployment in a SharePoint web arm.

Features in SharePoint

The *Features* "feature" in SharePoint is designed to make it easier to customize sites and deploy updates in a SharePoint farm. When building custom Web Parts or other functionality that's consumed by multiple sites hosted on multiple servers in a web farm environment, it is important to make sure that there is a consistent and robust method to group functionality and that it gets correctly deployed.

Features also provide a unit of functionality that administrators can enable or disable, ensuring that there is a consistent experience for a deployed Feature. For instance, if there is a schema for a list and an associated custom Web Part that should be deployed together, you can provide administrators with a vehicle to make sure those items are consistently deployed and enabled or disabled as a unit of functionality in SharePoint.

Feature definitions are deployed as folders under the directory `C:\Program Files\Common Files\Microsoft Shared\web server extensions\12\TEMPLATE\FEATURES`. The Features folder that you set up must contain a `Feature.xml` file that defines the elements bound to that particular Feature. A Features folder can also contain supporting elements including XML files, pages including `.aspx` and `.html` files, InfoPath schemas (`.xsn`), and Web Parts.

To better understand how Features work with SharePoint and how they can impact your ability to deploy Web Parts and other functionality to SharePoint sites, let's take a look at an example of a Feature that's implemented as part of Microsoft Office SharePoint Server 2007.

ExcelServerSite Feature

To understand how a simple Feature can be implemented, we'll look at the Feature that defines an Excel Services site. Since the Excel Web Access component is supported through one primary Web Part that lets you define an associated workbook and provide some definition for how the Web Part should display that data, this site has a simple scope included within the Feature.

To examine this feature, open the Features directory and navigate to the subfolder `ExcelServerSite` under Features. The directory contains a `Feature.xml` file, a `WebParts.xml` file defining supporting elements of the Feature, and the Excel Services Web Part, `Microsoft.Office.Excel.WebUI.dwp`.

To understand how this Feature is constructed, you first need to understand the syntax of the `Feature.xml` file. The Feature file is designed to specify all the dependencies and resource files associated with the Feature, including the XML files that define subelements of the feature. The supported attributes of the Feature element parent in `Feature.xml` are shown in the following listing, and the `Feature.xml` file for ExcelServerSite is shown in Figure 5-11.

```
<Feature
  ActivateOnDefault = "TRUE" | "FALSE"
  AlwaysForceInstall = "TRUE" | "FALSE"
  AutoActivateInCentralAdmin = "TRUE" | "FALSE"
  Creator = "Text"
  DefaultResourceFile =  "Text"
  Description = "Text"
  Hidden = "TRUE" | "FALSE"
  Id = "Text"
  ImageUrl = "Text"
  ImageUrlAltText = "Text"
  ReceiverAssembly = "Text"
  ReceiverClass = "Text"
  RequireResources = "TRUE" | "FALSE"
  Scope = "Text"
  SolutionId = "Text"
  Title = "Text"
  Version = "Text" >
</Feature>
```

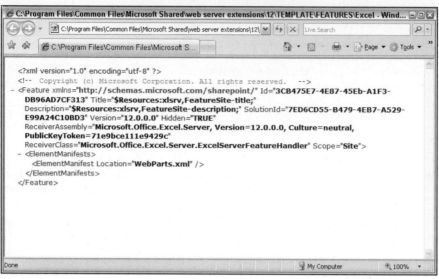

Figure 5-11

In addition to the attributes described for the Feature element, there are several child elements of the `Feature` parent element including `ActivationDependencies`, `ElementManifests`, and `Properties`. The `ActivationDependencies` element is really interesting because it prevents the Feature from being activated if a dependent Feature, identified by its ID, is not available. This allows you to provide robust links between Features and to ensure that once a certain Feature is disabled or removed from the SharePoint farm that all Features that are dependent on it explicitly will cease to function.

`ElementManifests` include specific modules that are required by that Feature. This is important because it provides the definition for the elements that are dependencies for a Feature. For instance, if a list, a custom content type, and a Web Part are all part of a Feature, you can ensure that the Feature will not be exposed if all of those elements are not present. The `ElementManifests` defines all of these components. `ActivationDependencies` then enforce dependencies based on these elements. This prevents you from having a partially deployed Feature that won't function correctly due to a missing element.

The `WebParts.xml` file for the Excel Services Site Feature is shown in Figure 5-12. In this case, the `Feature.xml` file for the `ExcelServerSite` specifically includes a reference to this `WebParts.xml` file.

As you can see in the figure, one of the elements included in this file is the relative file path to the Web Part supporting this Feature. The corresponding installed Web Part in the Web Part gallery is shown in Figure 5-13.

The reason that Features is important is that it provides the framework for defining the custom elements of a site, includes list definitions, content types, Web Parts (both custom and out of the box), and helps you as a developer to package these elements into a unit, or Feature, that can be assigned to a specific site or set of sites. This way, as you want to modify a site and change how the content elements and Web Parts are exposed to a user, you can simply redefine the elements of the Feature and deploy that update to your SharePoint server. The ability to define dependencies makes it much easier to ensure that you don't break existing sites of functionality by having interdependent elements of functionality that fail to deploy to a site.

For a deeper look at how Features work in the context of some broader examples, check out the Web Content Management and document workflow scenarios in Chapters 9 and 10.

Just as Features are important for packaging together key elements such as Web Parts, list definitions, and other resource files to make sure that users have a consistent user experience across multiple SharePoint sites in a farm, the master page concept is important to deliver a consistent user experience in terms of page appearance, consistent navigation elements, and key content elements across multiple pages in a site. In the next section, we'll examine master page functionality and how this works in SharePoint. We'll also look at what it takes to modify elements in master pages to update content across a number of pages.

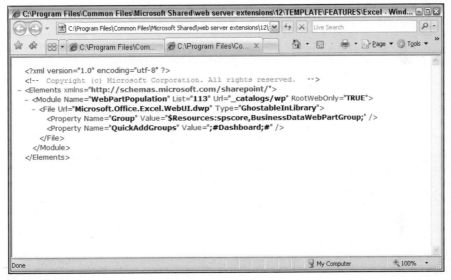

Figure 5-12

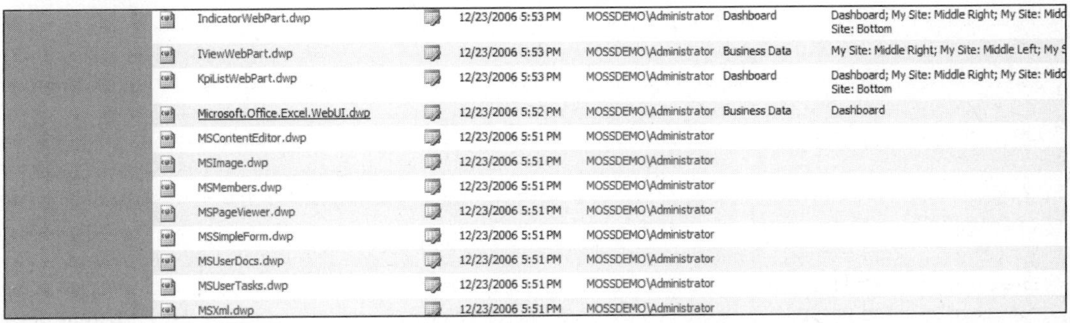

Figure 5-13

Master Pages in SharePoint

Master pages are part of the fundamental page model supporting Web content publishing and management in SharePoint. The architecture of master pages and how they fit into the overall page publishing and processing model was discussed in Chapter 2. In this chapter, we'll look at an example of the master page as component of the SharePoint user interface supporting content presentation and navigation.

Updating a Master Page in SharePoint Designer

To understand how master pages work and how they can be updated and modified, we'll walk through an update to a master page in SharePoint Designer. To follow this example, you need to have SharePoint Designer and a SharePoint site to access the master page for purposes of changing the navigation elements and updating the master page for the site. To update a master page, follow the steps listed below:

1. To get started, go to the Site Settings for your test site. Under the Galleries section, select Master Pages.

2. Under the Master Page Gallery, you should see at least one master page in the list named default .master. Select the Edit In Microsoft Office SharePoint Designer option for the master page when you click the arrow to the right of the master page name in the list.

3. Once the page is open in SharePoint Designer (shown in Figure 5-14), you need to select the Code tab at the bottom left portion of the design surface. This will take you from the Design view of the master page to the Code view, which lets you edit the code associated with this page.

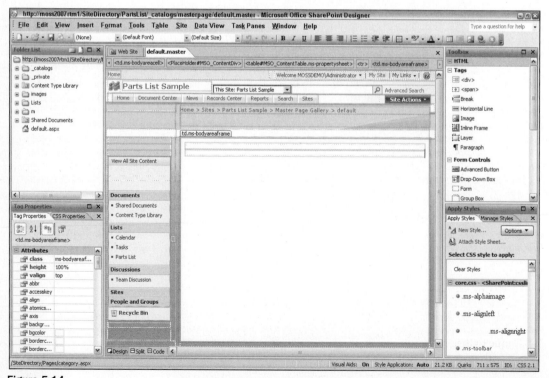

Figure 5-14

4. In this master page, you'll update the left-hand navigation menu, or the Quick Launch Menu. The menu is shown in Figure 5-15 prior to being updated.

Figure 5-15

5. In the code, find the section for the Quick Launch Menu. By default, the `StaticDisplayLevels` attribute is set to "2" and the `MaximumDynamicDisplayLevels` attribute is set to "0". Adjust these to "2" and "1", respectively.

6. Save the master page. Upon reloading the main page for the site, the Quick Launch Menu will appear as shown in Figure 5-16 with flyout menus.

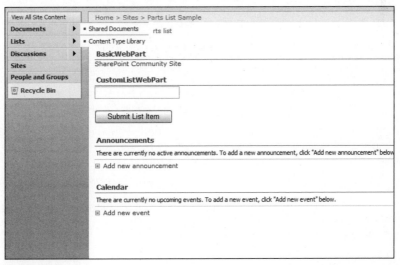

Figure 5-16

In this master page example, the `StaticDisplayLevels` and `MaximumDynamicDisplayLevels` are properties of the `Menu` control, one of the ASP.NET web controls that you can customize in SharePoint. This update to the master page can include specific code elements related to the SharePoint object model as well.

In addition to the modification of the master page for use in the SharePoint site, other ASP.NET web sites can also refer to these master pages. This enables you to leverage master pages across standard SharePoint sites that are published out of the box under the web content management features along with custom pages or applications that are published to the site through other means. To leverage the master page shown in the previous example with a custom ASP.NET web site, use the following code:

```
SPWeb customSite = SPControl.GetContextWeb(Context);
customSite.MasterUrl =
"http://moss2007rtm1/SitesDirectory/PartsList/_catalogs/masterpage/default.master";
```

The benefit of using master pages in this fashion is that it provides developers with a simple way to keep a consistent look and feel across a range of SharePoint sites and pages. For example, if you have a master page that provides a consistent set of navigation elements across a large number of sites, you can then update that master page and provide the latest changes in navigation elements to all sites at one time. Being able to do this in one place provides a more reliable and consistent approach to ensure that web site updates are applied effectively. This master page concept is based on ASP.NET, meaning that the concept applies for both your SharePoint pages hosted in this environment along with other web content targeted at ASP.NET. This provides flexibility for having custom ASP.NET web pages and code running side by side with SharePoint, both taking advantage of master page concepts to promote a consistent set of elements across those pages.

Creating a Custom Master Page in SharePoint Designer

In addition to modifying existing master pages, you can use SharePoint Designer to create a custom master page and then use that page as the basis for new pages or sites created within SharePoint. To create a new custom master page, you'll take the following steps:

1. Launch SharePoint Designer. From the File menu, select New and then click the Page option in the list. This will launch a dialog box designed to help you create new pages, sites, or SharePoint content items.

2. From the New dialog box, select Master Page in the middle of the form. Click OK.

3. Within SharePoint Designer, you'll see the basic outline for a master page in the Code view. Select the Design view option toward the bottom-left portion of the SharePoint Designer application surface.

4. In the content placeholder on the master page, type in **Test Content for Master Page**.

5. Now save the master page. You'll need to save this to the master pages gallery for one of your SharePoint sites. This directory is located at `_catalogs/masterpage/` under your site's directory. The code for the new master page is as follows:

```
<%@ Master Language="C#" %>
<%@ Register tagprefix="WebPartPages" namespace="Microsoft.SharePoint.WebPartPages"
assembly="Microsoft.SharePoint, Version=12.0.0.0, Culture=neutral,
PublicKeyToken=71e9bce111e9429c" %>
```

```
<%@ Register tagprefix="SharePoint" namespace="Microsoft.SharePoint.WebControls"
assembly="Microsoft.SharePoint, Version=12.0.0.0, Culture=neutral,
PublicKeyToken=71e9bce111e9429c" %>
<html dir="ltr">

<head runat="server">
<meta http-equiv="Content-Language" content="en-us">
<meta http-equiv="Content-Type" content="text/html; charset=utf-8">
<SharePoint:RobotsMetaTag runat="server"></SharePoint:RobotsMetaTag>
<title>Test Content for Master Page</title>
<asp:ContentPlaceHolder id="head" runat="server">
</asp:ContentPlaceHolder>
</head>

<body>

<form id="form1" runat="server">
    <WebPartPages:SPWebPartManager runat="server" id="WebPartManager">
    </WebPartPages:SPWebPartManager>
    <asp:ContentPlaceHolder id="ContentPlaceHolder1" runat="server">
            <p>Test Content for Master Page</p>
    </asp:ContentPlaceHolder>
</form>

</body>

</html>
```

Note that on its own, SharePoint Designer will prompt you to add the SharePoint Web Part Manager to allow you to modify the Web Parts on the page based on this custom master page and the SharePoint robots meta tag to enable indexing of the page for SharePoint search. To create a new page on your web site based on this master page, follow these steps:

1. Open SharePoint Designer. Select File, then click New, and select the Create From Master Page option.

2. This prompts a dialog box with the title Select Master Page. Select the radio option for Specific Master Page. Click the Browse button that is enabled by this option. From the file menu, browse back to the custom master page you created in the previous example. Remember, the URL for this master page will be `http://yoursite/subsite/_catalogs/masterpage/yourcustom .master`. Select that master page from the menu and click OK.

3. Click OK to close the Select Master Page dialog box.

4. This opens the design surface for the new SharePoint page based on your master page. The page name should be `Untitled_1.aspx`. Select the Save option to save this page back to your site. Name the page `CustomMPTest.aspx` and save it under your SharePoint site.

5. Navigate to this page in your web browser. You should see a single page with a white background. The page reads "Test Content for Master Page".

While this custom master page example is very basic, it should illustrate a few key concepts. First, this page was created based on the master page with no changes. This means that each instance of a page that is based on this master page will include a content placeholder with the text "Test Content for

Master Page". In addition, these pages will also include the Web Part Manager. One thing you'll note though is that no navigation elements or controls to edit this page within SharePoint were included. To ensure that custom SharePoint master pages are created with all of the key elements that are recommended to ensure a fully functional page within a SharePoint hosting environment, Microsoft has created a minimal master page template that you can locate at the following web address: `http://msdn2.microsoft.com/en-us/library/aa660698.aspx`.

Master pages provide a developer with the framework for creating sites that can be maintained and managed at a large scale. Traditionally, web developers wrote custom code or leveraged a content management system to create web sites. In most cases, these sites have leveraged CSS to help the developer to maintain a separation between the basic look and feel of the site and the logic behind individual pages. With master pages, a developer can separate some of the core presentation and navigational elements of the page or site. From there, you can use CSS to style the master pages themselves, offering a greater level of separation from the page style, the elements and features included within the page, and the ability to reference that master page from any number of SharePoint sites or applications.

In addition to the web parts and master pages functionality in SharePoint 2007, developers need to understand how to build some custom user interface elements in their Office applications, since SharePoint and Office are commonly used in integrated scenarios for data- and list-centric applications. In the next section, we'll look at the Office Ribbon and what is involved in customizing its appearance and functionality.

Customizing the Office Ribbon

In Office 2007, there is a new UI concept called the Office Ribbon or the Office Fluent Ribbon. The ribbon (for short) is designed to make it easier for users to find functions in Office in menus that are relevant to the context of the particular function that a user is performing. The aim of the ribbon is to provide users with easier access to more of the functionality in the Office tools. In order to understand how to incorporate custom ribbon elements into your applications, let's take a look at building a custom menu action into a Word document.

Adding elements to the ribbon in Office involves the use some custom XML elements to define the new UI elements. In Office 2007, there is a new file format based on the Open XML standard. While the implementation details of Open XML are beyond the scope of this book, we will look into Open XML a bit as we need to understand how to add custom elements into a Word file to affect our ribbon UI changes.

For more information on Open XML, check out `http://openxmldeveloper.org`.

To add the new ribbon elements, you'll need to create a sample Word document. Then you'll walk through the XML that's needed to add the custom user interface changes to the Word document. Along the way, you'll use a new tool designed to support the editing of Open XML for these types of customizations. Then you'll save the document and look at the results.

To build the sample document with a customized Ribbon, use the following steps:

1. First, create a Word document to update. Any sample Word document will work fine. Save a copy of that document so you can get back to it.

2. To create the new ribbon element, you use the `customUI` element supported through Open XML to support the ribbon extension. The XML is shown in the listing that follows.

```
<customUI xmlns="http://schemas.microsoft.com/office/2006/01/customui">
  <ribbon>
    <tabs>
      <tab idMso="TabHome">
        <group id="NewGroup" label="New Group"
               insertAfterMso="GroupClipboard">
          <button id="BlankPageButton" label="New Function Here"
                  size="large" image="Boxen"/>
        </group>
      </tab>
    </tabs>
  </ribbon>
</customUI>
```

The `<tabs>` element contains the individual tab children that you insert for the ribbon customization. As you can see in the listing, you are adding a new group to the Home tab. The group is named "New Group" and is inserted after the Clipboard group. In that group you add one new element, a button, that's labeled "New Function Here". You also specify a graphic for this button with the image attribute `Boxen`. For the sample in the book, I've created my own graphic using Paint with an orange and gray square box named `Boxen.png`. To follow along, you can create your own graphic or find another graphic to use.

3. Now that you have a defined element for this Ribbon UI change, we need to include this as an element of the document to be processed when the document is rendered upon opening. Simply put, you need to get this XML code from the listing into the document and embed your PNG graphic file so you can add a new icon to the menu. To do that, you can use a tool to help get the job done. On the `openxmldeveloper.org` site there is a tool you can download called the Microsoft Office 2007 Custom UI editor (catchy!). You use this tool to insert your XML definition of the new ribbon elements. The tool is shown in Figure 5-17.

You can use the UI editor to take your XML and to attach your `Boxen.png` image to insert into the file. Once you've pasted in the XML and attached the PNG file (remember, any image should work fine), you can save your Word `.docx` file. The result of this change to the document is shown in Figure 5-18.

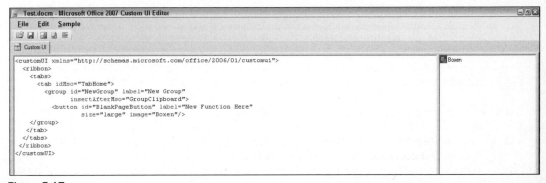

Figure 5-17

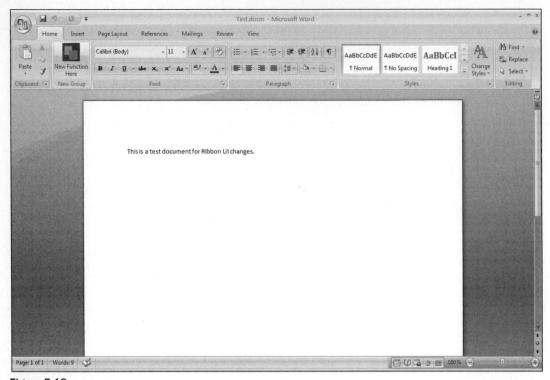

Figure 5-18

To understand what's going on, you can open up the Word document in a Zip editor tool, since the new Open XML Office file formats are based on Zip containers that contain XML files and resources to render the file. When you open the file as a `.zip` file, you'll see that the Custom UI tool created the XML file at the path `\customUI\customUI.xml` and the corresponding button image at `\customUI\images\Boxen.png`. The open file is shown in Figure 5-19.

Now that the custom UI element is created in the ribbon, you can also wire this button to existing Word macros or write custom code to invoke here. In order to do this, you'll need to further modify the `<button>` element in the `customUI` XML as shown in the following code:

```
<button id="BlankPageButton" label="New Function Here"
    size="large" image="Boxen"
    onAction="NewFunctionAction" tag="YourMacro"/>
```

The `onAction` attribute specifies a macro that you can create as part of a new VBA module. From there, you can invoke other existing macros or introduce other custom functionality. Take a look at some of the Word customization examples back in Chapter 4 and think about how that functionality could be surfaced more easily to the user with the addition of a customized ribbon user interface.

The Office Fluent Ribbon user interface is designed to make the Office applications more intuitive to the average user. The advantage to you as a developer is that this UI model is very easily extended using Open XML, as shown in the previous example. Now, you can easily modify the Office user interface and extend custom hooks into Word, Excel, or PowerPoint. For example, if you want to take some commonly used macros and expose them along with some eye-catching and intuitive icons on the main ribbon menu, you can do that very easily. If you want to take it a bit further and deploy custom code, such as a .NET assembly, that can call a web service that posts the content of your document to a custom application or repository, you can do that as well with relative ease. Any solution that can benefit from Office macros or other custom code along with an intuitive way for the user to initiate the functionality will benefit from the new ribbon UI in Office.

While the ribbon provides for some interesting user interface customization scenarios, one of the more common user interface elements that was introduced into Office 2007 is the Document Information Panel, which provides metadata support in the Office client. In the next section, we'll look at the Document Information Panel and its relationship to content in SharePoint.

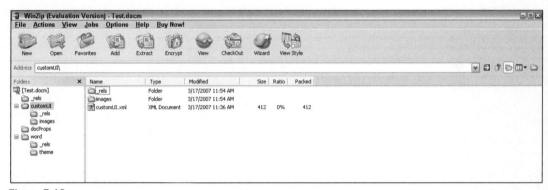

Figure 5-19

Document Information Panel

The Document Information Panel feature in Office 2007 is designed to integrate metadata collection and the binding of metadata data to internal documents and to external repositories, to make it easier to implement enterprise content management features. From a developer perspective, business processes that rely on documents to collect and process data or that produce document output for end users will likely take advantage of the Document Information Panel to help provide these features. In order to understand how the Document Information Panel can facilitate the collection of metadata, we'll create a content type in SharePoint and bind this content type to a document to enforce this metadata collection function.

In order to implement a Document Information Panel that's bound to all documents uploaded to or created within a Document Library, you first need to create a content type associated with your library. A *content type* is an extension to a SharePoint Library adding specific data elements to the list with requirements for data collection. One-to-many content types can be associated with a given library, enabling you to take a granular approach to managing how this data is collected. To create a content type and associate it with a library, take the following steps:

1. Start with a Document Library with no content types enabled. Select the Document Library Settings menu.

2. Under the General Settings header column, click the Advanced Settings link. This will launch the Document Library Advanced Settings: Content Type Library page.

3. Select the radio option to allow the management of content types, as shown in Figure 5-20. Leave the rest of these settings on the page as they are defaulted and click the OK button at the bottom of the page. This takes you to the Customize Content Type Library page.

4. Remaining on the same page, click on the Document link under the section Content Types. This launches the List Content Type: Document page.

5. Under the Settings header on the List Content Type: Document page, click the Document Information Panel Settings link. This launches the Document Information Panel Settings: Document page.

6. Select the Always Show option to ensure that the Document Information Panel is shown when documents are opened, as shown in Figure 5-21. Click OK at the bottom of this page. This will bring you back to the List Content Type: Document page.

7. Click the "Add from existing site or list columns" link at the bottom of the List Content Type: Document page to add content type elements to the document.

8. From the available columns list, select the Owner column, as shown in Figure 5-22. Click the Add button to add this to the list with the header Columns To Add. Click the OK button at the bottom of the page. This will return you to the List Content Type: Document page.

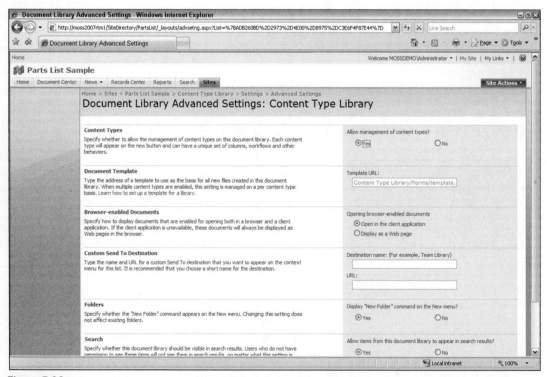

Figure 5-20

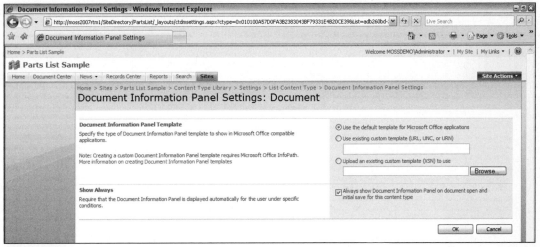

Figure 5-21

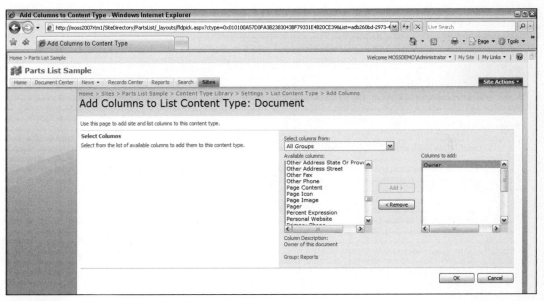

Figure 5-22

9. From the "List Content Type: Document" page, click the Title and Owner columns and mark those fields as required, as shown in Figure 5-23. To make sure they are required, click each link and select the Required option on the Change List Content Type Column: Document page. Be sure to click the OK button after making each change.

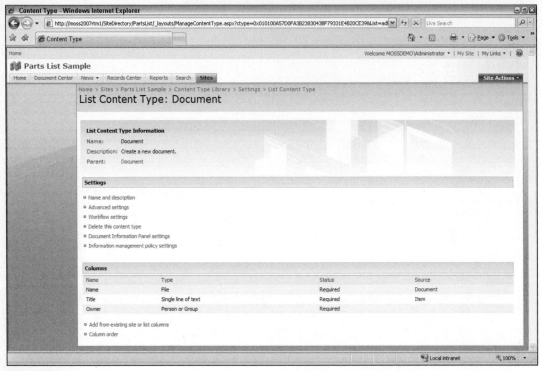

Figure 5-23

10. After making these changes, navigate back to the document library you started at in Step 1. Confirm that the Owner column has been added to your library's list of columns.

11. Create a new document in the library. When you open the document, confirm that the Document Information Panel appears with a Title and Owner field required.

12. Attempt to save the document without filling out the data to confirm that the fields are required, as shown in Figure 5-24. Since the Content Type specifies that these fields are required, Word will prompt you to complete the required fields in order to save the document.

13. To save the document, add a title and select an owner for the content and then save the document. (For this example we are using **Sample Content Type** for the title and **MOSSDEMO\ Administrator** for the owner.) Close the document and return to the original document library.

14. Confirm that the document content appears in the library as shown in Figure 5-25. When the document is saved to the library, this confirms that the content type requirements to include the title and owner fields were enforced and that the user included those fields when saving their document.

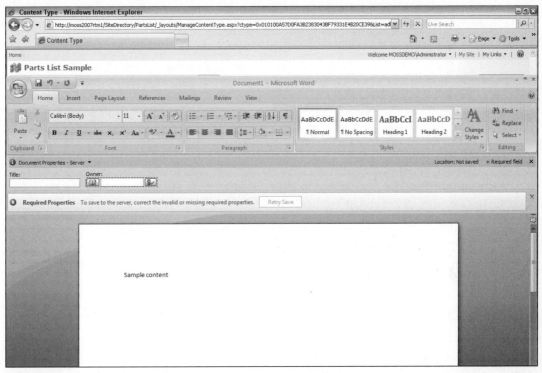

Figure 5-24

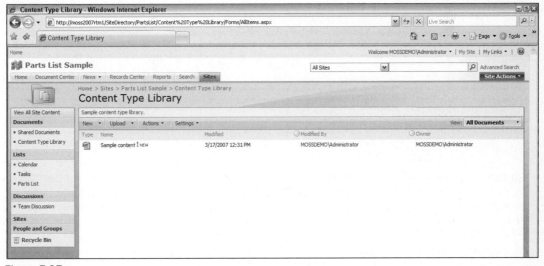

Figure 5-25

In addition to the content binding supported through the SharePoint content types and Document Information Panel fields, you can also extend the document using VBA to tie specific data elements within a document to the Document Information Panel fields as well. This ensures that you have integrity of the document metadata with the document content for specific applications.

The use of the content type in SharePoint to link document metadata from an Office document to a SharePoint library has many applications. First, this functionality can be used to make sure that documents are correctly classified according to business rules at the time they are created. If a document contains sensitive information or is associated with a specific department, that content can be included with the document, bound to the list in SharePoint, and acted upon appropriately. For instance, documents corresponding to a specific department or project can be more easily searched for within SharePoint. In addition, you can tie workflow to these documents and have rules to evaluate a specific field associated with this content type. This lets you use an Office document as an input to a business process or a workflow in a more effective manner, by ensuring that the content is correctly formed and meets the requirements as soon as it reaches the repository. Traditionally, document workflows that tried to implement this kind of functionality would often fail without the correct metadata, since users do not often understand or remember to add specific data elements to their documents. In older versions of Office, relying on users to enter data for custom properties in a document was subject to consistent failure.

Summary

This chapter on the user interface development aspects of Microsoft Office SharePoint Server 2007, Windows SharePoint Services v3, and Office 2007 has demonstrated several techniques to extend Office and SharePoint applications and to deliver new user interface elements for applications.

The SharePoint user interface services reviewed include components of the ASP.NET architecture integrated into the web content management infrastructure, including master pages. You also examined the use of Web Parts to build custom user interfaces and to leverage existing web controls in the ASP.NET programming model. In addition, you also examined the Features construct in SharePoint to determine how disparate functional elements can be combined into a group of Features that can be managed, deployed, and linked as dependencies to ensure a consistent and robust user experience across a farm of SharePoint servers.

In Office, you examined the new ribbon user interface design and the Open XML file format's support for UI extensibility in Office applications. Through an XML markup facility you can define new UI elements for an Office application and implement functionality that's tied to the new interface in a simple and straightforward manner.

Another user interface element that delivers a high level of functionality is the Document Information Panel. The process of binding content types in SharePoint to the Document Information Panel for new and existing documents uploaded to a library helps to provide a robust way to collect document metadata and to bind document data elements to a SharePoint list for report, search, and organizational purposes.

In the next chapter, you'll take a look at some of the services that SharePoint provides that serve as infrastructure for application developers to build business process, content management, and business intelligence applications. You'll also get to consider how these applications can be leveraged to deliver a mix of custom and out-of-the-box services to meet application requirements.

6

SharePoint Application Services

Microsoft's SharePoint technology architecture delivers a foundation of APIs and functionality that makes it easier for developers to build and deploy applications for a variety of scenarios. Workflow-enabled forms solutions, business intelligence and reporting aggregation tools, Excel-based publishing tools for Web reports, and records management tools are a few of the many functions that SharePoint supports out of the box, with hooks for customization and extensibility built in. These products have been packaged into Microsoft Office SharePoint Server and are primarily delivered through SharePoint's Enterprise edition product.

In order to leverage these existing application services to build new solutions, we'll look at the following components in the SharePoint application services stack:

❑ Forms services publishing, management, and administration

❑ Dashboard and reporting applications development in the Reports Center

❑ Document management solutions with the Records Center

In each of these areas, we'll look at the architectural components of the SharePoint products and technologies that enable developers to build out these services and leverage them in their applications. In addition, we'll look at some scenarios on how each of these feature areas can be leveraged and identify integration and extensibility points for custom applications. With an understanding of these features and services, you should be able to build InfoPath-designed forms that can be delivered through a browser to support rapid application development and simple integration with existing Web services and databases, to build dashboards and reporting solutions off existing Excel-based reports, and to implement records retention through SharePoint that can be integrated with external applications.

At the end of this chapter, you should have a good understanding of the out-of-the-box application services provided by Microsoft Office SharePoint Server 2007. This chapter will also serve as a preview for the business intelligence and workflow chapters, Chapters 7 and 8, which demonstrate custom business intelligence applications, Excel services, and SharePoint Designer and Visual Studio developed workflow applications.

InfoPath Forms Services

Microsoft SharePoint's InfoPath Forms Services are based on the InfoPath form's development tool. In SharePoint 2007, SharePoint provides a forms publishing capability to deliver forms in a web browser from SharePoint, enabling you to integrate workflow solutions into those forms to support business process automation.

The benefits of the InfoPath Forms Services are that a rapid application development tool like InfoPath can be leveraged to build applications that integrate workflow, can connect with external data sources such as web services and databases, and then can be quickly and easily deployed to a SharePoint environment for delivery and management purposes. While the full feature set of the InfoPath forms tool is not available in Forms Services, this can become a very quick and easy way to let power users or developers deploy small applications to support business process automation. To get a list of which features in InfoPath 2007 are and are not supported in InfoPath Forms Services, see the following article on the Microsoft Office web site: `http://office.microsoft.com/en-us/infopath/HA102105871033.aspx`.

While forms and workflow are covered in Chapters 9 and 10, we'll take a look at the Forms Services infrastructure and some of the functions available to you that enable you to turn on Forms Services features within Microsoft Office SharePoint Server and to build and expose forms that connect to external data sources.

Deploying Forms Services

SharePoint's Central Administration tool provides the management and administrative functions for most of SharePoint's key feature sets. One of the functions supported through this tool is the use of Forms Services, which includes the ability to upload new forms templates and to provision the various Forms Services features. To manage the Forms Services features, navigate to the Application Management portion of the Central Administration tool. In that tool you'll see a section for InfoPath Forms Services, as shown in Figure 6-1.

One of the options for Forms Services is the uploading of new templates. While the InfoPath client also supports forms uploads, forms that need to be provisioned across a broad infrastructure can be efficiently delivered from this Central Administration tool for Forms Services. To upload a form, select the Upload form template option. The Upload Form Template page will appear.

The form upload feature validates the form to determine if features in the form can be supported within the web-based hosting model in SharePoint and provides instructions on how to remediate any issues with forms. This validation is initiated when you click the Verify button on the Upload Form Template page.

To upload a form to InfoPath Forms Services, you select a form using the Browse button on the Upload Form Template page. Once your form is selected (and verified), you click the Upload button to upload it to SharePoint. Once a form is uploaded to SharePoint, the Upload Form Template page will refresh to indicate whether or not the form successfully uploaded. Once you click the OK button, you're directed to the Manage Form Templates page. Your new form will appear in the Manage Form Templates list as shown in Figure 6-2.

Uploaded forms can be assigned to workflows or can have a category assigned to them for other functions. Once forms are uploaded and exposed to users, the forms are rendered in a Web Part designed for InfoPath forms.

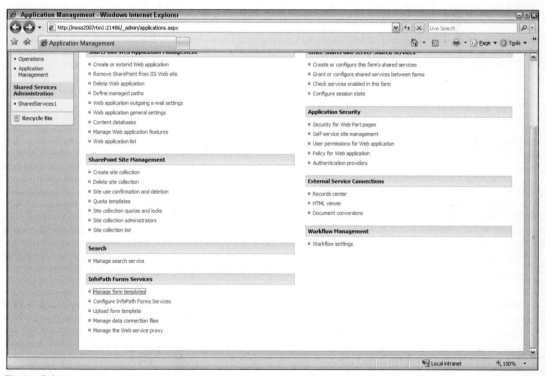

Figure 6-1

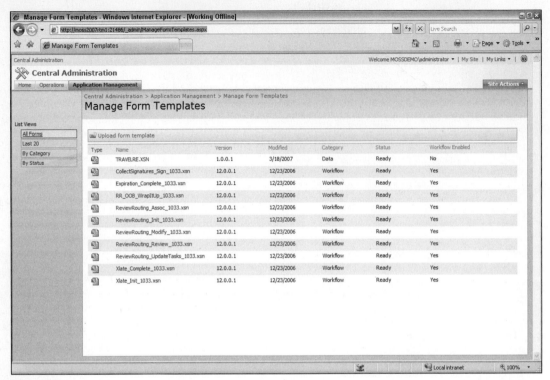

Figure 6-2

Creating a Custom Form

In order to leverage the InfoPath Forms Services in Microsoft Office SharePoint Server, you need to understand what's involved with basic form design in InfoPath. InfoPath form design is based on the use of XML schema to define the format of the form along with the operations that are supported by the various control elements of the form. InfoPath can support scripting, managed code, and direct hooks into web services or databases that are defined by InfoPath data connections.

When you are using InfoPath as a development client for web-based Forms Services, there are compatibility options that are designed specifically to enable the conversion of forms into a mode that is supported by the Forms Services browser-rendering features. For example, use of the InfoPath rich text edit control is not supported by Forms Services. A number of other features are not supported, and by following the validation procedures mentioned in the previous section you can easily determine what forms features are supported upon upload to the Forms Services library.

To understand what is involved in creating forms to deploy to the Forms Services environment, you will walk through the creation of a form template that consumes a SharePoint web service. You'll create the form, connect it to the data source, and then see a demonstration of the basics of the form design process. Later on you can then publish that form to Forms Services and see a demonstration of its functionality. Here are the steps to create a form template designed to call a SharePoint web service:

1. With the InfoPath 2007 client open, select the Design A Form Template link on the Getting Started menu. You can also select this option from the File menu. Once the Form creation wizard is launched, select the option to create a form template using a web service option. Then click OK.

Be sure to select the "Enable browser-compatible features only" check box to best ensure compatibility with the Forms Services rendering engine.

2. Next, you'll need to specify how the form works with the web service with the Data Connection Wizard. The options you can select include Receive And Submit Data, Submit Data, or Receive Data. In this case, select the Submit Data option. Click Next.

3. At this point you'll need to enter a URL for the web service. The SharePoint web services listing in Chapter 3 contains a list that you can review for your sample applications, or you can use any custom web service that you may want to test with. For this example, select the Lists service to enable the creation of a new list. The format of the URL should look something like this: `http://[servername]/[site]/[subsite]/_vti_bin/lists.asmx?WSDL`, as shown in Figure 6-3. Once you've entered this URL, click Next.

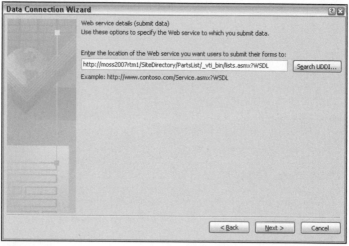

Figure 6-3

4. Now you'll need to select a method of the web service to connect the form functionality to a supported operation of the web service. The Data Connection Wizard exposes options based on the web service semantics as defined in the Web Service Definition Language (WSDL). For this example, select the AddList method. Click Next.

5. To complete the data connection definition, you need to enter a name for the data connection. The name for this sample data connection is Main Submit. Click Finish.

At this point, a new form template has been created. This form definition includes the fields defined by the web service to support data binding of these fields to the web service for submission.

Now that you have a basic template created, you can customize the form in the InfoPath client. In the case of the sample form, you can add custom text or formatting to the Form and Add controls. For this sample form, you'll need to add controls that map to the fields supported by the web service method binding to the AddList method of the SharePoint Lists service.

To further develop the sample form, select the fields from the web service and drag each one onto the body of the form where indicated. The sample should look like Figure 6-4. The data source includes all of the fields associated with the AddList method, and when dragged onto the form, each field creates a label and a corresponding text box to support data entry. Where the form labels read "Click to add a title" and "Click to add form content," you can add your own descriptive text or form controls.

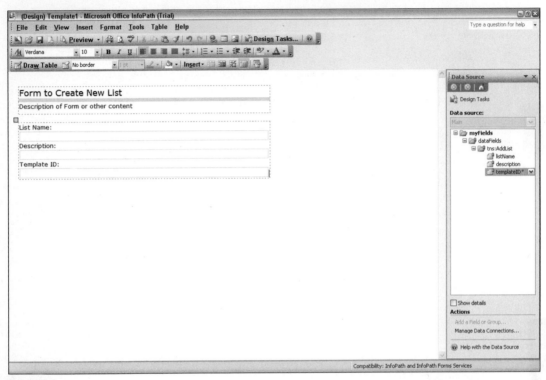

Figure 6-4

To ensure that data entered into the form is formatted correctly or includes the right default data, you can select properties for each field and leverage data validation routines or create rules associated with the data. For example, you can call a web service to populate a list or drop-down on the form and then bind those values to a particular field. This is particularly useful when constructing InfoPath applications, since you can rely on the existing logic that already resides in your web services. When binding that functionality to InfoPath, you can reduce (or eliminate entirely) the code required to expose those web services in new applications.

In this sample, you need to make sure that the Template ID associated with the AddList method is set to an appropriate integer that corresponds to the SharePoint list template IDs for standard list templates.

This will ensure that the new list that's created corresponds to the appropriate list template and supports the fields and formatting requirements of the list users. As shown in Figure 6-5, you can set the default Template ID to be 106, the 32-bit integer corresponding to the Events list type in SharePoint. Other commonly used list types include the following:

- Announcements: 104
- Contacts: 105
- Document Library: 101
- Discussion Board: 108
- Events: 106
- Form Library: 115
- Issues: 1100
- Links: 103
- Picture Library: 109
- Survey: 102
- Tasks: 107

Many other list template types are supported and can be specified through this API. In addition, there are APIs that can enumerate the different list types to support dynamic selection of the list type to be created. We'll explore this a bit further in the next section.

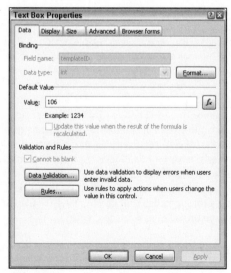

Figure 6-5

When creating this type of form template, there are many other functions that can be considered. While this form only submits data, you can configure the form to dynamically receive data and render it on the form based a web service or a database source. In addition, these data sources can be bound to form controls to support a more dynamic experience for rendering specific data elements or default values for entry. Just as you can customize and extend the function of the user interface of your applications written using the forms and supporting controls in Windows or ASP.NET pages and controls, you can also extend InfoPath scenarios. While InfoPath doesn't offer all of the same flexibility, much of the customization in InfoPath can be accomplished with much less code.

Once the form has been created, it needs to be saved as a file to preserve the formatting, data source definitions, and other form configuration elements. From there, the form needs to be published, so that it can be rendered within InfoPath Forms Services. Next, we'll walk through some extensions for this form, using the controls customization features and the addition of new data sources to create a richer form application.

Extending a Custom Form

One of the most important features in the InfoPath client with Office and the Forms Services features in SharePoint is the ability to extend the form's functionality to validate fields, connect to multiple data sources, and implement custom logic. As mentioned earlier in the chapter, the full functionality of the InfoPath 2007 client isn't available within the InfoPath Forms Services environment. This limits your ability to execute custom code when forms are saved, to use specific forms controls, and to use certain scripting techniques. At this point, you may want to refer back to the article on InfoPath 2007 and InfoPath Forms Services compatibility.

In this section, we will focus on the features that do work well in InfoPath Forms Services such as field-level validation, the addition of multiple data sources to the form for receiving and submitting data, and the customization of control behavior to help users interact with the forms application. Adding these elements to a form is essential to creating forms-based applications that are similar in functionality to what you might develop in an ASP.NET or other Windows application.

In the last section, you created a form that was bound to a web service that can create a new list for a SharePoint site. In that form, the list was bound to the Lists web service for a particular SharePoint site, and the form could submit data to the `AddList` method of the service. This is useful for creating new lists, but the functionality is a bit limited. For example, the form as created doesn't provide any features to validate the user input. The form also requires that the user know what the list name should be and supports free form entry on each field. Finally, the list template ID must be entered as an integer, and there is no guidance to the user on how to select the ID or what IDs are preferred. To improve upon the design of this form, we'll add custom formatting to the data input fields, bind the form to an existing SharePoint library containing project names that can be used for the list, and provide some options to the user when selecting the type of list they want to create.

Conditional Formatting in InfoPath Forms

One important component of a forms application is to help users understand what fields are required and to draw their attention to fields that need data input. For this sample form to create a new list, we can call attention to the fields that are not filled out and perform data validation on a field to make sure that the field is filled out correctly.

To get started with the data input fields and user interface formatting, you use the following steps:

1. Double-click the List Name text field to bring up the Text Box Properties. Click the Display tab and click the Conditional Formatting button, shown highlighted in Figure 6-6.

2. From the Conditional Formatting dialog box, click the Add button.

3. From the Conditional Format dialog box (not to be confused with Conditional Formatting), you can create conditional formatting based on the data included in a form field, in a hidden data field associated with the form (more on those later), or based on an expression or evaluation of a data element. Set the conditional formatting to be based on the listName value and select the "is blank" expression to evaluate the listName field. As shown in Figure 6-7, set the Shading to yellow. When the text field is blank, this conditional formatting option will shade the text field to yellow.

4. Click OK for each of the three displayed dialog boxes.

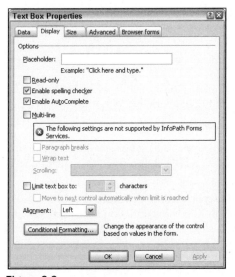

Figure 6-6

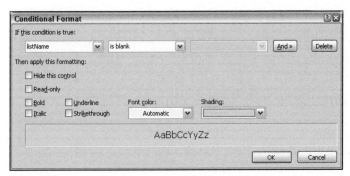

Figure 6-7

Now if you click the Preview button on the InfoPath toolbar, your form will be presented in a preview mode, and you should be able to confirm that the conditional formatting is working. For the Description text field, follow the same procedure just mentioned to have the field shaded yellow when the field is blank.

Now that these steps are completed, the InfoPath form will highlight the blank fields to bring the user's attention to the fact that the List Name and Description are not filled out. Once text is entered in those fields and the user changes the focus to another control on the form by using the Tab key or the mouse, the field will return to its default white shading.

The third field on this form is the Template ID field. This field is denoted as required based on the data bindings performed when the form was created, and a data connection was established with the Lists web service's `AddList` method. Since the template ID value is required for SharePoint to create the right type of list based on an existing template, when the form is active, either in preview mode or in normal launch mode by a user, the template ID field will include an asterisk to denote the field is required. However, for your application you may want to also make sure that the user enters the correct type of data, such as a string or an integer, to make sure that the web service functions correctly when the form is submitted. To make sure that the data entered in the form is valid, you can create a conditional formatting or data validation routine on the form.

To create a conditional formatting routine for the template ID, you use the following steps:

1. Double-click the Template ID text field to bring up the Text Box Properties. Click the Display tab and click the Conditional Formatting button as shown previously.

2. From the Conditional Formatting dialog box, click the Add button.

3. From the Conditional Format dialog box, set the conditional formatting to be based on the `templateID` value and select the "does not match pattern" expression to evaluate the `templateID` field.

4. In the third drop-down box on the Conditional Format dialog box, click the Select A Pattern option. It is the only option listed until you have created a customer pattern. This launches the Data Entry Pattern dialog box shown in Figure 6-8 to help you create a custom pattern. Note that you can also select an existing pattern to format a number such as a ZIP code or phone number.

5. For the example form, the template ID field needs a three- or four-digit integer. To create a custom pattern, which is based on regular expressions, for a three-digit number enter **\d\d\d** in the Custom pattern field. Click OK.

Figure 6-8

6. When you return to the Conditional Format dialog box, click the And button next to the third drop-down list. This lets you create another condition to check for when formatting the field. In this case, create another condition following steps 3-5, but this time use the **\d\d\d\d** custom pattern when entering the pattern matching text. Click OK.

7. As shown in the figure in the previous example, select the Shading option for the template ID field. This time, select the color red. Now, when the template ID field contains a value other than a three- or four-digit number, this conditional formatting option will shade the text field red.

8. Click OK for each of the three displayed dialog boxes to get back to the main form design surface.

Note that the same expression evaluation used for conditional formatting can be used to create a data validation rule on a field. The data validation rules are available from the Control Properties dialog box, such as the Text Box Properties shown in the previous example, in the Validation And Rules section of the Data tab.

Creating a Dynamic Lookup on a Form

Now that we've looked at customizing the form's interface to help the user with inputting data, we want to turn to creating a dynamic lookup to populate one of the fields. For many forms applications, you'll need to populate some of the fields or controls with data from a web service or another data source. In the next part of this example, we'll use a dynamic lookup from another SharePoint list to show how we can leverage existing list data and Forms Services to create new lists with other relevant pieces of information.

To get started with creating this dynamic lookup on the existing form, go through the following steps:

1. Find a list that you want to use for the data source in SharePoint. For this example, we created a simple list of projects that includes three items with the following Title fields — Construction Project, Data Center Project, and Facilities Move Project. Any list should work for this example.

2. From the InfoPath form design surface, make sure you're in the Data Source section of the Design Tasks explorer. From there, there is a Manage Data Connections link. Click this link.

3. From the Data Connections dialog box, you should see the Main Submit data connection created when the form was first created. Click the Add button to add a new data connection to this form.

4. From the Data Connection Wizard, click the radio button that reads "Create a new connection to:" and then select the radio button option to "Receive data." This will create a data connection designed to read data from another source. Click Next.

5. The next radio button option offers the option to select a data source, including an XML document, a database, a web service, or a SharePoint list or library. Select the SharePoint list or library option. Click Next.

6. Input the location of the SharePoint list you're using. The format of the list URL will be constructed in a similar fashion to the following: `http://[site]/[subsite]/Lists/[YourList]/AllItems.aspx`. This URL will be used to determine which lists associated with the site you want to select. Click Next.

7.   Select the list or library from the list box that corresponds to your sample list, as shown in Figure 6-9. Click Next.

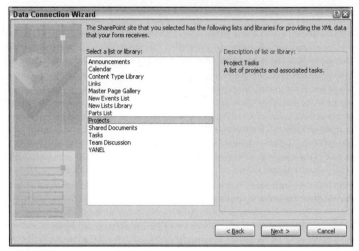

Figure 6-9

8.   Select the check boxes for the fields you want to use in your list. For this example, just select the Title field of the list and uncheck the other options. Click Next.

9.   On the next step of the wizard related to offline data caching, click Next. (This offline option provides support for offline forms but is not relevant to this example.)

10.   Enter a name for this connection. The default name will be the name of your list, but you can change it to something else. For this example, enter **Projects** and click Finish.

This new data connection is now part of the data connections associated with the form. But since the list is going to include multiple items and the form contains text boxes for data entry only, you need to convert one of the controls to a drop-down list. To do so, right-click the List Name text box, select the Change To option in the menu, and then select Drop-Down List Box, as shown in Figure 6-10.

When you double-click the Drop-Down List Box control, you are presented with the Properties dialog box for the control. On the Data tab, there is a List box entries section, as shown in Figure 6-11. You can select the option to use an external data source for the source data for the list box. Your data source created for Projects should show up in this list. If you select the data source and then click the button next to the Entries text box in that section, you can select the Title field from the list of fields or groups. Click OK to close that dialog box and OK again to close the Properties window for the control. If you select the Preview option in the InfoPath toolbar, you should see that the drop-down list box control is connected to the SharePoint list data dynamically, as shown in Figure 6-12.

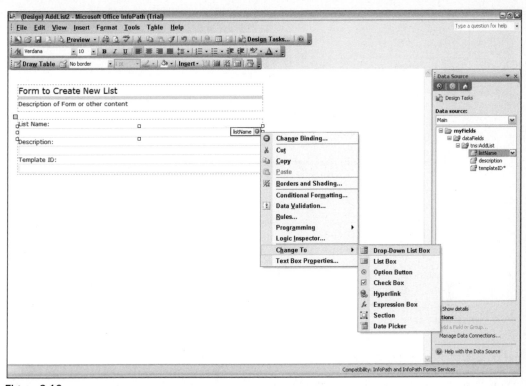

Figure 6-10

Figure 6-11

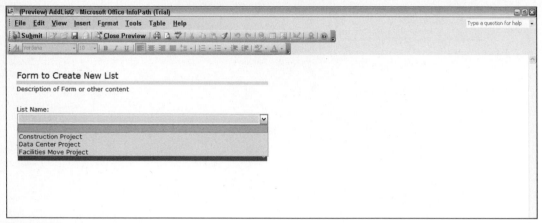

Figure 6-12

Now we've constructed a form that includes conditional formatting, data validation, and dynamic lookups to provide a more dynamic and rich application experience for the user. In the next section, we'll walk through the forms publishing and rendering process to show how this form can be delivered through a Web browser in SharePoint.

Publishing a Custom Form

Now that you've created a custom form, you need to walk through the process for publishing the form to InfoPath Forms Services and see the functionality to use the form to access SharePoint web services.

To publish the sample form template, you'll use the following steps:

1. From the File menu for the form template, select the Publish option to spawn the publishing wizard. This will initiate the process to publish the InfoPath form to SharePoint and make it accessible to the forms server rendering engine.

2. The publishing wizard includes several publishing options. Select the option to publish to a SharePoint Server with or without Forms Services. Click Next.

3. Next, you'll need to enter the URL to a site that you want to publish the forms to. This URL will need to be a site that supports permissions to all users that will need to access the form. Click Next.

4. Select the option to publish the forms to a document library and enable the forms publishing option within a Web browser, as shown in Figure 6-13. Click Next.

5. Next, pick the option to create a new document library and add the name and description for the library to host the form templates and supporting list. Click Next.

6. As shown in Figure 6-14, add columns to the list to support metadata related to the form or to expose data elements within the form contents for list rendering. Click Next.

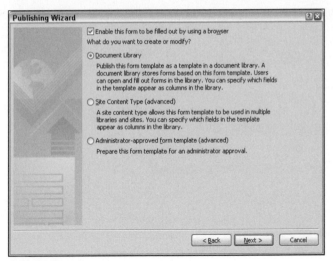

Figure 6-13

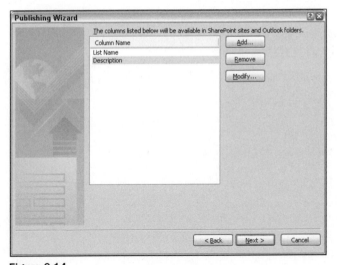

Figure 6-14

7. After the "Connecting to InfoPath Forms Service" message, the form will be published. Select the "Open this form in the browser" link to render the form, as shown in Figure 6-15. This is the form based on the underlying form template that you've created and stored as a template. The Forms Services rendering engine takes a URL parameter to link to the form template and perform the browser-based rendering features.

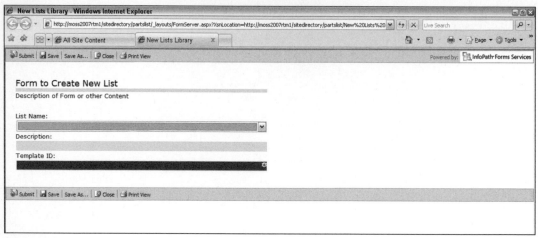

Figure 6-15

The new document library for the form template is shown in Figure 6-16, and the corresponding rich InfoPath client form is shown in Figure 6-17.

Figure 6-16

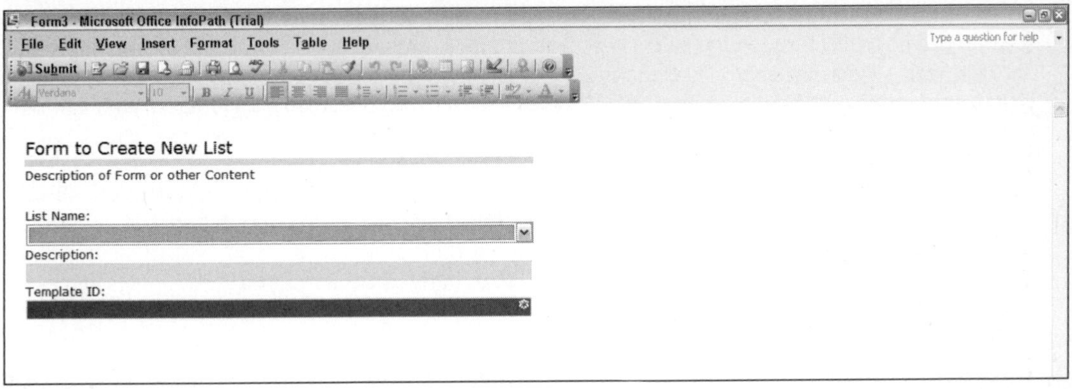

Figure 6-17

This form template can be accessed via URL through the Forms Service rendering engine or can be created as a new form from within the document library. From both of these modes, you can edit the form and save an instance of the form back to the library as well as submit the form to call the `AddList` method of the Lists web service.

When you load the web-based form through the browser, you'll note the structure of the URL that processes the request. The format of the URL is shown below:

```
http://moss2007rtm1/sitedirectory/partslist/_layouts/FormServer.aspx?XsnLocation=ht
tp://moss2007rtm1/sitedirectory/partslist/New%20Lists%20Library/forms/template.xsn&
OpenIn=browser&SaveLocation=http://moss2007rtm1/sitedirectory/partslist/New%20Lists
%20Library
```

The format of this URL is based on your server and site configuration supporting InfoPath Forms Services. You may see slightly different results, but the basic components of the URL will likely be consistent. As you can see, the URL consists of the following elements:

❑ `http://[server name]/[site]/[subsite]/_layouts/FormServer.aspx`: This portion of the URL includes the server name, site, and sub-site that contain the *FormServer.aspx* rendering engine for web-based forms.

❑ `?XsnLocation=http://[server name]/[site]/[subsite]/[document library]/forms/`
`[template].xsn&OpenIn=browser`: This portion of the URL includes the server name, site, and sub-site location that contain the forms library location. In this example, the file name with the `.xsn` extension is referenced with another parameter that specifies that the form be opened in the browser.

❑ `&SaveLocation=http://[server name]/[site]/[subsite]/[document library]`: This portion of the URL includes a parameter to specify a save location for the rendered form. In this case, this parameter causes the form to be rendered with an option to save the form to a library.

If you fill out the sample form and submit it as shown in Figure 6-18, the AddList method of the Lists web service is invoked and will create a new list item for your site based on the site location specified for the Lists web service. When you execute this form and then review the View All Site Content page for your sample site, you should see an update to the list based on your form data entered. An example of this update is shown in Figure 6-19.

Another option with Forms Services is to save the results of the form directly to a document library when the form is completed. The end user can select the Save or Save As option and will be prompted to save the document to a library, as shown in Figure 6-20. From there, a workflow can be used to trigger activities based on the submission of the new form to the document library, and logic can be created to act on the form submission based on the form's content for specific fields.

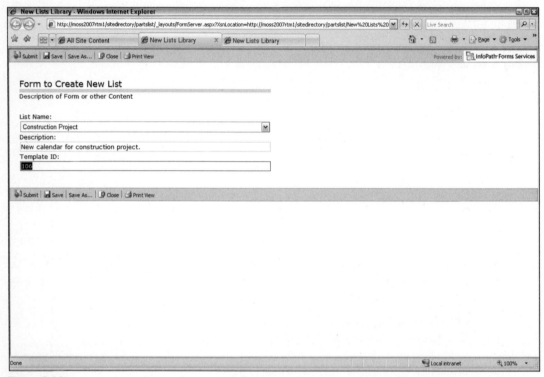

Figure 6-18

Lists			
Announcements	Use the Announcements list to post messages on the home page of your site.	1	3 months ago
Calendar	Use the Calendar list to keep informed of upcoming meetings, deadlines, and other important events.	0	3 months ago
Construction Project	New calendar for construction project.	0	1 minute ago
Links	Use the Links list for links to Web pages that your team members will find interesting or useful.	0	3 months ago

Figure 6-19

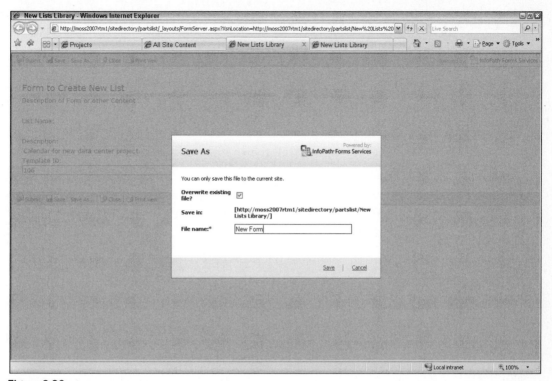

Figure 6-20

As shown in the previous sections on InfoPath Forms Services, the InfoPath client can be used to author forms that can be used in the rich client to connect to data in SharePoint and use web services to automate forms-based processes while integrating with back-end systems. Examples where the InfoPath Forms Services can be used include new project request forms for an IT department, electronic forms that can used by human resources departments to collect data from employees, or any other requirement to collect data and integrate it with SharePoint or with external web services or data sources.

The value of the InfoPath Forms Services for developers is in its ease of use in creating new front-end data collection applications that with little or no code can be integrated with existing SharePoint and web services data sources. For applications that require reach, the InfoPath Forms Services option makes it easy to create these applications in the InfoPath client and then quickly publish them to users through the browser without regard for what tools (or even browser platform) that they may be running.

In addition to the functionality of Forms Services that can support data collection and business process automation through workflow, SharePoint also delivers an extensive set of reporting and business intelligence capabilities. Chapter 7 will go into the business intelligence scenarios in more detail, but to get started, the next section takes a look at the features of the Reports Center and some of the basic tools available for publishing and managing reporting data.

Services for Reporting and Dashboards

Another key set of functionality in Microsoft Office SharePoint Server 2007 is the support for delivery of reports, particularly Excel-based reports, through the Web browser. The Report Center in SharePoint 2007 is designed to provide a solution to integrate Excel-based reports, dashboards, Key Performance Indicator (KPI) lists based on SharePoint lists, and SQL Server Analysis Services and manually entered data to develop reporting and dashboard sites that can be deployed within SharePoint and managed like other SharePoint sites.

While the SharePoint platform has delivered integration with reporting tools in the past through Web Parts for Reporting Services and a variety of Web Parts published by other business intelligence tools vendors in previous versions, SharePoint Server 2007 builds the concept of business intelligence (BI) into the core of the product with a site template for the Reports Center, the introduction of Excel Services, and integrated support for KPI dashboards. To understand how the BI features in the Report Center work, the following presents an example of publishing reports and building a dashboard with Excel-based data with an integrated KPI dashboard.

Publishing a Report

The Report Center is a site template in SharePoint that serves as an aggregation point for various reports, dashboards, and Web Parts to construct reporting solutions. To get started with the Report Center, this example starts with an Excel-based report that you can publish to the reports library. The following steps deploy the report:

1. First, create a sample Excel workbook for your report. An example of a simple worksheet with conditional formatting (heat map) is shown in Figure 6-21.

   To understand how the publishing model in Report Center works, any basic spreadsheet should provide a good starting point for demonstrating the functionality.

2. The Report Center site template includes a library for reports that will automatically render Excel workbooks through Excel Services. To upload the workbook to the Reports Library, click the Upload option to upload a single report document or multiple report documents in the Reports Library. This will launch a page that will prompt you for the location of the report file and for Version Comments. Click OK to initiate the upload.

3. Once the report document has been uploaded to the report library, you'll be prompted to provide additional metadata about the report. Fill out the form for report data as shown in Figure 6-22, and click OK.

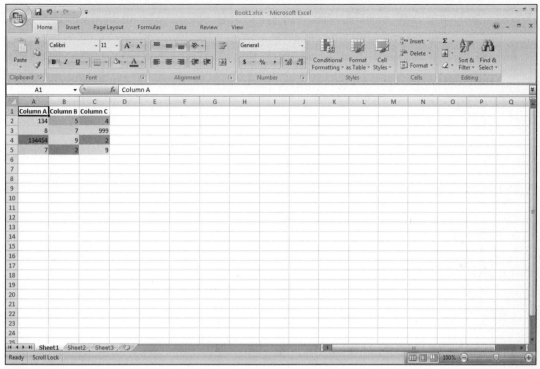

Figure 6-21

You can select a content type that will determine what metadata will be collected for that report. Remember that in addition to the default content type, such as Report, you can create custom content types and deploy them to your report libraries to collect additional data that's unique to your organization or reporting site requirements.

4. To extend a content type to include options that are relevant to your organization or site requirements, such as defaults for the Report Category element of the Report content type, click on the Document Library Settings option from the main page of the Report Library. From the resulting Customize Reports Library page, follow the Content Type link for the Report Content Type by clicking on its link.

5. On the List Content Type: Report page, follow the link for Report Category and click the "Edit column in new window" link, as shown in Figure 6-23.

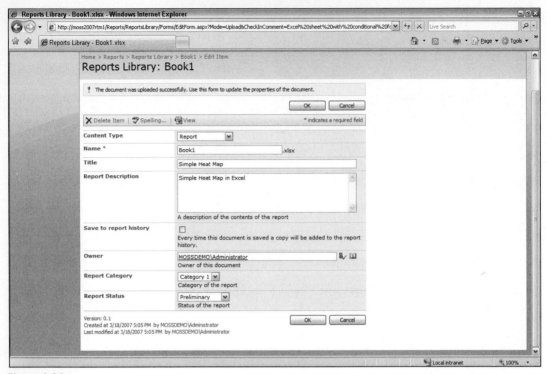

Figure 6-22

Figure 6-23

6. Add or customize choices for report category, as shown in Figure 6-24. These categories can be used to provide descriptive metadata for search, list filters in the report library, and workflows that are associated with reports. When creating your own categories, be sure to provide clear and descriptive categories that are likely to be reused.

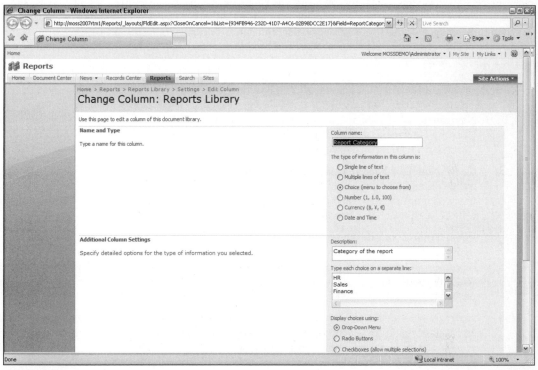

Figure 6-24

After these steps are completed, a new report is published and can be assigned to the appropriate business category, department, or other organizational element used to categorize reports. This level of categorization and the resulting ability to effectively search for, filter on, and assign workflow actions to reports based on these categories deliver a substantial portion of the benefit to developers building their reporting applications on SharePoint Server 2007.

Now that you have a basic report published, you can take a look at the functionality for dashboards and further customization of reports.

Creating Dashboards for Reports

Microsoft Office SharePoint Server 2007 delivers dashboard support for BI applications within your portal and collaboration web sites. These dashboards are supported in the Report Center and are designed to help you create reporting dashboards featuring links to reports, text-based content related to the reporting data, lists of Key Performance Indicators (KPIs), and data from Excel-based reports using the Excel Services Web Parts. This enables you as a developer to create reporting applications targeted at a department or specific organizational function with all of the key reporting elements required for business managers and decision makers to use when running the business.

From the Reports Center site, you can select the Dashboards link to see a library of dashboard sites. The Dashboards link takes you to a specific view of the Reports Library that you worked with earlier when publishing reports. There are several default filters applied to this library to allow you to view only the dashboard pages, reports, or other combinations of the data within the library when building your applications.

To create a new dashboard site containing Excel-based reports and a KPI list, follow these steps:

1. Navigate to the dashboard view of the Reports Library within the Report Center site template. Click the New menu item and select New Dashboard Page. This launches a new page called New Dashboard, where you can begin configuring the dashboard application.

2. On the New Dashboard page, select a layout for the dashboard page. The default options include a multi-column horizontal layout or two different vertical column layouts. For this example, select the two-column vertical layout. There is also an option on this page to create a KPI list automatically. Select this option to create the KPI list on the new page.

   When naming your dashboard application, think about how your reporting users will navigate to the application and whether or not they'll need to browse the dashboard library to find their reporting content. Use descriptive terms in the dashboard page name to help the users identify with the content of the dashboard reporting data. To create the new page after configuring the dashboard options, click OK.

3. Now you should see the new dashboard page in its configuration mode. By default, you'll see a KPI Web Part and two Excel Services Web Parts where can you configure connections to Excel workbooks or data connections. To assign an existing Excel workbook in the reports library to both Web Parts, you can use the workbook published earlier in the first example on report publishing to get started. Select one of the Excel Services Web Parts on the dashboard page and click the "Click here to open the tool pane" link. This will launch the configuration menu on the right side of the screen for configuring the Excel Web Part. This configuration panel is shown in Figure 6-25.

There are a number of configuration options for connecting a workbook to the Web Part in Excel Services. To get started with this dashboard, you'll need to select the location of the workbook using the Workbook field and the number of rows and columns to display. Advanced options, such as which toolbar menu commands to expose, what level of interactivity is supported — such as PivotTable support — filtering, and sorting can be configured for each workbook connection. In addition, the Audiences feature in SharePoint can be used to target this Excel reporting content to specific audiences based on group membership or organizational role. This enables you to create more dynamic dashboard applications that can be targeted to a large range of users with data delivery targeted toward specific organizational membership or role functions.

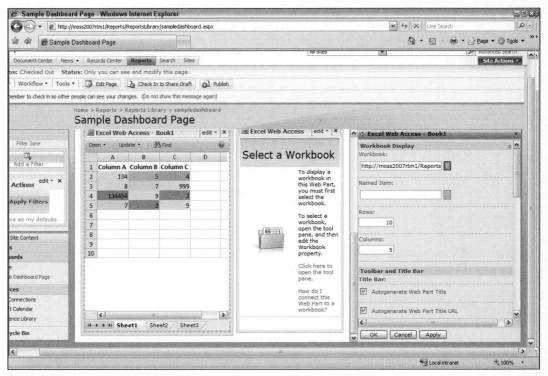

Figure 6-25

Another component of the new dashboard page that has been created is the KPI Web Part. The KPI Web Part is used to present manually entered data, list data, or data from other sources such as Analysis Services in a banded fashion based on rules and configuration settings. These settings can be used to identify which values are "better," which icons should be used as indicators, and what calculations might be displayed based on underlying KPI data. To configure the KPI Web Part on the dashboard page, go through the following steps:

1. Select a new KPI based on an existing SharePoint list. For this sample, you can use the Parts List that was demonstrated in Chapter 3. As you may remember, that list contains a Part Name field, a Part Number field, and a Price field. If you need to recreate a similar list for this example, go ahead and do so. To get started with creating a KPI based on the list, click the New button on the Key Performance Indicators Web Part, as shown in Figure 6-26. Select the "Indicator using data in SharePoint list" option. This will take you to the Dashboard KPI Definitions: New Item page.

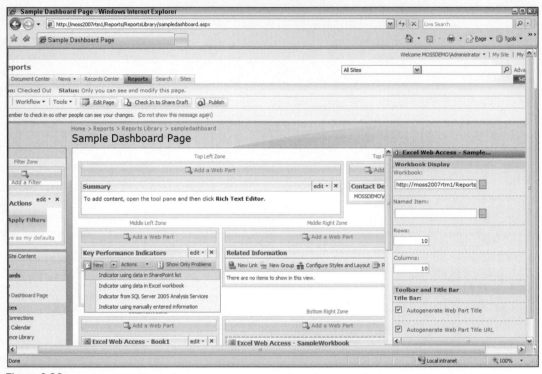

Figure 6-26

2. On the page for defining the KPI items, you'll need to provide a name, description, and comments about your indicator. Note that there are multiple sections of this page, and each subsequent configuration step for the KPI will occur on this same configuration page. Since we're using the parts list from the earlier Chapter 3 example, we can establish a KPI to determine how many of the parts are being sold for $3 or more. If $3 is the minimum price a business would need to sell at in order to stay profitable, it is important to make sure that you can provide an indicator tied to the price list to a manager responsible for profitability.

3. Also on the page for defining the KPI items, there is a section called Indicator Value. In this section, you need to enter the List URL for the SharePoint list.

4. Since we want to set our KPI to monitor profitability of parts, we need to configure the indicator value to be based on the price of the items in the parts list. There are a number of ways to do this. We can use an average calculation or track this by the percentage of the parts that are on the list that are priced at $3 or more. For this example, we'll use a percentage calculation of the items in the list for the indicator. To configure this, select the "Percentage of items in the list where" option under Value Indicator. Under that section, there are two drop-down boxes to select the column of the list and the operator to use, and one text box for a value. Select the Price column of the list, select the "is less than" operator from the drop-down list, and enter **3** in the value text box. This defines a value indicator based on the percentage of items that are less than $3 in the SharePoint list for tracking parts.

5. Finally, you need to configure the status icon for the KPI. This is where the "banding" occurs. Under the Status Icon section of the page, we'll need to configure the icon associated with this

KPI. Since we're tracking profitability and want to look for the percentage of items priced at less than $3, you'll need to use the drop-down list to set the rule that better values are lower. In our case, the lower the percentage of items sold for less than $3, the higher the profitability margin should be.

6. Green, yellow, and red indicators are defined for each band. You need to define the threshold percentage for which the green, or good, indicator should be shown and the value for which the yellow indicator should be shown. Anything greater than that percentage, and you'll see the red indicator. In this example, we use 10 and 50, meaning that less than 10 percent of the parts priced at less than $3 is our goal, and less than 50 percent of the parts priced at less than $3 causes us to get a yellow warning. Anything more than 50 percent of the parts being priced at less than $3 will cause the red indicator to show for this KPI.

7. At the end of the configuration page for this KPI there is a Details link, where you can provide a link to a more detailed document explaining how this KPI is calculated within your web site. There is also an Update Rules setting that enables you to recalculate the KPI for each viewer, which is important if your KPI is based on a dynamic list, or to manually initiate the recalculation periodically. To save this KPI, click the OK button at the bottom of the page.

8. Once configuration of the dashboard page is complete, select the highlighted Publish option, as shown in Figure 6-27. Since the dashboard publishing page is integrated into the overall Web Content Management publishing model in SharePoint, which is discussed in greater detail in Chapters 9 and 10, it requires that the page be put into an editor mode and subject to workflow and approval for all content updates if required.

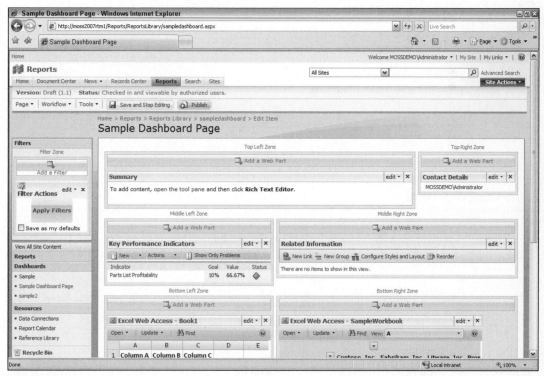

Figure 6-27

9. Once the completed dashboard is published, you'll note that the KPI list shows a percentage of 66.67 (or 2/3) of the products in the parts list are priced below $3. As a result, the red indicator is showing on this KPI. This result will be different for you based on the number of items in your list and the prices that you used. To get a sense of how the KPI calculation and banding works, you can play around with adding items to your list and adjusting the percentage thresholds as defined for the Status Icon. To go back and edit the properties for the KPI or change its configuration, you can click on the arrow next to the KPI name and select Edit Properties, as shown in Figure 6-28.

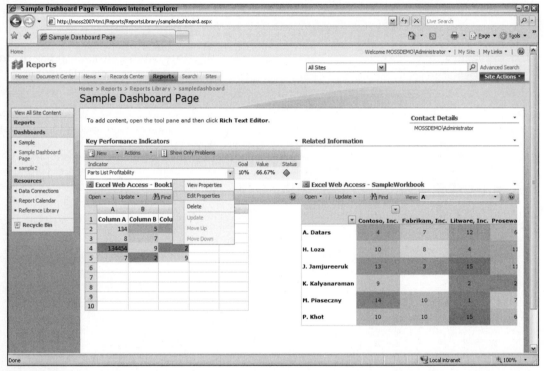

Figure 6-28

The completed dashboard application includes a KPI list and Excel-based reports based on Excel workbooks that were published to the Reports Library. Of course, report documents posted to other document libraries or Excel data connections to data sources outside of SharePoint are also supported as part of SharePoint's BI features. For more detail on KPIs, Excel Services, and configuring BI applications in SharePoint, take a look at Chapter 7, which dives deeper into the BI capabilities of Excel 2007 and SharePoint.

Note that additional Web Parts that are connected to other data sources could be incorporated to capture business data exposed through the Business Data Catalog or other list data needed. For example, if your SAP portal application includes IViews, a type of Web Part supported by SAP that exposes key business data, you can use the IViews Web Part in SharePoint 2007 to connect the SAP functionality into a dashboard that's surrounded by other SharePoint BI features. In addition, the KPI lists can be based on existing data from Analysis Services or can be defined and updated manually as required. Again, Chapter 7 goes deeper into connecting these SharePoint BI applications to Analysis Services.

In addition to the reporting functionality delivered in SharePoint, there is also extensive support for document and records management functions. In the next section, we'll look at a few of the basic features designed to support records management and advanced document management scenarios.

Records Management in SharePoint

An important feature in Microsoft Office SharePoint Server 2007 is the Records Center, supporting records management, document holds, and other policy-based management of document content. This is a key function that's required for an enterprise content management system, and can also be part of the underlying infrastructure for building document management processed on SharePoint or for integrating SharePoint content with another document management solution. To better understand the Records Center in SharePoint, the following sections take a look at a scenario involving setting a hold on an unclassified record and the tools that are available to place documents on hold and report on status.

Records Hold Scenario

Part of the Records Center functionality that's important to developers and content administrators is the ability to collect unclassified records and to route them as required to ensure proper classification, and to place holds on the documents for specific processes. Examples of holds that might be required are legal holds to support ongoing legal discovery processes, or a human resources hold request to perform an investigation of personnel involved in an incident at work. Supporting these processes in a consistent way across a variety of applications is critical to making sure that the business users of IT services can consistently manage their organization's data and classify it appropriately for governance by the appropriate policies and technical tools. To better understand how the records hold process works, let's walk through the following scenario:

1. Upload a sample document to the Unclassified Records library in the Records Center site. Select the Manage Holds menu option related to that document by clicking on the arrow to the right of the document name in the library list.

2. This action will bring up the Item Hold Status page. Select the "View all of the holds in this site" link on the left side of the page to create new Holds types. These should correspond to the types of document holds that would apply to your organization. For this example, we'll focus on the human resources hold for personnel issues or investigations and the legal holds for pending litigation.

3. Create a new Holds Item, as shown in Figure 6-29. An example would include the HR investigation hold or general legal hold. Again, remember that these holds will be specific to an organization's requirements for document retention and hold policies. After entering the new hold item, click OK at the bottom of the page.

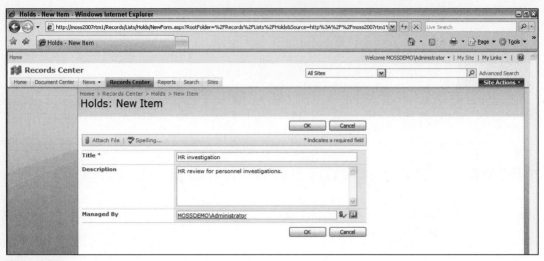

Figure 6-29

4. Navigate back to the Item Hold Status page in your web browser. On the Item Hold Status page, select the "Add to a hold" radio button and select the Hold type from the drop-down menu, as shown in Figure 6-30. Click the Save button to save the item.

5. From the main Records Center site, select the Holds link under Lists. Under Holds, select the holds type assigned to your item (for example, HR investigation). This will open the Holds: HR Investigation page.

6. On the Holds: HR Investigation page, select the "View hold report..." option, as shown in Figure 6-31.

Figure 6-30

7. A report summary is produced, as shown in Figure 6-32. The individual documents on hold are listed on the Items On Hold tab of the spreadsheet, as shown in Figure 6-33.

Note that this report is dependent on the proper administration and configuration of your Records Center functions on your Microsoft Office SharePoint Server 2007 site. The Holds Processing and Reporting job must be scheduled and running. To verify the status, check the SharePoint 3.0 Central Administration Console on your server.

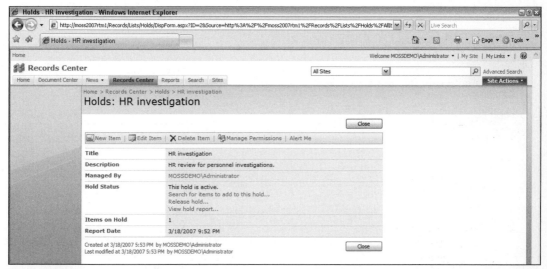

Figure 6-31

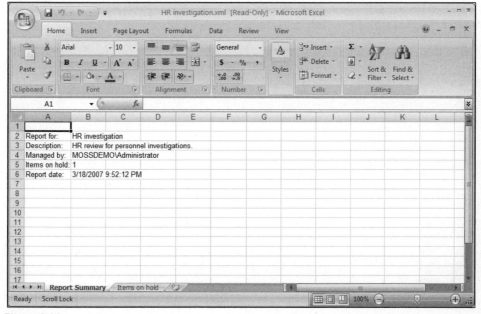

Figure 6-32

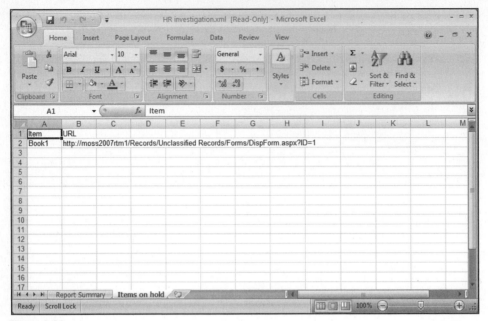

Figure 6-33

In this records management scenario, you have options to assign workflow to items on hold and to create other processes that are triggered by the hold status of the item. The Records Center also provides a repository web service, to make it easy to pass documents from third-party applications and to support integration with other document management tools or records management applications. This web service is based on underlying SharePoint object model features that are implemented to support the SPFile class and the handling of file storage within SharePoint.

Automation of Records Retention

In order to support the automation of records management and retention features in SharePoint from non-SharePoint applications, you can use the existing SharePoint object model support and web services extensions to assign records to be retained and to automatically create the copies of those records, along with the associated metadata, for retention purposes.

While the SharePoint user interface enables a user to send a document to the Records Center for purposes of retention, this level of user interaction is not often reliable enough to build into a business process. However, with the use of content types and workflows that are automatically invoked when documents are stored in SharePoint document libraries, an invocation to a web service or SharePoint API that supports a records retention feature can bridge the gap between the manual process and the information retention needs of your organization.

To understand how to use the SharePoint APIs to automate record retention and handling procedures, the following example walks through a sample implementation of a custom application to create official copies of an existing file stored in a document library. To build the sample application, you'll go through the following steps to write the code shown in the listing that follows:

1. Create a Windows Forms application with a button and label. The button will be used to assign a record to the SharePoint records center, and the label will be used to capture the status of the records retention procedure.

2. In the code, create an instance of the SPSite class for the site collection and then create an instance of SPWeb to return the target site within the collection.

3. The SPWeb class supports a GetFile method that takes the file location and name as an input parameter. This is used to create an instance of the SPFile class.

4. The SPFile class supports the SendToOfficialFile method. This method sends a copy of the file and associated data to the SharePoint Records Center.

```
//String parameters to enable site and list access
const string siteUrl = "http://moss2007rtm1/";
const string siteName = "SiteDirectory/PartsList/";
const string sourceFileName = "Shared%20Documents/TestUpload.doc";

OfficialFileResult result;
string fileResult;

//Freeze UI
Cursor = Cursors.WaitCursor;
buttonAssignRecord.Enabled = false;

//Return a site collection using the SPSite constructor providing the
site URL
SPSite siteCollection = new SPSite(siteUrl);

//Return the target web site based on site name
SPWeb site = siteCollection.AllWebs[siteName];

//Access file from within site based on path
SPFile file = site.GetFile(sourceFileName);

result = file.SendToOfficialFile(out fileResult);

//UI clean-up
labelStatus.Text = "Records Hold Status: " + result;
Cursor = Cursors.Default;
buttonAssignRecord.Enabled = true;
```

The OfficialFileResult enumeration provides a set of statuses to indicate the results of the file retention action. These status results include the following:

- ❏ FileCheckedOut
- ❏ FileRejected
- ❏ InvalidConfiguration

- ❑ MoreInformation
- ❑ NotFound
- ❑ Success
- ❑ UnknownError

In the case where there is a configuration problem with the Records Center, the file can't be found, or there is some other error, you can implement a notification to inform a records retention administrator of the issue. For more common errors such as the file being checked out, you can maintain a list of files to be sent to the Records Repository that are checked out and then have a service monitor the list that can either send the file to the Records Repository on check-in or force an undo of the checkout to gain control over the file in the retention system.

When a file is sent to the SharePoint Records Center Repository via the object model, a copy of the file is created and sent to the repository. The original file stays in its location and isn't directly linked to the file that's maintained in the Records Center site. The copy of the file that is retained in the repository does include the content type metadata for the file, audit history, and original location as metadata.

Summary

This chapter on the architecture of Microsoft Office SharePoint Server 2007 and the supporting applications services was designed to cover a broad range of services to support business process automation, dashboards, and reporting and document management.

The SharePoint application services reviewed include InfoPath-based forms solutions, business intelligence and reporting tools, Excel-based publishing tools for web reports, and records management tools for developers. The combination of these services with SharePoint's existing object model and support for web content management and publishing provide a platform that enables a range of applications, some requiring very little or no code. These tools can be used to develop simple business process automation, reporting, and document retention solutions.

For example, with a single form that connects to a web service or database, you can create an application and host it within a browser in the SharePoint environment. The functionally equivalent application written in ASP.NET would require administrative access to a Windows Server to deploy the solution and would require more code to implement. In the case of the reporting solutions, an existing Excel workbook with data source connections can be delivered as a web-based report and support the alerts, security features, and other content management capabilities that are inherent in the SharePoint environment. This data can also be combined with KPI scorecards and other content to provide more context for the reports, helping business users to better understand the data and make better decisions. In the case of the records management scenarios, a single invocation of the SharePoint object model can be used to automate the retention for an existing document without user intervention. More advanced scenarios with workflow can be implemented to support routing and retention of content based on content types in SharePoint.

The next chapter takes a deeper look at the business intelligence and reporting features in SharePoint to understand what level of customization is possible for reports, and how Excel Services can be used to leverage existing reports with connectivity to external data sources to deliver a reporting portal, along with individual KPI dashboards and reporting components to integrate into existing SharePoint sites.

7

Building Business Intelligence Applications with SharePoint and Office

As businesses collect data at a constantly increasing rate, the problem that most face isn't not getting enough data, but making the best use of what is available. In the past, business intelligence tools have been targeted at a narrow range of analyst users, but in a modern business, people in almost every role need these kinds of tools to analyze data and make good decisions.

A business intelligence (BI) platform is made up of several components, as shown in Figure 7-1.

Extract Transform Load (ETL) systems take data from source *Line-of-Business systems* to standardize and consolidate that data into a BI data store, which can be any combination of a *data warehouse*, *departmental data mart*, or *OLAP (Online Analytical Processing) cube*. In these reporting-focused data stores, the data is restructured to make reporting easier and more consistent. Key business calculations are performed and related data from disparate line-of-business systems are combined to enable business users to answer questions that span more than one application:

- ❑ A single enterprise *data warehouse* store can be an elusive goal due to the complexity and cost involved in taking on every perspective on the data.

- ❑ Therefore, it is common to take a more focused approach with *departmental data marts* targeted at specific types of questions and often a smaller subset of source systems.

- ❑ *OLAP cubes* are complementary storage technology designed to simplify and accelerate queries across large data sets.

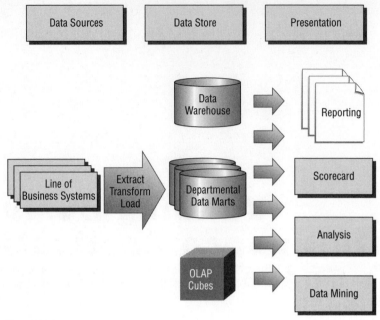

Figure 7-1

The *data sources* and *data stores* of a BI system are invisible to end users. For them, the *Presentation Services* are the face of the BI system and provide the real value to its use:

❑   *Reports* are the traditional vehicle, either printed or through electronic distribution. Electronic reports can be generated in various formats, including Excel spreadsheets that can then be further manipulated.

❑   *Scorecards* are an increasingly common way to present performance data with trends, especially when aligned with goals.

❑   *Analysis* can be done through several clients, including Microsoft Excel's built-in ability to consume OLAP data. This chapter shows this enhanced capability in Excel 2007.

❑   Finally, *data mining* is a computer-driven analysis of data. It is designed to find trends and correlations in the data that may be impossible for users to find manually.

This chapter covers the business intelligence features of Microsoft Office SharePoint Server and Microsoft Office Excel, focusing specifically on the following:

❑   Data visualization in Excel that makes it easier to make sense of large amounts of data

❑   Relational and dimensional data access improvements in Excel that make it easier to create and maintain data-connected spreadsheets

❑ Online spreadsheet publishing with Excel Services and Excel Web Access

❑ Automating and controlling server-hosted spreadsheets in custom applications to reuse business logic and automate processes

Excel Business Intelligence Features

Business intelligence was a major focus area for Microsoft in designing the Office 2007 wave of applications. Excel, perhaps the most popular data analysis tool on the market, has received major enhancements in the way data connections are managed and in how data is presented. Even limits on Excel's already formidable maximum spreadsheet size were expanded in response to a broadening of how users view and use Excel, which is now being used for more general-purpose data analysis and list management.

Data Visualization in Excel 2007

Business users are frequently faced with an overwhelming mass of numbers, with the task of finding the patterns and value in these numbers. Excel has helped that effort with increasingly powerful graphing tools and the capability to set up rules to highlight data values based on user-specified criteria. Excel 2007 has advanced features in this area by supporting several new or expanded capabilities in the area of conditional formatting, which we'll explore here.

Conditional formatting makes it easy to identify the trends and outliers in data quickly without reformatting a spreadsheet. By selecting a range of cells and performing one click to the Conditional Formatting menu in the ribbon (see Figure 7-2), you can choose a variety of predefined visualization styles.

In this example, we're working with a sales report from the Adventure Works sample database, but we need to analyze that data. You can show data bars based on the range of values in the range (see Figure 7-3) with the Data Bars feature, enabling you to identify patterns in the data.

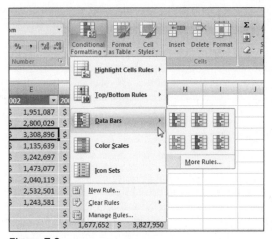

Figure 7-2

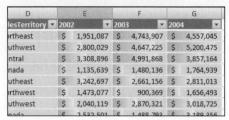

Figure 7-3

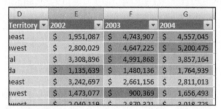

Figure 7-4

These bars are adjusted automatically based on the values in the selected range. Color Scales work the same way, but uses color scales instead to highlight values (see Figure 7-4).

Another useful conditional formatting feature is the ability to highlight cells based on top/bottom rules. This formatting rule makes it easy to spot outliers and mistakes. This rule supports top/bottom x, top/bottom x percent, and variations from the average. While manual conditional formatting rules were supported in prior versions of Excel, the new conditional formatting capabilities in Excel 2007 make this feature much faster and more approachable. In prior versions, this kind of formatting would have required creating custom formatting conditions or even VBA code.

Conditional formatting is useful on its own, but combined with some of the other data capabilities we'll explore with Excel 2007, this feature becomes even more powerful.

Excel also includes an enhanced graphics engine. Charts and graphs now support rich effects such as glossy reflections, realistic shadows, and 3D lighting. Moreover, complementing the Office emphasis on making better-looking documents, these features are enabled automatically, and many prebuilt styles are available in the product to quickly create cohesive and attractive layouts.

Accessing Data Sources

By providing an easy way to analyze data stored in back-end systems, Excel provides tremendous capabilities to business users. The challenge is that setting up these connections has typically been out of reach for most nontechnical users who don't understand database drivers, OLEDB, or connection strings. Here, we'll look at how Excel can connect to both relational and multidimensional data sources, as well as how Microsoft Office SharePoint Server (MOSS) provides the capability to centralize the management of these connections so users have a pre-defined catalog of data sources.

Relational Data Sources

Excel 2007 adds an improved database connection management system that makes it easier to understand what data sources a workbook uses. It also makes it easy to manage those connections. We can start off with a simple situation in which we need to pull a table from a SQL Server database.

We'll start by using Excel's Data ribbon, specifically the Get External Data and Connections panes shown in Figure 7-5.

Figure 7-5

The first three options (From Access, From Web, and From Text) provide a quick interface for importing data from systems typically used by end users.

The web interface is especially useful because it automates *web scraping,* pulling data from a table on a web site and enabling users to refresh that data at any time. Excel handles generating the request and parsing the HTML to get to the correct data.

For developers, though, the most important capabilities are in the From Other Sources button, which exposes options for connections to various data types, including SQL Server, Analysis Services, and XML. Figure 7-6 shows the list.

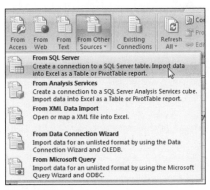

Figure 7-6

For our simple relational example, we'll use a connection to a SQL Server database:

1. After specifying a server name and security parameters, you are presented with the dialog used to select the source database and table (see Figure 7-7).

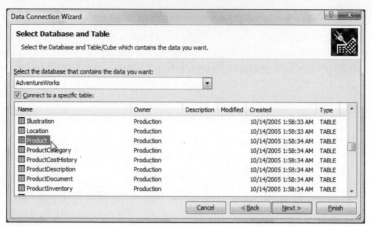

Figure 7-7

In this case, we'll select a simple table from our database, but views are also available in this dialog. This makes it possible to use a pre-defined SQL query that joins in reference tables or filters data as a data source.

2. After this, Excel gives us the ability to save this connection for future reuse in other queries, as well as keep it around for refreshing that Excel workbook data in the future (see Figure 7-8). It also enables sharing this connection with other users, but we'll explore that more later in this chapter. For now, we'll just save this connection with the default name, provide a simple description, and finish.

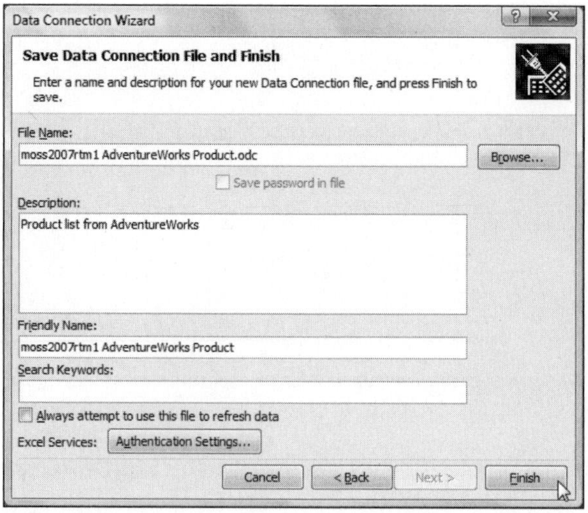

Figure 7-8

3. The last step in the process is the Import Data dialog (see Figure 7-9), which prompts for the target of this new data, either in Table, PivotTable Report, or PivotChart and PivotTable report formats.

This time, we'll choose a table as a target and put the data in an existing sheet in the workbook. We now have the data in an Excel table, which provides some rich interactivity such as sorting, filtering, and formatted layout (see Figure 7-10).

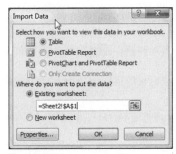

Figure 7-9

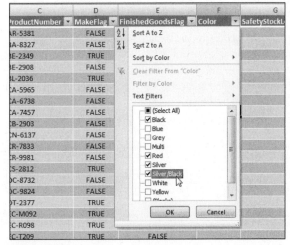

Figure 7-10

With this simple spreadsheet, any user familiar with Excel now has the capability to work with this relational data, using the power of the local workstation to set filters, sort data, and even produce graphs or use the data visualization tools covered earlier in this chapter. This data is also refreshable without losing a user's customizations, so the refreshed view will retain filters, sorting, and conditional formatting.

A key enabling feature for this user scenario in Excel 2007 is the ability to support over 1 million rows in a worksheet, instead of the 64,000 limit in previous versions. This means that users can download larger data sets and manipulate them with less risk of hitting this hard limitation.

Centralizing Connection Management

In the previous section, we created a connection and a worksheet with a refreshable data set, but now we have a spreadsheet with its own embedded connection settings, including server name and security parameters. If this database were to move or somehow change, this connection would break and each user of this spreadsheet and any derivatives would need help reconfiguring it with the new connection settings. This isn't ideal, especially in the vision of service-oriented architecture whereby location and implementation details are abstracted from users and applications.

MOSS 2007 has the ability to centrally store and manage database connections for an enterprise. This gives administrators the ability to publish a set of connections for broad use, with the capability to update these connection settings as changes occur. The whole process is seamless to users — they work with spreadsheets and refresh them as before, but now Excel in the background keeps its connections up-to-date with the published version in MOSS. Instead of storing the connection itself, the spreadsheet holds a URL reference to the master copy of that connection (as well as a cached copy of that connection in case the user is working offline).

The other benefit of centralized management of connections is that it gives end users a starting point for new worksheets when they need to access a corporate system. Instead of treating every data request as a one-off, most user requests can be directed to a catalog of prebuilt and tested connections. Even though in these examples we're building connections with Excel, these same connections can be used by the other Office applications as well, including Microsoft Access.

We'll take the connection we created in our previous example and publish it to our enterprise so it can be used more broadly and managed centrally. Before we begin, we need to have a Data Connection Library on our MOSS site. These are automatically created as part of the Report Center template but can also be manually created elsewhere. A MOSS site can have any number of Data Connection Libraries, each with its own permissions and location in the site.

1. To start, we'll find the connection definition file we created in the Data Connection Wizard. By default, Excel creates a My Data Sources folder under the current user's Documents folder to store database connections. In that folder, we find the .ODC file (Office Data Connection) for our prior example (see Figure 7-11). The .ODC file contains general metadata and an XML section containing the connection information, including a connection string and a query definition for this data set.

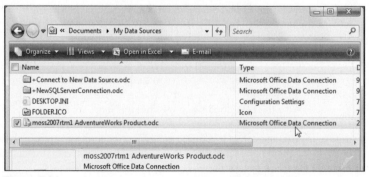

Figure 7-11

2. Now that we have located our data connection file, we'll go to the MOSS Data Connection Library and upload this connection file just as we would upload any other file to SharePoint, using the Upload button in the library toolbar (shown in Figure 7-12).

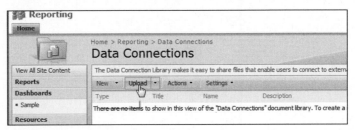

Figure 7-12

3. After going through the standard SharePoint file upload page and choosing the ODC file we found previously, we are presented with the properties dialog shown in Figure 7-13.

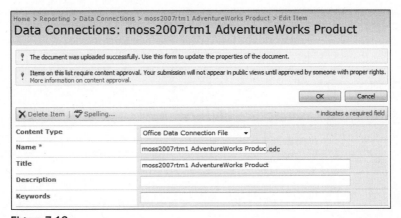

Figure 7-13

Notice the warnings in this dialog regarding approval. This is part of the administrative capabilities of shared data connections in MOSS. A broad range of users might have the capability to upload connections, but only specified users can approve these connections for broad use by other consumers of this shared library. Again, as with any other MOSS library, this approval is handled in the same way that an appropriate user would approve a document in a document library.

4. In this case, click OK on the properties dialog without any changes and open the item drop-down menu for the new connection to approve it, as shown in Figure 7-14.

Figure 7-14

This opens a form on which the user can approve or reject the connection, or just leave it in a pending state if desired.

Once approved, the connection is available for users. This works the same way as the document approval process, so the connection can be revised, but users will see the original version of the connection until the revised connection is approved.

Publishing the Data Connection Library

To make it easier to discover enterprise resources, the Office applications support a centrally defined list of resources, including shared connection libraries. In addition to storing the connection libraries themselves, Microsoft Office SharePoint Server is also used to keep a list of these published links.

Here are the steps for setting up a published connection library:

1. Start in Shared Services Administration for the appropriate site (from Central Administration, click the Shared Services Administration link in the left navigation panel and choose the appropriate shared services instance).

2. In the User Profiles and My Sites section, click the link for Published links to Office client applications.

3. Click the New toolbar button to access the form to create a new item entry (see Figure 7-15). Complete the URL (use the URL of the data connection library you created earlier), specify a user-friendly name for the library, and select the Data Connection Library type.

Now that this link is published to MOSS's shared services, Office clients associated with these shared services will receive these links during a regularly scheduled sync. Users are associated with a MOSS shared services instance when they select a default My Site. Once this is done, that Office client is associated with the shared services associated with that My Site store. This is done through a scheduled sync process so it may take a few hours for all clients to be updated with the new links.

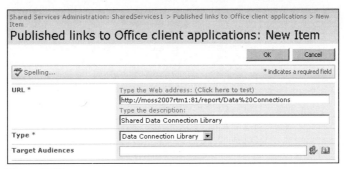

Figure 7-15

With the increasing tightness of integration between the Office client and server applications, Shared Services are a way to keep the Office client application configuration in sync with the server without going through a manual configuration process. Some organizations may push the My Sites (and therefore Shared Services) settings to workstations using software and patch distribution systems.

Analysis Services Integration

Reporting from relational data sources is important, but SQL Server includes another way to structure and query data — SQL Server Analysis Services. Analysis Services is an Online Analytical Processing (OLAP) engine and it provides some advantages over relational reporting:

❑ **Support for business-level description of data:** When creating an Analysis Services database, technical details such as table relationships, field names, and calculation formulas are defined and stored along with the data. Users see an abstracted business-level view of the data.

❑ **Efficient queries over large data sets:** Whereas relational databases often start to experience performance problems when summarizing large data sets, Analysis Services has been designed from the ground up to provide quick results on large summary queries.

❑ **Granular security controls:** Analysis Services can secure data in many different ways, and includes the capability to restrict the level of detail a user can employ on a query. This capability would need to be custom built in most relational sources.

Microsoft Excel has been a client for Analysis Services for several versions through an extension of Excel's Pivot Table and Pivot Chart capabilities. The 2007 release adds a streamlined pivot user interface, support for many new SQL Server Analysis Services 2005 features, and server-side capabilities through Excel Services (which are covered in more detail later in this chapter).

Data in an Analysis Services database is stored in a cube. A *cube* is a multidimensional data structure that contains summarized data (*measures*), broken out by a number of criteria (*dimensions*). For example, a sales analysis cube may have measures for total dollar sales, unit count, and total discount dollars. Dimensions for this sales cube may include sale date, product, store, and salesperson.

Connecting to Analysis Services from Excel is similar to connecting to a relational data store, but this time you choose the From Analysis Services source from the From Other Sources menu, as shown in Figure 7-16.

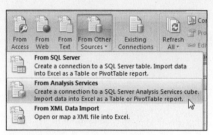

Figure 7-16

Analysis Services will prompt with several dialogs for the target server and cube and then display the same finish and save dialog shown earlier at the end of the relational data connection wizard. In addition, as with the relational connection, you are prompted to specify how this data should be added to the spreadsheet, as shown in Figure 7-17.

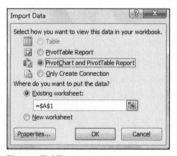

Figure 7-17

For an Analysis Services connection, there is no way to simply show the multidimensional Analysis Services data in a simple table, so notice that option is disabled. However, you do have the option of creating just a connection to be used later.

Excel's user interface for pivot tables works in two primary modes: PivotTable mode, a worksheet format with dynamic features, and PivotChart, an interactive graphical format. Figure 7-18 shows Excel with a PivotTable and PivotChart bound to the same data source (in this case, the sample AdventureWorks Analysis Services cube).

The following UI elements are exposed when working with PivotTables and PivotCharts:

Element	Function
Pivot Table Field List task pane	Lists the measures and dimensions available in the PivotTable and the different sections that these dimensions and measures can use (Report Filter, Column and Row Labels, Values)

Element	Function
PivotTable Options ribbon	Commands to manipulate the PivotTable data and the formatting of the selected PivotChart
PivotTable Design ribbon	Changes the look of the PivotTable with either a custom layout or one selected from a set of predesigned templates

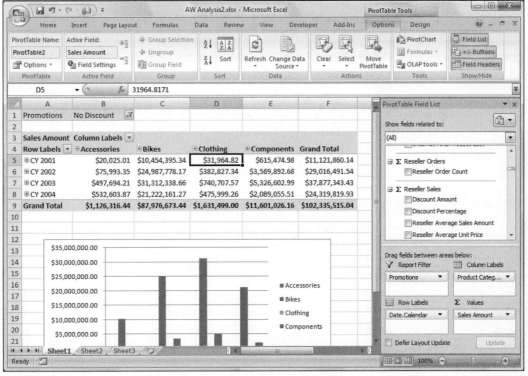

Figure 7-18

The PivotChart also shows a floating filter pane for making changes to the filter, category, or series data.

PivotTable or PivotChart reports are created by dragging data fields to one of the four data areas in the Field List pane. The Values area defines the actual data values that will appear in the body of the report. Only value fields are allowed in this list. The other three areas (Report Filter, Column Labels, and Row Labels) only accept dimension values, which are ways the data can be grouped or filtered. Each grouping can also expose its own hierarchy — date is a common example, whereby the hierarchy typically may appear as follows: year ⇨ quarter ⇨ month ⇨ day. Excel exposes this hierarchy as a nested set of expanding rows or columns, as shown in Figure 7-19.

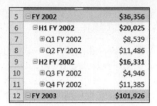

Figure 7-19

Excel also supports putting multiple fields in a single data area. These also appear as nested expanding rows or columns in the PivotTable or PivotChart.

XML for Analysis (XMLA) Support

SQL Server Analysis Services 2005 uses XML for Analysis (XMLA) as its native protocol for queries, data definition, and management. Anytime Excel is working with SQL Server 2005 Analysis Services, it is using this standardized XML interface. This makes it easier to write a custom application that can consume these web services either directly at the web service level or using the ADOMD.NET object model Microsoft provides for accessing XMLA data sources in .NET applications.

ADOMD.NET is a standalone download available from Microsoft. You can find it at www.microsoft.com/ downloads/details.aspx?FamilyID=790D631B-BFF9-4F4A-B648-E9209E6AC8AD&displaylang=en.

Offline Cube Support

Having access to data when connected to a corporate network is great, but many users work away from the office with limited or no connectivity. This is why Excel and Analysis Services support offline cubes, which are local copies of Analysis Services cubes that can be a subset of that cube or sometimes the entire cube.

Creating an offline cube in Excel is done through these steps:

1. Start with an Excel workbook with a Pivot Report or PivotChart pointing to a SQL Server Analysis Services cube. Ensure that macros and interactive content is enabled for this workbook (if it isn't, you will see a warning bar at the top of the workspace, with an Options button to enable it).

2. With the Pivot Report or Chart in focus, go to the Pivot Table Tools ⇨ Options ribbon, click the OLAP Tools drop-down menu button and select Offline OLAP.

3. Click the Create Offline Data File button, which will start the Create Cube File wizard.

4. Click Next to continue through the wizard introduction page.

5. The next page lists each dimension (a way of filtering and grouping data) of the cube in a checklist. Check any dimensions that should be built into the cube. Note that any dimensions used in the current view on the cube will be checked by default. Click Next to continue.

6. The following page looks similar to the last, but this time you are selecting members within the selected dimensions, including the measures you want to build into the offline cube. Make the desired selections and click Next to continue.

7. The final step prompts for a local filename for this offline cube. It may default to the desktop, but this can be changed by typing in a new path or using the Browse button. Click Finish when satisfied with the offline cube filename.

8. Excel will now build the offline cube. This may take a few minutes, depending on the size on the cube, the selections made in the Offline Cube Wizard, and the speed of the network connection.

9. When the wizard is finished, it returns to the Offline OLAP dialog, which now has the Offline Data option selected. Clicking OK returns to the Excel workbook, but now the pivot task pane only shows the dimensions and members available in the offline cube. As long as the supporting data is available in the cube, the user can now modify the existing cube view without a connection to the server.

If the users later decide that they want an additional dimension or other element that is not in the cube, going to the same menus and repeating the same process again enables them to change the offline cube parameters.

Key Performance Indicators (KPIs)

MOSS has built-in support for creating a dashboard, pulling data together from many different systems. Quite often, this data ends up being rolled up to Key Performance Indicators, or KPIs. A KPI is a way to measure the actual value of something against a goal. Sometimes these KPIs roll up into broader KPIs that are also tracked against a goal. Figure 7-20 shows the default sample KPI list with a few additional PKIs, one with a hierarchy.

Figure 7-20

KPI data can come from many different places. MOSS's built-in KPI feature supports data from the following:

❑ SharePoint lists

❑ Excel workbooks

❑ SQL Server Analysis Services

❑ A manually edited list

We made our simple KPI list in Figure 7-20 by building a site with the Report Center template, but a KPI list can be created elsewhere by creating a list of that type. Now we will add another KPI from Analysis Services to this list:

1. Make sure you are drilled into the list so that you have the menu bar across the top. Then select the drop-down list next to the New menu and pick "Indicator using data in SQL Server 2005 Analysis Services" (see Figure 7-21).

2. Now you are presented with a form to add the KPI. The first box is a prompt for the data connection. This will come from the Data Connection Library we defined earlier in this chapter. Remember that you are pulling data from Analysis Services, so the connection has to be an OLAP/Analysis Services connection.

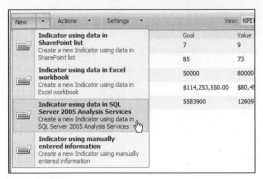

Figure 7-21

3.   Once the data connection is established, the KPI folder and list are populated with KPIs defined in that cube. Select a folder and a KPI from the list. The form will show a preview of the KPI's value under this selection (see Figure 7-22).

4.   The only remaining required field is the KPI display name. You can optionally also add comments to the KPI and make the KPI a link to another page on the site.

5.   Click OK. The KPI is now added to the list and refreshed from the Analysis Services cube when displayed.

That was the process for using an Analysis Services KPI in a MOSS dashboard. Using an Excel spreadsheet as a source is similar but takes a few additional steps because the user must define the cells to use for the data value as well as the warning and goal levels for that KPI. (With Analysis Services, this was built into the cube when the cube was designed.)

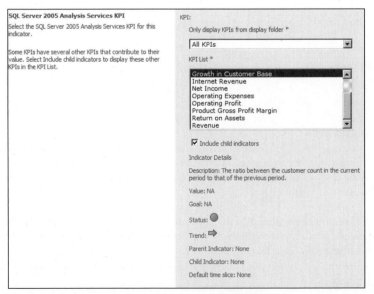

Figure 7-22

Another way to refresh cube data is by using manual lists and SharePoint web services to update those lists from anther application. This "push" model can be used when an application needs to update the dashboard directly instead of updating a spreadsheet and then uploading that spreadsheet to a SharePoint library, or updating a database and then refreshing the Analysis Services cube to reflect the changes.

Dashboards and Key Performance Indicators are popular features of intranets and project sites. By using MOSS's support for these features, the amount of time required to support these capabilities is reduced dramatically, and the KPI dashboards can pull from the same data sources that users may later connect with to do detailed analysis on the same data. This makes it easier to maintain one of the goals of BI — a "single version of the truth."

Excel Services

Excel Services is a new feature of SharePoint in the MOSS 2007 release, and this is the first time Microsoft has published the Excel calculation engine for developers to reuse beyond the client-side Excel application. Excel Services is not designed to be a general-purpose server-based replacement to the desktop Excel product. Instead, it is designed to better extend Excel's strengths for analyzing and manipulating data into new scenarios, such as in web dashboards and custom applications.

This is not the first time Excel has been reusable in these scenarios, but the previous client-side only model is restrictive. Loading client-side Excel ActiveX controls in a web dashboard to visualize data only works for clients that have these controls installed and the appropriate browser and security settings to make them work. For analyzing data, however, these controls did not support an efficient client-server model, so the client was forced to download large amounts of data to support interactive features such as drilling and filtering because all of this processing was being done on the client.

In contrast, Excel Services supports several usage scenarios:

❑ As a set of SharePoint Web Parts for interaction with server-side spreadsheet data without client-side controls

❑ As a rich, interactive spreadsheet viewer for clients that do not have Excel installed or that choose to stay in the browser (including support for setting calculation parameters and refreshing data)

❑ As an API for developers to use for building custom applications that need to work with data or business logic built into Excel workbooks

Although Excel Services supports rich consumption of spreadsheet data, including the capability to refresh data and perform calculations, it is a read-only system and does not store any of its own data. It is purely a server-side engine that reads spreadsheets posted in network locations such as SharePoint libraries or network shares.

Excel Services gives developers and architects some new options when planning for the BI aspects of enterprise systems. It can be used as part of a security plan so users can get data-driven spreadsheets, but these spreadsheets can be refreshed and published outside of Excel Services instead of giving users direct access to refresh spreadsheets. Excel Services also makes it possible to restrict user access to only portions of a spreadsheet.

Using a distributed architecture, Excel Services can protect the intellectual property contained in a worksheet. For example, users can be restricted from modifying underlying data or even seeing the formulas within a workbook.

Architecture

Excel Services is modular and built from three core components:

- **Excel Calculation Services:** Loads, modifies, refreshes, and calculates Excel spreadsheets
- **Excel Web Access:** Renders Excel spreadsheets to browser clients
- **Excel Web Services:** Gives custom applications access to Excel Calculation Services through web services

Figure 7-23 shows the relationships between these components in a typical deployment.

A user can access Excel Services through Excel Web Access, Excel 2007, or through a custom application through web services. Excel Web Access and Excel Web Services both expose services from Excel Calculation Services.

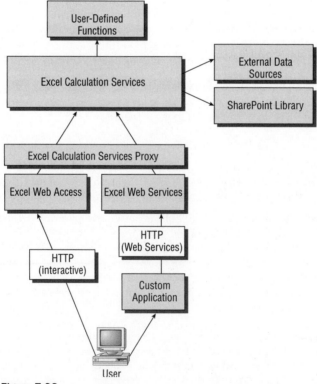

Figure 7-23

Excel Calculation Services is the core data engine of Excel Services. Because it does the work of opening and manipulating spreadsheets, the processor and memory requirements for an effective Excel Calculation Services server are higher than they are for the other Excel Services components. Because Excel Calculation Services spends most of its time performing calculations and refreshing the data in spreadsheets, adding CPU cores and expanding RAM can improve its performance. Excel Calculation Services also supports a farm deployment, enabling additional Excel Calculation Services servers to be added as needed to increase capacity with growing demand.

Excel Web Access and Excel Web Services share a smart proxy that handles the overhead of user requests, manages sessions, and routes user requests to a pool of Excel Calculation Service instances in the farm. Excel Web Services is a thin layer over this proxy intended to give developers access to Excel Calculation Services outside of SharePoint.

Excel Web Access is implemented as a SharePoint Web Part with a tool pane that enables nondeveloper users to publish spreadsheets in SharePoint pages. It also supports connections to other filter Web Parts so Excel Services can be used in a consolidated dashboard with other controls that use the same filters.

Limitations

Excel Services comes with some limitations in the spreadsheet features that are supported in Excel Calculation Servers and rendered through Excel Web Access. These are some of the more common unsupported features (A full list is available at `http://msdn2.microsoft.com/en-us/library/ms496823.aspx`):

❑   VBA (Visual Basic for Applications) macros

❑   External references to other Excel workbooks

❑   Most graphical elements such as WordArt, embedded images, and image backgrounds (although Excel chart graphs are supported)

❑   SmartTags and ActiveX controls

In most cases, Excel Services will still work with spreadsheets containing these items; they will simply not function (for example, VBA macros will not execute, ActiveX controls will not appear).

A limitation of Excel Services' ability to refresh data sets is that it can only refresh PivotTables and Pivot Reports. Other data elements such as the query table we created against our SQL table in Figure 7-9 will not refresh on the server. (See the "Troubleshooting Excel Services" section at the end of this chapter for more details about this.)

Configuring Excel Services

In this section, we will go through the process of configuring Excel Services using settings for a standard configuration. In a large-scale farm deployment, there are additional options and tasks that do not come into play with our configuration.

Because Excel Services is a server-side application that does work on the behalf of remote users, most of the work in configuring Excel Services involves setting up its security parameters. Excel Services reads spreadsheets from SharePoint libraries and refreshes data in spreadsheets from external databases, so configuring Excel Services security properly is important. An incorrectly configured environment can lead to errors for users attempting to access data they should be able to see. Similarly, it can give users access to unauthorized data.

Excel Services is configured mainly at the Shared Services instance level. The administration page can be found by starting at the SharePoint Center 3.0 Administration site, following the Shared Services link in the left navigation bar, and selecting the desired shared instance provider (there is only one by default with a standard MOSS install) by clicking on its name in the list. This will open the Shared Services instance configuration page, which includes a menu of Excel Services configuration options (see Figure 7-24).

Figure 7-24

We are going to use most of these in our configuration process. Here's what they do:

❑ **Edit Excel Services settings:** General settings for Excel Services, including security, memory, session management, and load balancing

❑ **Trusted file locations:** Defines a list of locations from which Excel Services is allowed to open spreadsheets

❑ **Trusted data connections:** Defines the trusted list of data connection libraries that can be used to load and refresh spreadsheet data

❑ **Trusted data connections:** Defines a list of data connections Excel Service may use

❑ **Trusted data connections:** Manages the list of allowed database drivers used in a data connection

❑ **User-defined function assemblies**: User-defined functions (UDFs) are custom code that can be called from spreadsheets; this tool manages the list of allowed UDF assemblies.

In this configuration process, we'll start by configuring Excel Services' general settings and then configure security by adding items to the trusted lists.

Excel Services Settings

Excel Services has several global configuration settings that are managed from the Excel Services Settings page. The configuration page has a description of these parameters and their values, but we'll take a look at them from a developer's perspective here.

Parameter	Possible Values	Comments
Security: File Access Method	Impersonation (default) Process account	Excel Services can load workbooks not only from SharePoint libraries, but also from non-SharePoint sources such as file shares. By default, SharePoint impersonates the user making the request, but depending on how security is configured in the MOSS environment, that user account may not have access to the necessary resources. Configuring this to Process Account makes Excel Services use the Shared Service Provider account, or the Excel Services unattended service account (if configured).
Security: Connection Encryption	Not required Required	This setting lets an administrator force an encrypted (HTTPS) connection between a client and the front-end web server.
Load balancing: Load balancing scheme	Workbook URL Round Robin Local	Determines which load balancing algorithm will be used for requests between the web front-end servers and the Excel Services servers. If only one sever is running Excel Calculation Services, then this setting does not matter.
Load balancing: Retry interval	(Integer value in seconds)	After a server connection times out, this is how long the web front end will wait before retrying its connection to the Excel Calculation Services.
Session management: Maximum sessions per user	-1 (unlimited) or integer value	Because a user can have multiple windows and Web Parts active, this sets an upper limit to contain excessive system resource use.
Memory utilization: Maximum private bytes, Memory cache threshold, Maximum unused object age		These are memory tuning settings that can be adjusted in advanced deployments such as dedicated Excel Calculation Services server(s).
Workbook cache: Workbook cache location	(local file path)	This is a location for Excel Services to store a cached copy of recently used workbooks. This improves load times for popular workbooks.

Continued

Parameter	Possible Values	Comments
Workbook cache: Maximum size of workbook cache	(MB value for cache size)	Sets the location of the workbook cache; the default is a subdirectory from the TEMP directory.
Workbook cache: Caching for unused files	(checkbox)	Toggles whether Excel Services keeps cached workbooks in cache after they are closed.
External data: Connection lifetime	-1 (no recycling) or integer value (seconds)	Excel Services caches connections; this value determines how long these connections are allowed to stay open.
External data: unattended services account	Username/password	The unattended services account is a fallback account for connecting to data sources when a Single Sign-On (SSO) client cannot map to a Windows account for impersonation.

Trusted File Locations

Because Excel Services does not act as a primary data store for spreadsheets, Excel Services uses the trusted file location list as a whitelist to define the safe locations where it can open spreadsheets, regardless of whether they are stored in SharePoint, on a network share, or on a non-SharePoint web site.

Not all file locations are alike, so each trusted location is defined with its own security and performance settings. A public network drive may not be trusted with the same level as a managed document library that is restricted to a handful of contributors.

The Add Trusted File Location form prompts for the path and type (WSS, UNC, or HTTP) and for the following parameters:

- **Session management:** Session and request timeouts
- **Limits on properties**
- **Calculation:** Mode and timeout
- **External data connections:** Enabling/disabling, data caching parameters, concurrent query limits, and error/warning handling
- User-defined functions: Enable/disable

Because Trusted Location parameters are defined at the file location level, they provide you with the flexibility to adjust these settings based on the use and permissions on these locations. It's best not to set these permissions too restrictive, but rather at a realistic level. Use more selective storage locations for spreadsheets that need more advanced rights.

Trusted Data Connection Libraries and Providers

Just as with file locations, data connections must be secured properly to prevent access to unauthorized data or possibly open an attack on the SharePoint server infrastructure. In addition, just as with Trusted

File Locations, the Trusted Data Connection list ships empty and needs to be populated to support data connections in Excel Services.

The Add Trusted Data Connection Library form has no configuration settings other than the path to the Data Connection Library. For instructions on creating data connection libraries, see the section "Centralizing Connection Management" earlier in this chapter. Once a connection library is added to the trusted list, Excel Services can use that connection to refresh data within spreadsheets.

Because data access providers essentially is code executing on the server, Excel Services uses a separate whitelist for data providers. This list is prepopulated with the common providers from the major database vendors. If a data refresh is requested with a provider not on this list, that refresh request will fail, even if the Data Connection Library storing that connection profile is in the Trusted Data Connection Library list.

Trusted data providers are managed in the Trusted Data Providers administration page in the Excel Services configuration menu. Adding a new data provider to this list only takes the identifier for the data provider and the provider type (OLEDB, ODBC, or ODBC Data Source Name). The provider identifier is the short provider name used in connection strings (for example, "SQLOLEDB" for SQL Server's OLE DB driver or "MSOLAP.3" for the OLE DB SQL Server Analysis Services driver).

We will return to these data connection whitelists in our upcoming section on troubleshooting because they are a common cause of errors when working with a server-side data connection in Excel Services.

Using Excel Services

Now that Excel Services is configured with the proper permissions, we'll publish a spreadsheet through Excel Services so users can view and refresh the data in this spreadsheet on the server. Because Excel Services only refreshes Pivot Reports and PivotCharts, the spreadsheet you upload here should have one of those data structures if you want to test server-side data refresh. (If this doesn't work, try following these steps again with a simple spreadsheet with no refreshable data — many of the errors in publishing to Excel Services stem from data security settings. See the section "Troubleshooting Excel Services," at the end of this chapter.)

Excel 2007 has a feature for publishing into Excel Services found under the Office button ⇨ Publish ⇨ Excel Services. Actually, the important difference between Publish to Excel Services and a normal Save As to a SharePoint library is a button in the Save dialog.

The Excel Services Options button (see Figure 7-25) exposes a dialog with options to control how Excel Services will expose the spreadsheet. (This same dialog is available on a SharePoint-hosted spreadsheet under the Office button ⇨ Server ⇨ Excel Services Options.)

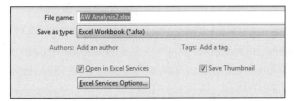

Figure 7-25

The first tab in this dialog controls which workbook objects will be available. The Entire workbook can be made available (see Figure 7-26).

Figure 7-26

Conversely, the contents can be restricted by worksheet (see Figure 7-27).

Figure 7-27

Contents can also be restricted by item (see Figure 7-28).

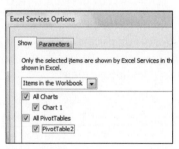

Figure 7-28

The Parameters tab of the Excel Services Options dialog (see Figure 7-29) enables named cells to be exposed in Excel Services so you can set parameters on server-side calculations.

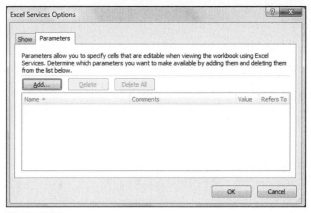

Figure 7-29

The reporting worksheet we're uploading here does not have any parameters, but this dialog supports publishing spreadsheet parameters to be used in server-side calculations. Excel Services can use these defined parameters to create prompts for users to fill in these parameters for interactive recalculations.

Once the spreadsheet is published, it is available through Excel Services. The easiest way to view the spreadsheet in a browser is to navigate to the document library where the Excel worksheet is stored, expand the document library item drop-down list, and select View in Web Browser (see Figure 7-30).

By default, SharePoint will always attempt to render content to the client in the highest fidelity that the client supports. This means that if the server detects a full-blown Excel 2007 installation, then it will default to showing that content in Excel. The server will fall back to lower-fidelity methods such as Excel Web Access as needed to support the widest reach.

There are situations in which a developer may want to render in Excel Web Access regardless of whether Excel 2007 is installed on the client or not, perhaps as a matter of consistency to avoid user confusion and to simplify training. Whatever the reason, the document library configuration page has an Advanced Settings configuration page that defines the behavior of that library. The Browser Enabled Documents option on that page defaults to "Open in the client application," which uses the rich client if it is detected, but the "Display as a Web page" option will render in HTML even if the rich client is available.

Figure 7-30

At this point, if the trusted locations setting was configured properly in the farm's Shared Services, then the workbook will appear in the browser window. It should retain most formatting and several interactive features in this mode.

Expanding data-driven sections and refreshing data from the database server are supported only if all of the following are true:

❑ The location where the worksheet is stored is listed in the trusted locations for the farm's Shared Services and has the Allow External Data set higher than None.

❑ The connection used by this worksheet was linked from a trusted connection library for this farm's Shared Services.

❑ Security is configured properly in the SharePoint farm for accessing data source.

There are several things to check to make sure this will work properly — check the "Troubleshooting Excel Services" section at the end of this chapter for some assistance.

Note that in Figure 7-30, users can also open this spreadsheet on the client using Excel, but there is an option to view the snapshot in Excel. Instead of downloading the spreadsheet in its original form, opening a snapshot performs all spreadsheet calculations on the server and sends only the results to the client. This means that the user sees none of the underlying formulas in the original spreadsheet. This is another way to retain control of the intellectual property in a spreadsheet.

Now that we have published Excel worksheets for a broad range of users, both in full-screen document views and as Web Parts, you can proceed to look at how this infrastructure can be leveraged to reuse business logic and automate the processes of refreshing data.

Excel Web Services

Excel Web Services exposes the server-side spreadsheet engine in Excel Calculation Services to any client capable of calling a published SOAP interface. Excel Web Services is referenced by the path `/_vti_bin/excelservice.asmx?WSDL` relative to any site root. For example, the following are valid references:

```
http://myserver/anysite/_vti_bin/excelservice.asmx?WSDL

http://myserver/_vti_bin/excelservice.asmx?WSDL
```

Although any site in an Excel Services–enabled server will support the Excel Web Services interface (including the root site, as the second of the preceding examples shows), the best practice is to create a dedicated blank site specifically for Excel Web Services and use that so a future site reorganization doesn't break established web service references.

A second way to access Excel Web Services is directly through the `Microsoft.Office.Excel.Server.WebServices` namespace declared in the `Microsoft.Office.Excel.Server.WebServices.dll` assembly, which is located in the following path on a server configured with Excel Services:

```
Program Files\Common Files\Microsoft Shared\web server extensions\12\ISAPI
```

This direct namespace reference works mostly the same way as Excel Services references through the server-based web service, but it is intended for code running SharePoint itself, such as custom workflows or custom Web Parts.

API Overview

Both the Web Service interface and the direct .NET reference define the following classes (some additional types are declared, but they are not designed for external use):

`ExcelService` (web service)	Primary service with methods for session management, reading/writing cell values, calculating worksheets, refreshing data, and reading out an entire workbook
`RangeCoordinates` (type)	A data structure for defining ranges in worksheets
`Status` (type)	Status messages from the server are returned in an array of this type.

A typical Excel Web Services session using `ExcelService` may have the following steps:

1. The client starts a session by making a request to Excel Services to open a specific workbook. That session is now associated with that workbook instance. If other clients (or the same client in a different session) make a request for the same workbook, they are isolated because Excel Services gives each client a separate instance of that workbook.

2. Using the established session, the client can read and write workbook values in Excel Services in that session's workbook. Note that these changes are stored on the server only for the length of the session and are not seen by other users or persisted back to the original workbook location. (Excel Services does not write data back to persistent stores.)

3. If required, the worksheet in an established session can have its external data refreshed. This is restricted by the same policies and limitations on trusted data connections. Again, changes live only as long as the session, as nothing is persisted.

4. The one way to keep changes made to a workbook in Excel Web Services is for the client to read that entire workbook out of the Excel Web Access session. This session copy of the workbook includes any changes made up to that point in the session.

5. When the client no longer needs to use this Excel Services session, it closes the workbook, thereby closing the session.

Because a client can hold multiple sessions on a single Excel Web Access instance, sessions are keyed using a server-assigned session ID that is returned from the `ExcelService.OpenWorkbook()` method. This session will stay alive on the server and terminates when one of the following occur:

❑   The client calls the `ExcelService.CloseWorkbook()` method

❑   The session times out on the server (this is configured in Edit Excel Services Settings in Shared Services Configuration)

Every Excel Services method (except `OpenWorkbook`) takes the session ID parameter to access the appropriate session workbook. Every method also includes an output parameter or return value named `status` (type: **ExcelServices.Status**) that can return informational alerts from the request.

In the next section, we will implement a Windows Forms application to leverage existing business logic.

Example: Reusing Business Logic Built into Excel Worksheets

In this scenario, we start with a workbook that contains some important business logic that we also need in a custom Windows Forms application. In this case, we're using a workbook that performs a what-if calculation for an auto loan based on three input parameters: financed amount, interest rate, and term in years (see Figure 7-31).

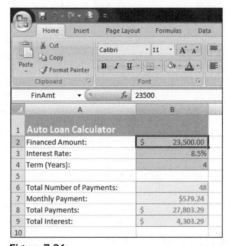

Figure 7-31

Our Windows Forms application user interface, shown in Figure 7-32, has a corresponding set of controls and a Calculate button that initiates the call to Excel Web Services.

Figure 7-32

The form includes three labels for the output values that are not visible here because they are empty.

Before we can access Excel Web Services, we need to add a web reference so Visual Studio can build a proxy assembly (by using the Project menu and selecting Add Web Reference). The Add Web Reference dialog prompts for a Web Service URL that is the relative path `/_vti_bin/excelservice.asmx?WSDL` from any site on a SharePoint server as long as Excel Services is enabled. Enter the reference URL and `ExcelSvc` as the web reference name and click Add Reference to generate the proxy assembly.

Now we'll open the form code and start implementing the calls to Excel Web Services. First, we need a few namespace references:

```
using ExcelServicesCalcClient.ExcelSvc; // web service proxy assembly
using System.Web.Services.Protocols;
using System.Net;
```

`ExcelServicesCalcClient.ExcelSvc` is the namespace generated by Visual Studio for our Excel Web Services reference (`ExcelServicesCalcClient` is the VS project namespace). We use `System.Web .Services.Protocols` because it declares an error class we want to catch, and `System.Net` provides support for managing credentials.

Now we'll start the button `Click` event handler by declaring some local variables for the web service:

```
private void calcButton_Click(object sender, EventArgs e)
{
    // set up the Excel Web Services
    ExcelSvc.ExcelService ews = new ExcelSvc.ExcelService();
    ExcelSvc.Status[] status;
    string sessionID = "";

    ews.Credentials = CredentialCache.DefaultCredentials;
```

We'll call the `ExcelService` class to do most of the work here. We declare a variable for an array of `Status` to catch any status codes that might be returned from our Excel Web Services calls and use `sessionID` to maintain our session handle. The web services call will need authentication, so we use `DefaultCredentials` to pass the credentials active in the current user session:

```
    try
    {

        sessionID =
ews.OpenWorkbook(@"http://moss2007rtm1/bi/Documents/CarLoanCalc.xlsx",
            "en-US", "en-US", out status);
```

Now the code goes into a `try` block for error handling. The first call to Excel Web Services is the `OpenWorkbook` method, which returns a session ID. The main parameter for the `OpenWorkbook` method is the workbook location, which, depending on the trusted location settings in Excel Web Services, can be a SharePoint library URL, a UNC network path, or a URL pointing to a non-SharePoint web location. The second and third parameters are for UI and data culture, respectively.

```
        status = ews.SetCellA1(sessionID, "Sheet1", "FinAmt", finAmountBox.Text);
        status = ews.SetCellA1(sessionID, "Sheet1", "IntR", intRateBox.Text);
        status = ews.SetCellA1(sessionID, "Sheet1", "Term", termBox.Text);
```

Now we use the `SetCellA1` method to set cell values in the workbook using the named cells in the worksheet. (There's also an alternate version called `SetCell` that uses numerical coordinates covered after this example.) Like every request for the rest of the session, we pass in `sessionID` to associate the request with the workbook we just opened. The second parameter is worksheet name, and the third is the cell reference. We could just as easily have used the cell coordinates B2, B3, and B4 instead of their names, but using named cells makes the solution more resilient because these cells can be moved and still maintain their names.

We're passing the string values in the text boxes directly into the worksheet. The server parses these values, so special characters in Excel are treated as they would be if they were typed into the spreadsheet — for example, passing in "8.5%" formatted as a percent on the server and divided by 100 before it is stored.

Because our workbook is set to calculate automatically, the calculation was done after every call to `SetCellA1`. Later in this chapter we'll look at ways of controlling this more efficiently.

```
monthlyPaymentOut.Text = ews.GetCellA1(sessionID, "Sheet1",
    "MonthlyPayment", true, out status).ToString();
totalPaymentsOut.Text = ews.GetCellA1(sessionID, "Sheet1",
    "TotalPayments", true, out status).ToString();
totalInterestOut.Text = ews.GetCellA1(sessionID, "Sheet1",
    "TotalInterest", true, out status).ToString();
```

Reading data from the worksheet is done in mostly the same way using the `GetCellA1` method, again passing the sheet name and using named cells for robustness. The Boolean parameter determines whether the cell should be returned formatted. The cell value is returned from the method call so the status is returned as an `out` parameter.

At this point we are finished with the web service so we perform error handling and cleanup:

```
    }
    catch (SoapException)
    {
        errorBox.Text = "Error attempting refresh of server-side spreadsheet";
    }
    finally
    {
        ews.CloseWorkbook(sessionID);
    }
}
```

First, the `try` block is closed and a `catch` block for `SoapException` traps any web service exceptions. An application could inspect the `SoapException` for more information.

The `finally` block will be executed regardless of whether an exception was raised in the try block or not. This ensures that the `CloseWorkbook` method will be called to free up server resources.

In this example, you saw a simple application for a server-side calculation consumed by a client application. Not only are we reusing Excel's calculation abilities, but we also get a level of abstraction from the business logic. You can think of this model as a web service that can be built, tested, and maintained by business users. Once the service contract is defined (named ranges for input and output parameters), the actual calculation in the spreadsheet can be changed without any changes to the client application.

More on Excel Web Services

The car loan calculator example provides an overview of what's involved in building a simple Excel Web Services client application. Excel Services provides additional ways to set and read cells, as well as methods for controlling calculation and refreshing data. We'll look at these in more detail here.

Setting and Reading Values

The following Excel Web Services methods support reading and writing cell values:

Method (Numerical coordinates)	Method (Excel range syntax)	Description
GetCell	GetCellA1	Reads the value of a single cell
GetRange	GetRangeA1	Reads the values of a range of cells
SetCell	SetCellA1	Sets the value in a single cell
SetRange	SetRangeA1	Sets the value in a range of cells

When reading or writing cells, individual values are treated as object types, and ranges are referenced as object arrays. Depending on the data type in the actual worksheet cells, the underlying .NET type will change accordingly.

Whenever possible, it's better to use the range methods over the single-cell equivalent because this cuts down on request round-trips.

Because Excel Web Access is an XML-based web service interface, it carries XML's limitations regarding handling special characters in string values going in either direction, and Excel Web Access does not have built-in provisions for quoting special characters in strings.

Controlling Calculation

In the previous example, the code did not call a method to trigger a worksheet recalculation before reading the data out. Just as in desktop Excel, by default, the workbook is recalculated when the it is loaded and then when data changes. In this case, that happened every time `SetCellA1` was called. In a performance-sensitive system, desired behavior would be to make all necessary `SetCell` calls and then have the workbook calculate once before the cell values are read out.

The calculation behavior on a workbook is set in Excel in the Formulas ribbon using the Calculation Options drop-down menu (see Figure 7-33).

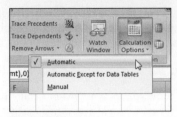

Figure 7-33

Setting this to manual calculation on a worksheet also has an impact on calculation behavior in Excel Services. Excel Services provides `Calculate`, `CalculateA1`, and `CalculateWorkbook` methods to support explicitly recalculating workbook data. These recalculations can be as granular as a single cell, a range of cells, a single worksheet, or an entire workbook. If a worksheet is being built to run on the server and it has multiple inputs, the best practice is to set this calculation to manual in Excel and call the `.CalculateWorkbook` or `.Calculate` methods when all parameter values have been set. This eliminates the overhead of unnecessary interim calculations, the results of which will not be read.

This refresh setting can also be overridden in the Trusted File Location administration page. The Calculation Behavior setting defaults to the calculation settings in the file, but can also override the file's settings with manual or automatic recalculation.

Refreshing Worksheet Data

Excel Services supports refreshing worksheet data from external sources, and Excel Web Access exposes this through a `Refresh` method that takes an optional connection name parameter to specify a single connection to refresh. Otherwise, it refreshes all workbook connections if this parameter is left out. Excel Services uses the connection from a trusted connection library to access the data sources and update the workbook.

Because refreshing the database connection may be a time-consuming process, an interactive application may be a good candidate to use the .NET Framework's built-in support for calling web service methods asynchronously. See `http://msdn2.microsoft.com/en-us/library/55xs7d7f(vs.71).aspx` for more information about this.

The capability to automate the refresh of a spreadsheet through a simple client application or script can come in very handy. Organizations that publish data through Excel to a wide range of users can choose this capability in a scheduled job. The biggest challenge is that using only the techniques we have covered so far, this refreshed workbook disappears when the session ends, as Excel Services does not persist data. Next, you'll learn how you can save data from Excel Services back into a persistent document store.

Saving Workbooks from Excel Services

Excel Services is a read-only environment, but there is actually a way to manipulate a worksheet in an Excel Web Access session and save those changes back out to an Excel document. Excel Web Access exposes a

`GetWorkbook` method that returns a binary copy of the workbook as it is in the current session. This method takes a `WorkbookType` enumeration as a parameter, with the following possible values:

FullWorkbook	All workbook contents including any changes in the current session
FullSnapshot	Workbook values *including any changes in the current session*, but data references are flattened out. Any interactive content, including PivotTables, data connections, metadata, and other noncell content, is removed.
PublishedItemSnapshot	Workbook values *of the original version of the workbook*, but data references are flattened out. Any interactive content, including PivotTables, data connections, metadata, and other noncell content, is removed.

To get a full-fidelity copy of the manipulated workbook, calling the `GetWorkbook` method with `WorkbookType.FullWorkbook` will return a binary stream that can then be saved to an `.xslx` file on a local file system or back to a SharePoint library through SharePoint web services.

Troubleshooting Excel Services

As with any technology, the more flexibility a product offers regarding configuration and deployment, the more ways it can be broken. In Excel Services, most problems come down to configuration settings in the multiple types of trust a worksheet must have to perform various tasks. This sections covers some of the common error messages and describes a few settings to check.

❑ **Is Excel Services enabled for this web application?** Under Central Administration ⇨ Application Management ⇨ Manage Web Application Features, make sure that "Office SharePoint Server Enterprise Web application features" is enabled.

❑ **Is the location where the workbook is being open from a trusted file location?** In Central Administration ⇨ Shared Services Administration ⇨ Excel Services Trusted File Locations, add the HTTP or UNC location to the trusted locations list if it is not already there. Don't forget to check the "Children trusted" box if the workbook is in a subdirectory of this trusted location.

❑ **When using a workbook with a refreshable option, is the connection definition stored in a trusted data connection library?** Make sure that the data connection library is listed in Central Administration ⇨ Shared Services Administration ⇨ Trusted data connection libraries. (There are no additional parameters for this configuration other than the path itself.)

❑ **If this workbook uses a nonstandard data provider, is that provider listed in the trusted data provider list?** Check Central Administration ⇨ Shared Services Administration ⇨ Trusted data providers to determine whether that entry is listed. (Note that even some common providers such as the MS Access driver are not trusted by default.)

❑ **Make sure you are using a server-refreshable data structure in Excel if attempting to perform a server-side refresh.** Excel Services can only refresh PivotTable reports and charts, not query tables like those we created earlier (refer to Figure 7-9). (Some developers on the Excel Services product team have built add-ins to convert query tables into PivotTable reports, retaining many of the interactive features of query tables and adding the capability to refresh them on the server. You can find this add-in by searching MSDN for "query table excel services," without the quotes.)

For additional help troubleshooting Excel Services, especially with external data access, see:
`http://technet2.microsoft.com/Office/en-us/library/7e6ce086-57b6-4ef2-8117-e725de18f2401033.mspx`.

Summary

The Office System can be an integral part of an organization's business intelligence architecture. Users already familiar with Excel can use its new capabilities in data visualization to identify trends and patterns in data. These visualization capabilities adapt to the data being analyzed so they are easy to use even for people without experience customizing worksheet styles in Excel.

Excel's improved data connection management helps centralize the more difficult tasks of configuring data connections and setting security parameters, enabling users to simply choose a connection from an enterprise library of managed connections. This centralization improves users' ability to find the data they need and makes it possible for administrators and developers to change the way these connections work without forcing users to rebuild their workbooks.

Excel Services enables several new scenarios. In situations where users need a focused set of data elements from an Excel workbook, that data can be presented through Excel Web Access in a SharePoint site through a set of Web Parts. These parts can be configured independently or used alongside SharePoint's data filters and lists to provide a rich dashboard experience.

Excel Services also gives developers a way to use the knowledge built into existing spreadsheets to automate calculations, and makes it possible to automate server-side spreadsheet recalculation and refreshing to automate processes that may have previously been done manually.

8

Creating Custom Workflows for Windows SharePoint Services and Office SharePoint Server

Businesses are placing greater demands on their employees to make better decisions faster in order to remain competitive in the marketplace. When attempting to come to a decision, we typically follow a series of steps. The steps can be formal in terms of a standard operating procedure or informal in the sense of an implicitly understood way of operating, but collectively they represent a business process. Because these business processes fundamentally require human interaction, making better decisions faster may be limited by human interactions. Therefore, by increasing the effectiveness of human interactions, we can improve the overall efficiency and quality of the outcome. Business processes can be modeled using flow charts and represented using workflow terminology. Software that facilitates and manages this "human workflow" can provide a way of automating the interaction among the people participating in the process and thereby deliver improved effectiveness.

In previous chapters, you have seen how Windows SharePoint Services 3.0 (WSS) and Office SharePoint Server 2007 (MOSS) provide a robust, customizable, and extensible collaboration environment that enables team members to share business information contained in SharePoint data repositories (for example, documents and lists).

Business processes such as document approval can be automated by associating a workflow with the SharePoint data, notifying the approver of work to be done and the submitter of the outcome of the approval process. Other workflow examples include coordinating the interaction of a group of individuals (such as preparing a PowerPoint presentation for a conference or a new project proposal) or tracking a set of issues, including the tasks, the responsibilities, and the status, on an ongoing basis.

WSS can execute workflow applications through the use of technology called Windows Workflow Foundation (WF). People interact with these workflows through the web browser and Office 2007 applications such as Word 2007. WSS provides the framework for utilizing WF, and MOSS extends the capability by delivering several workflows that automate common business processes. Custom workflows can be created using tools such as SharePoint Designer and Visual Studio 2005, and then associated with SharePoint data.

Therefore, workflow is not only part of the collaboration process, but also represents and automates the collaboration process. Workflows expedite the decision-making process by helping to ensure that the right information is made available to the right people at the right time. Workflows also help ensure that individual workflow tasks are completed by the right people and in the appropriate sequence (or in parallel).

This chapter focuses on an introduction to WF and the tools for creating custom workflows: SharePoint Designer and Visual Studio Designer for Windows Workflow Foundation.

The main objectives for this chapter are as follows:

❑   Understanding the architecture, concepts, and capability of Windows Workflow Foundation

❑   Understanding the concept of a workflow and the role that Windows SharePoint Services and Office SharePoint Server play in workflow creation, deployment, and utilization

❑   Creating custom workflows using SharePoint Designer and Visual Studio Designer for Windows Workflow Foundation

❑   Comparing the capabilities of SharePoint Designer and Visual Studio Designer for Windows Workflow Foundation for creating custom workflows

Windows Workflow Foundation Architecture

Windows Workflow Foundation (WF) is a component of the .NET 3.0 Framework, which resides natively inside the Windows Vista operating system and is available as a download for Windows XP SP2 and Windows 2003 Server. WF provides the capability to create an application that executes a multi-step sequence of work items logically formulated to model a business process. This application consists of one or more workflows, and each workflow is composed of one or more activities.

Workflow Activities

A workflow is composed of discrete work units that are organized into a predefined order or set of states. In the context of WF, each work unit is called an *activity*. Activities can be performed by the system (for example, computers and applications) or by individuals (users), and they represent the building blocks of any WF workflow. In WF, activities have the following possible characteristics:

❑   Activities represent actions, such as approving or rejecting a document, sending an e-mail, creating a list item, deleting a list item, etc.

❑   Activities can also control the logic and flow of execution in the workflow, such as looping, pausing, else-if branching, and so on.

❑ Activities can be composed of methods, properties and events. Activities can also be composed of other activities. These activities are called *composite activities*.

❑ WF provides a set of predefined activities that can be used during the workflow authoring process. Custom activities can be created using Visual Studio 2005 Designer for Windows Workflow Foundation.

❑ Activities are executed in a predefined order by the WF runtime engine.

Workflow Markup

WF workflows utilize several different files to describe and document their functionality. The number and composition of these files varies depending on the actual workflow functionality, but it typically includes an XML file, or markup file, and one or more code files.

XML File

The XML file includes a declarative description of the workflow and is written in Extensible Application Markup Language (XAML). XAML files can be created with any text editor because the schema is publicly available. SharePoint Designer and Visual Studio 2005 Designer for Windows Workflow Foundation both provide a graphical interface to create workflows and automatically generate the XAML description. XAML files have the extension `.xoml`.

Code File(s)

One or more code files describe the functionality and business logic of the workflow. These files can contain C# or VB.NET code or declarative markup (rules).

The workflow markup can be precompiled into .NET assemblies or compiled directly at runtime depending on how the workflow was created and deployed. We'll revisit the details of the markup later in this chapter when we create a custom workflow. Regardless of how it was created, execution is accomplished by the WF runtime engine.

Workflow Execution

The WF runtime engine is not a service but a class that is instantiated within a .NET process that is hosted by another application (that is, WF is not an executable that is launched by double-clicking the code file). The WF runtime manages workflow execution, including the loading and unloading of workflows, and it allows workflows to remain active (persistence) for extended periods of time, including a reboot of the computer. Workflow execution depends on a set of "pluggable" services that provide support for persistence, transactions, scheduling, and tracking. These services are discussed briefly in the following list:

❑ **Persistence:** The state of an executing workflow is temporarily stored to some medium and removed from memory. At some future point in time, the workflow state data is retrieved by the WF runtime, the workflow is brought back to life, and its execution continues until another point of suspension is necessary or termination. This capability is sometimes referred to as *dehydration* and *hydration* of workflows.

❑ **Transaction:** This service provides the capability to combine completed workflow steps into atomic units (*batch*) so that consistency is maintained between the internal state of the workflow and the external data. For example, SharePoint workflow services provide methods that perform actions, and event handlers that dehydrate and hydrate the workflow. Batched actions are not committed until the workflow is dehydrated.

❑ **Scheduling:** Workflow execution must be managed, just as it would be with any other application code. Specifically, this service controls and manages the number of threads, their availability, their priority, and the order of their execution.

❑ **Tracking:** The tracking service monitors the flow and execution of currently running workflow instances. This information is made readily available to the host and is capable of being persisted.

Workflow Fundamentals and Types

A workflow is a model or representation of a real-world process. Recall that the workflow is composed of individual work units or items called *activities*. The workflow describes how these activities relate to one another, their order of execution, and how data flows into and out of the workflow.

WF provides a library of prebuilt activities. Additional activities can be created and made available for use by other workflows. WF provides two types of built-in workflows, sequential and state machine:

❑ **Sequential:** This workflow executes as a series of steps in a predefined order until the last step is completed. Because the workflow can respond to external events, the actual execution order varies, including branching and multiple parallel paths.

❑ **State Machine:** This type of workflow is represented by a set of predefined states, transitions, and actions. In response to some action or event, the workflow will execute a transition and move to another state. A beginning state is necessary to launch the workflow, and a final state defines the termination of the workflow; however, a final state is not required.

It's important to reinforce the concept that WF is not a complete workflow application. It is a foundation for software developers to create workflow-enabled applications. WF functionality runs inside a workflow host, which can be any Windows application process or service. The WF runtime can be customized to accommodate specific host requirements and different application requirements by replacing or extending the native runtime services discussed previously in the chapter.

Windows SharePoint Services and Workflow Capability

WSS serves as the host for the WF runtime feature. It has the capability to execute WF workflows and includes one out-of-the-box workflow called the *three-state workflow*. As part of being the host for the WF runtime, WSS provides custom implementations of some of the key WF services. For example, the WF persistence service needs to be customized to accommodate SharePoint's requirements, as described in the following paragraph.

Human workflow interactions can extend into hours and even days. For example, in a document approval process, the approval or rejection of a document could encompass multiple approvers and include author revisions before the process is complete. Under practical conditions, the approval workflow cannot remain in memory over this extended time. Review by an approver introduces a natural delay whereby the runtime engine is waiting for input. At this point, the workflow can be dehydrated. Activities that have been successfully executed are batched into a single SQL Server transaction. If the transaction is successful, then WSS unloads the workflow instance from memory. The workflow can be hydrated in response to external events. For example, when the reviewer changes the approval status to accepted or rejected, WSS loads the persisted workflow data and recreates the workflow instance. The workflow instance receives the approver's input, and the workflow continues.

Persistence helps maintain a link between the state of the dehydrated workflow and the document or list item with which the workflow is associated. This link is critical to SharePoint and illustrates a key SharePoint requirement that is satisfied by customizing the persistence service. Therefore, despite the fact that numerous workflows may be executing at any time, only a small subset of those workflows may be resident in memory and consuming critical resources. The ability to customize the native WF services is a feature of the WF pluggable environment and is at the discretion of WSS as the WF host.

Workflow Templates

Workflow capability is delivered to WSS through the concept of templates. *Templates* describe the workflow and govern its action, are the result of the workflow authoring process, and must be deployed to the WSS web server. Workflows are enabled by creating an association between a specific template and a document library, list, or content type. This template is then loaded by the runtime to create an instance. Multiple workflows can be associated with any list or document library directly, or associated with a content type.

Workflow Instances

An *instance* of a workflow is created once a workflow is started. A single list or document may have several different workflows available, and instances of each of these workflows can execute concurrently. However, only one instance of any specific workflow may be active at any point in time for an individual list or document.

A workflow instance can be started using the web browser or an Office 2007 client. The workflow can be configured so that it starts whenever a new item is created, an existing item is changed, or a new major version has been published. Users that have been assigned tasks can be automatically notified once previous users have completed work on an item. Users can be assigned either in series or in parallel, and have the flexibility to assign tasks to someone else. While running, workflow instances can be tracked and their performance monitored. For example, the Activity Duration report describes the length of time each activity and workflow instance takes to complete. The Cancellation & Error Report shows you which workflows are being canceled or are encountering errors before completion.

There are several different phases to creating a custom workflow: authoring, deploying, associating, and instantiating. Within each of these phases, roles are played by different individuals who participate in the workflow: workflow author, approvers, users, and administrators.

Authoring and Deploying a Workflow

Each workflow consists of logic that defines the workflow and the user interaction with the workflow. For WSS, users will interact with workflows via web forms. Therefore, WSS relies on ASP.NET to display its forms, and aspx pages to represent those forms. The workflow logic is defined during the authoring of custom workflows by the tool being used.

SharePoint Designer (SPD) and Visual Studio 2005 Designer for Windows Workflow Foundation (DWF) are two different tools for creating custom workflows. Each has different capabilities and advantages, which are discussed in detail later in the chapter. For now we will cover only some of the key differences:

❏   SPD workflows cannot include custom .NET code, are limited to a predefined list of activities, and are created specifically for a single list or document library. The result is a workflow template that is automatically deployed to the WSS web server. Creating custom workflows using SPD is designed for the nonprofessional developer, such as an information worker or web designer.

❏   DWF produces a workflow template that can contain custom .NET code, can utilize custom activities, and can be utilized across sites, multiple lists, document libraries, and content types all at the same time. Custom activities can also be created without creating a custom workflow. Once the custom template or activity is created, it must be deployed to the WSS web server by a server administrator. Custom activities, once deployed, are also available for use by SPD.

Regardless of the tool, the outcome is a workflow template. SPD simplifies deployment while DWF requires assistance from a server administrator or equivalent with the proper server permissions. Once workflows have been created and deployed, they must be associated with a list or content type before they can be used.

Associating a Workflow

A workflow template must be "attached" or associated with a WSS list or content type before an instance can be created (instantiated) and executed. As you've already seen, SPD templates are associated with a list or document library during the design of the workflow (design time). For DWF, association occurs after deployment, and the scope of the association is not only broader but can also include WSS content types. Template association is accomplished via the web browser. ASP.NET web pages can be used to represent data entry forms during the configuration of the association.

The same template can be associated with multiple lists, multiple document libraries, and multiple content types at the same time. Each different association can be configured specifically for the attached item depending on the requirements. Once association has been completed, a workflow instance can be created and executed.

Instantiating a Workflow

A workflow *instance* is an executing workflow. For each specific association, a single workflow instance can be created and started. A workflow can be started in one of three different ways:

❏   The workflow can be started manually from the WSS list item or document and can be started from the Office 2007 client application.

❑   The workflow can be configured to start automatically when an existing document or list item is changed.

❑   The workflow can start automatically when a new document or list item is created.

Creating the workflow instance is the last step that launches an executing workflow. As part of creating an instance, the user who launches the workflow can interact with the workflow by providing customization or configuration data that will be passed to the workflow and used during execution. A user can also interact with an executing workflow and modify the executing instance if the workflow was designed to allow this capability. If these user interactions are necessary, the workflow author must provide workflow forms as part of the authoring process.

Workflow Forms

Workflow forms enable user interaction with the executing workflow and therefore make the workflows dynamic and more flexible. The form collects information from users at predefined times and enables users to interact with the workflow tasks. Workflow forms can be classified into three different types, each of which is briefly summarized in the following list: association and initiation forms, task forms, and modification forms.

❑   **Association** forms enable default values to be entered by the user, or configuration information to be set by the administrator. **Initiation** forms provide the opportunity to enter starting values for forms that launch manually.

❑   For workflows that require users to complete tasks, **task** forms can track actions, maintain status, and trigger events that influence the executing workflow.

❑   **Modification** forms can be used to adjust the flow of the executing workflow without the need to restart the workflow.

Regardless of the form type, all workflow forms need to perform two functions:

1.   Collect data from the user.

2.   Call an object model (OM) function to perform the desired action, create an association, start the workflow, edit tasks, or modify the execution. User data is passed to the OM function as a parameter, and the function passes the data into the workflow.

For WSS workflows, ASP.NET web forms are used to represent workflow forms because the user interacts with the workflow via the web browser. Therefore, the workflow author would need to create the web forms and write the custom code that collects the data and calls the OM.

The WSS primary responsibility as the WF host is to provide the capability to execute workflows, but it also includes one out-of-the-box workflow. SPD and DWF are two different authoring tools that can be used to create custom workflows. Both of these tools produce workflow templates. Workflow instances are created from these templates after the template is associated with the WSS list or content type.

Workflow forms can also be used as part of the custom workflow. Because WSS relies on ASP.NET web forms for workflow forms, these forms must also be created and deployed to the WSS web server, along with the workflow templates, prior to use. Additionally, custom code is necessary to move user data from the ASP.NET web form into the workflow. MOSS provides an alternative to creating custom work-flows because it provides several out-of-the-box workflows ready for use without writing any custom code. It enables Office clients to interact with workflows and simplifies web form creation and data processing.

Office SharePoint Server and Workflow Capability

MOSS extends the workflow capabilities of WSS in three different areas: support for the Office 2007 clients to launch workflows, the ability of workflows to use InfoPath 2007 forms, and a set of six prede-fined workflow templates. The MOSS templates supplement the Three-State workflow template pro-vided by WSS.

Predefined Templates

The predefined workflow templates can be utilized directly without writing any custom code and can be configured through the web browser interface. The WSS and MOSS out-of-the-box workflows include the following:

❑   **Three-State:** Used to track and move list items through a series of states. Can be configured to react differently at various stages.

❑   **Approval:** Routes a document for approval. Approvers can approve or reject the document, reassign the approval task, or request changes to the document.

❑   **Collect Feedback:** Routes a document for review. Reviewers can provide feedback, which is compiled and sent to the document owner when the workflow has completed.

❑   **Collect Signatures:** Gathers signatures needed to complete an Office document. This workflow can be started only from within an Office 2007 client.

❑   **Disposition Approval:** Manages document expiration and retention by allowing participants to decide whether to retain or delete expired documents.

❑   **Group Approval:** Similar to the Approval workflow, but uses a designated document library and offers a personalized view of the approval process. It includes other features such as a hierarchical organization chart from which to select the approvers and allows the approvers to use a stamp control instead of a signature. This solution was designed specifically for East Asian markets.

❑   **Translation Management:** Manages document translation by creating copies of the document to be translated and assigning translation tasks to translators.

The number of out-of-the-box workflow templates available varies depending on the site template selected to provision the site and the features that have been activated for the given site. Features represent a new SharePoint technology that enables the deployment and activation of SharePoint site components. The name can be misleading; it doesn't refer to the SharePoint feature set like Web Parts, lists, and so on. (Features are discussed in more detail in Chapter 5.)

MOSS workflows are deployed as Features. A list of the available workflow templates is shown in the Site Collection Workflows gallery, accessed from the Site Settings web page. An example is shown in Figure 8-1.

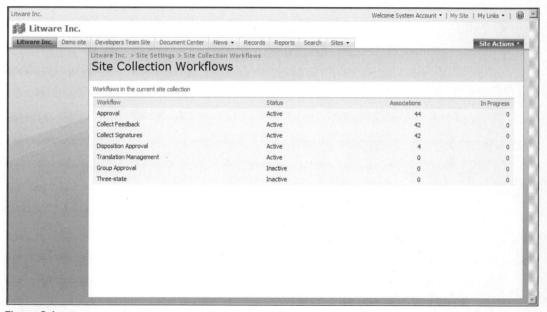

Figure 8-1

As you can see from Figure 8-1, the Approval, Collect Feedback, Collect Signatures, Disposition Approval, and Translation Management workflows are active and ready for use. The Group Approval and Three-State workflows are not active. These two inactive workflows can be activated individually as Site Collection Features from the Site Collection Features web page, as shown in Figure 8-2.

You'll notice in Figure 8-2 that the Group Approval workflow feature (fifth from the top) and the Three-State Workflow feature can be activated by clicking the Activate button. Once these features are activated, the corresponding workflows are visible in the Site Collection Workflows gallery. The process just described for identifying the available workflow templates and activating them also applies to custom workflows that have been created and deployed using the Feature process.

Many of the MOSS workflow templates use InfoPath 2007 forms to present data entry or configuration capability to the user. This enables workflow interaction for the Office 2007 clients, a feature not available for WSS workflows.

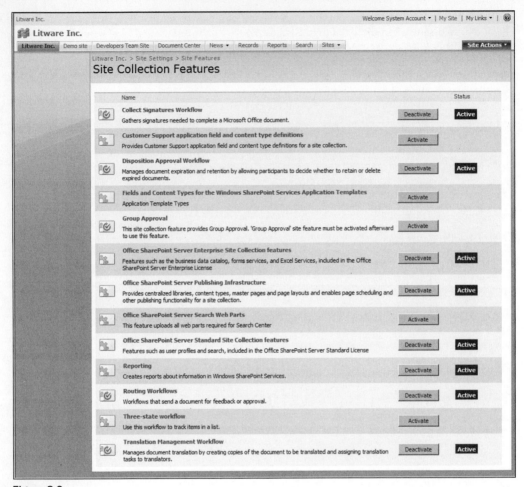

Figure 8-2

InfoPath Workflow Form Advantages

There are several advantages to using InfoPath 2007 forms as workflow forms:

❏ They can be used directly from Office 2007 client applications (Word, Excel, PowerPoint, Outlook, InfoPath). Office 2007 client applications and ASP.NET web forms can both host InfoPath forms for workflow. While using an Office 2007 client, the InfoPath task form will appear as a dialog directly in the application. This form is exactly the same form you would see in the browser, so a single InfoPath form can be utilized for both environments.

❏ InfoPath forms are designed in a visual designer similar to SharePoint Designer and are created using a drag-and-drop user interface. A form designer does not need HTML or ASP.NET development experience.

❑ InfoPath workflow forms don't require that custom code be written to process form data. MOSS web form pages that host the InfoPath forms make the OM function calls that handle data processing. If the out-of-the-box form processing does not meet your needs, you can write your own custom code as part of the web form or you can write custom code as part of your InfoPath form.

MOSS provides out-of-the-box workflows and enhances workflow functionality in general. The integration of InfoPath forms with the custom workflows enables both Office client and web browser interaction. A single InfoPath form can be utilized for both experiences. Unlike using WSS ASP.NET web forms for user interaction, you have no need to write custom code to move data between the InfoPath form and the workflow. As discussed earlier, this is different from WSS workflows, which can only utilize ASP.NET forms and therefore have no Office 2007 client capability and require custom code for forms processing.

> *The InfoPath forms that are utilized for the out-of-the-box workflows can be viewed outside of their associated workflows. To view these forms, navigate to* `c:\Program Files\Common Files\ Microsoft Shared\web server extensions\12\Template\Features`. *Perform a search for *.xsn and you will see a list of InfoPath forms associated with deployed features. Before viewing, copy any of the forms you are interested in looking at to a separate folder.*

If the out-of-the-box templates don't provide the necessary capability, custom workflow templates can be created using SharePoint Designer and Visual Studio Designer for Windows Workflow Foundation. That's the subject of the rest of this chapter.

Creating Custom Workflows with SharePoint Designer

It's time to look at an example to illustrate some of the concepts we've been discussing. One of the most common human workflow scenarios is related to content approval. Figure 8-3 shows a flowchart describing an approval and routing business process. This scenario can be represented by a custom workflow. Before we walk through creating the workflow, we'll discuss the overall process.

Content Approval Workflow Scenario

The approval process is initiated when a user or author creates a new document and saves the document to a SharePoint document library (or uploads a new document). Individuals responsible for approving the content are notified by e-mail that there is new content pending approval, and confirmation is also sent automatically to the author. At this point in the process, a delay is introduced into the workflow due to the time it takes for an approver to review the material and decide whether to accept or reject. The approver downloads and reviews the content. After review, the content is either accepted or rejected. Based on the reviewer's response, the workflow can follow two different parallel paths:

❑ If approved, then the document is moved from its current location to a different document library for published content. An e-mail is sent to the author and approvers notifying them that the document has been approved.

❑ If the content is rejected, then the document is moved to another document library and the author and approvers are notified electronically.

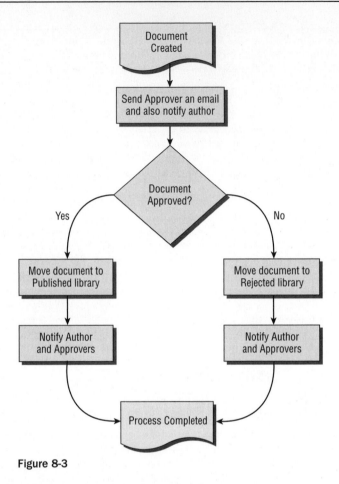

Figure 8-3

Regardless of the path, the workflow terminates once notification has completed.

Creating the Content Approval Workflow: Exercise and Tutorial

The following is a step-by-step walkthrough for creating a custom workflow that models the business process outlined in Figure 8-3. This walkthrough introduces you to the concept of creating a custom workflow using SPD. You are encouraged to walk through this exercise on your own:

1. Create a new team site called **Developers Team Site** or use one you have already created.

2. Create three new document libraries and name them **Submitted**, **Published**, and **Rejected**. For the Submitted document library, enable "Require content approval for submitted items" (from the Versioning Settings page).

3. Open SharePoint Designer. Close any site that may be open using the File ⇨ Close Site command.

4. Open your Developers Team Site by choosing File ⇨ Open Site. Inside of the Open Site dialog, enter the URL of your team site for the Site Name and click Open.

5. Open the Workflow Designer dialog by selecting File ⇨ New ⇨ Workflow. Assign the name **Publishing** to the workflow. Attach the workflow to the Submitted list. Enable "Automatically start this workflow when a new item is created" and disable all other start options. The result, with the Workflow Designer dialog showing the initial configuration for the Publishing workflow, is shown in Figure 8-4. Click Next.

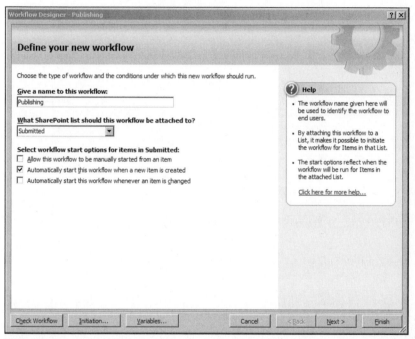

Figure 8-4

6. Assign the name **Launch and Wait** to the Step Name.

7. From Conditions, select Compare Submitted Field. Assign Field to Approval Status and Value to Pending. From the Actions menu, choose Send an email. Click this message hyperlink to open the Define Email Message dialog. Click the address book icon at the end of the To: line to open the Select Users dialog. Select the Approvers group and add it to Selected Users. The result should resemble what is shown in Figure 8-5. Add the user who created the current item to the CC: line using a similar process. Fill out the e-mail appropriately or leave it blank and then click OK. The e-mail message should resemble the one shown in Figure 8-6.

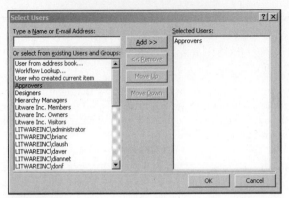

Figure 8-5

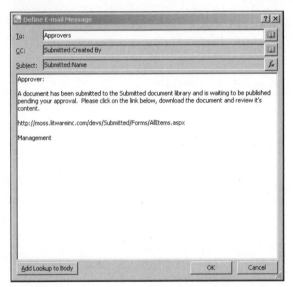

Figure 8-6

8. From Actions, select More Actions and then select "Wait for field change in current item" and click Add. Assign Field to Approval Status. Change the "to equal" phrase to "not equal" and assign Value to Pending. This completes the Launch and Wait step. Compare your result to that shown in Figure 8-7.

9. Under Workflow Steps, click the Add Workflow Step hyperlink. Assign the name **Review and Routing** for step 2.

10. From the Conditions option, choose Compare Submitted Field. Click the Field hyperlink and select Approval Status. Click the Value hyperlink and select Approved.

11. From Actions, select Copy list item. Click the "first this list" hyperlink and select Current Item. Click the "second this list" hyperlink and choose Published.

12. From Actions, select Delete Item. You may need to choose More Items in order to see this option. Click the "this list" hyperlink, select "current item," and click OK.

13. From Actions, select Send an email. Click the "this message" hyperlink and fill out the e-mail. Send an e-mail to the user who created the current item and CC: the Approvers group. An example of the completed e-mail for an approved document is shown in Figure 8-8.

14. Click the Add 'Else If' Conditional Branch hyperlink.

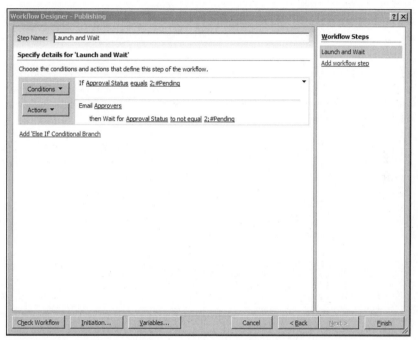

Figure 8-7

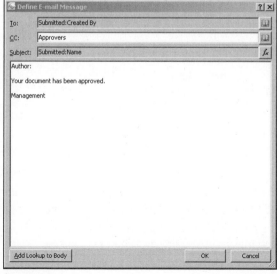

Figure 8-8

15. Add the Compare Submitted Field condition for the second condition. Assign Field to Approval Status, and Value to Rejected.

16. From Actions, select Copy list item. Assign the first "this list" to Current Item and the second to Rejected.

17. Add an action to delete the current item.

18. Add an action to Send an e-mail to the item creator stating that their document has been rejected and CC: the Approvers group. This completes the Review and Routing step; an example is shown in Figure 8-9.

19. Click Finish. As described previously, workflows authored using SPD are automatically deployed and bound to the specific list during creation. The dialog shown in Figure 8-10 confirms that the Publishing workflow has been deployed and associated with the Submitted document library and that no instances have been created.

If you are going to utilize the e-mail capability from SPD workflow, make sure that you have configured an e-mail server as part of your SharePoint deployment. If not, the workflow may have problems and not function as intended.

20. Navigate to the Submitted documents library. Create a new document and save it with the name **First Document** to the Submitted document library. Add some example text if you wish, such as **this is a test of a new workflow**. Close Word.

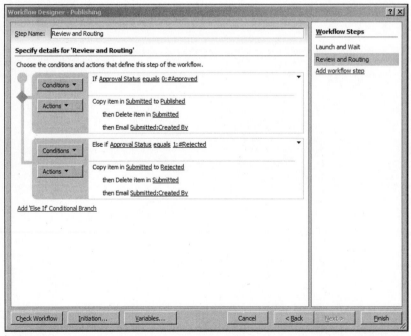

Figure 8-9

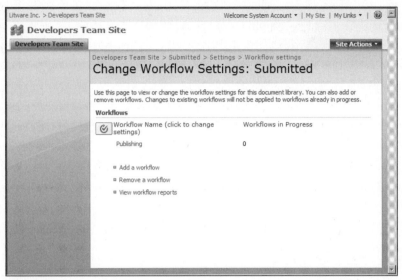

Figure 8-10

21. View the newly created document item in the Submitted library. Notice that the Approval Status of First Document is pending because you enabled content approval (back in step 2). Also notice that the Publishing column contains In Progress, indicating that an instance of the Publishing workflow has been created. Navigate to the Workflow Settings page of the Submitted document library to confirm that one instance has been created and is currently running.

22. Approve First Document by hovering the mouse over the item inside of the Submitted document library, clicking the drop-down arrow to reveal the context menu, choosing Approve/Reject, and then approving the content in the subsequent pages.

23. Once the document has been approved, confirm that the document is moved to the Published document library.

24. Repeat the process with a second document but this time reject the content to demonstrate that the created document is moved to the Rejected document library.

25. As an exercise, view the e-mails sent to the author after the document is submitted and approved or rejected. In addition, you can add users to the Approvers group and review the corresponding e-mail communication.

Creating the Content Approval Workflow: Review

The previous exercise demonstrated several key points, advantages, and limitations of the SPD environment for creating custom workflows:

❑ A workflow is a model that represents a real-world business process. Therefore, a workflow consists of a series of one or more steps.

❑   SPD creates custom workflows without writing any custom .NET code. The workflow author associates a sequence of conditions and actions with a single WSS list or library. Changes to the WSS item can trigger actions in the workflow.

❑   The conditions and corresponding actions chosen by the workflow author are formulated into a set of rules. Each rule is comprised of zero, one, or more conditions and one or more actions. These rules apply conditional logic so the workflow performs the associated action only if that condition is true. Multiple actions can be associated with a single condition, and actions can be subject to compound conditions. The actions in our custom workflow were performed serially but could have been executed in parallel.

❑   SPD takes this information and creates a rule-based template, based on XAML, that represents the custom workflow. For those interested, you should note that our custom workflow produced three separate files. These files are described as follows and can be viewed from the SPD Workflows folder, as shown in Figure 8-11:

   ❑   **Publishing.xoml:** The workflow markup file. This is only needed if the workflow uses conditions.

   ❑   **Publishing.xoml.rules:** This is the workflow rules file.

   ❑   **Publishing.xoml.wfconfig.xml:** This is the workflow configuration file.

Figure 8-11

A detailed discussion of the contents of each of these files is beyond the scope of this chapter. SPD publishes these documents to a hidden document library called Workflows for each WSS site. The Workflows document library is hidden from the web browser by default. To enable viewing of this document library, right-click the Workflows folder in the SPD folders list and select Properties. Select the Settings tab and uncheck the Hide from Browsers check box and click OK (see Figure 8-12).

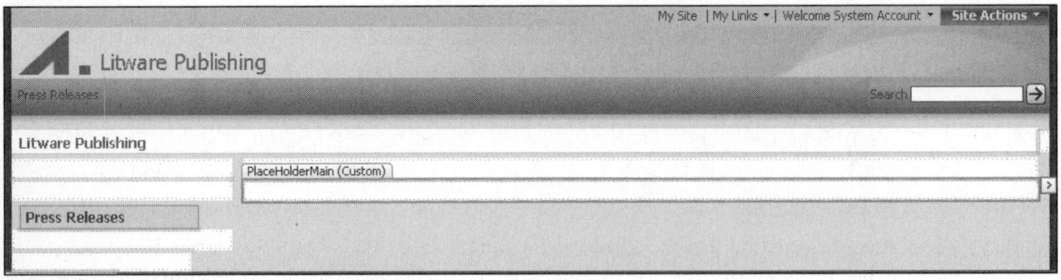

Figure 8-12

This example illustrated the use of SPD to create a custom workflow. Note a few key points:

❑ During initial definition of the SPD workflow, it must be associated with a specific list or library.

❑ The workflow template is automatically deployed to WSS. No additional administration or configuration is necessary.

❑ The workflow was initiated automatically in response to a new item being created.

❑ The SPD workflow, once initiated, executes in the security context of the logged-on user. This has implications for any resource that the workflow is accessing. For example, from our scenario, if the logged-on user did not have permission for one of the document libraries accessed by the workflow, then the workflow would not work properly.

SPD has the capability to create custom workflows that are configured to be manually started by the user. Additionally, SPD can automatically create forms that are displayed to the user so that the workflow can collect information from the user. These forms are .NET aspx pages. In the next scenario, we will create another SPD workflow that illustrates these and other capabilities.

Quality Assurance Testing Scenario

Imagine the following scenario: Your company wants to track the types of quality assurance testing the development team is performing, including the results of that testing. Developers and QA testers will submit and store testing documents to a document library, and these documents will contain the details of testing that is to be performed. This library needs to show the status of this testing (Pending, Successful, Failed) on an ongoing basis. After submitting the document, the user needs to manually launch a workflow that tracks the status of the testing. The workflow needs to create a task in the Tasks list on the WSS site. This task maintains the status of the testing. Supervisors of the technical personnel submitting the test need to be notified that a new test has been submitted. QA testing personnel will complete the testing and then update the Tasks list to indicate the result of the testing. To complete the process, the document library status needs to be updated, and the task in the Tasks list needs to be deleted.

Creating the Quality Assurance Testing Scenario: Exercise and Tutorial

1. Create a document library called **QA Testing** on the Developers Team Site used previously. Create a column called **Test Result** for the QA Testing library. Select Choice for the information type and enter **Successful**, **Failed**, and **Pending** as the possible options. Set Pending as the default value.

2. Start SPD and open the Developers Team Site.

3. Create a new workflow called **QA Test Assessment**. Attach QA Test Assessment to the QA Testing list. Ensure that the workflow is configured to start manually only.

4. Click the Initiation button to reveal the Workflow Initiation Parameters dialog. This dialog enables the workflow designer to automatically generate an ASP.NET web form that will collect data from the user prior to execution of the workflow.

5. Click the Add button. Enter **Supervisor's Email Address** as the field name. Keep Single line of text as the Information type. Click the Next button. Leave the Default Value field blank and click Finish. Click the OK button and then click Next.

6.   Enter Create Task and Notify for the Step Name. Leave the first condition blank because you always want the first action to execute.

7.   From Actions, select Collect Data from a User from the drop-down menu. Click the Data hyperlink to reveal the Custom Task Wizard dialog. This dialog provides the capability to enter details of a custom task that will be created and assigned to a user in order to collect data from the user. Click Next.

8.   Enter **QA Test Result** for the Name and **Result from the quality assurance assessment of the submitted test** as the Description. Click Next.

9.   Click the Add button to add custom fields. Enter **QATestResult** for the Field name, **Field to hold the result of the assessment** as the Description, and select Choice as the Information Type. Click Next.

10.   For options, enter **Successful** on one line and **Failed** on the next line. Uncheck the Allow blank values check box. Click Finish. Click Finish one more time to return back to the Workflow Designer dialog.

11.   Click the This User hyperlink. Select "User who created current item" and click the Add button. Click OK.

12.   At the end of the collect data action sentence, click the Variable:collect hyperlink and select Create a New Variable from the drop-down menu.

13.   Enter **QATestResultID** as the Name and click OK. This completes the first step of the QA Test Assessment workflow.

14.   Click the Add Workflow Step link and name the step **Process QA Result**.

15.   Click Conditions and select Compare Any Data Source. Click the Value hyperlink and then click the fx button that subsequently appears. In the Lookup Details section, select Tasks from the Source drop-down and then select QATestResult from the Field options. Within the Find the List Item section, select Tasks:ID in the Field selection and then click the fx button. Select Workflow Data for the Source in the new dialog box that appears and Variable:QATestResultID for the Field option. Click OK and then click OK again.

16.   Click the second Value hyperlink and select Successful from the list.

17.   From Action, select "Set field in current item." Click the Field hyperlink and select Test Result from the drop-down menu. Click the Value hyperlink and select Successful.

18.   From Action, select Delete item. (If the Delete item is not visible, click More Actions.)

       Click This List and select Tasks. Select Tasks:QATestResult in the Field drop-down menu and Successful for the value. Click OK. If a dialog box displays asking if you want to continue, click Yes.

19.   In the upper-right corner of the Condition/Action step, click the inverted triangle, and select "Run all actions in parallel."

20.   Click the Add 'Else If' Conditional Branch link.

21.   Add a second condition that tests whether QATestResult equals Failed. If so, set the Test Result column to Failed. Delete the task from the Tasks list. You can refer to steps 15–19 if you need help in creating the condition.

22. Feel free to also add e-mail notification for the item creator, just you did in the previous exercise, and e-mail notification of the submitter's supervisor.

23. Click Finish to complete the workflow and deploy the template to WSS.

24. Review the QA Test Assessment workflow in the Workflows folder in SPD. Note that there are five separate files, similar to what you saw previously for the Publishing workflow but two additional aspx files that represent ASP.NET web forms that were automatically generated to accept user input.

25. Navigate to the QA Testing document library and create a new document. Launch the QA Test Assessment workflow from the document's context menu. Enter an e-mail address for the supervisor and click Start.

26. Navigate to the Tasks list. Select Edit Item from the context menu of the QA Test Result item. Assign the QATestResult to Successful and click Complete Task.

27. Confirm that Test Result = Successful for the document in the QA Testing document library, and that the item in the Tasks list has been deleted.

Creating the Quality Assurance Testing Scenario: Review

SPD automatically generates ASP.NET web forms whenever user input is required for the executing workflow. The SPD workflow author can create two different types of workflow forms: an initiation form and a task edit form:

❏ **Initiation form:** This form gathers information from users when they start the workflow. Initiation forms are displayed to users when they manually start a workflow on a given WSS item. Recall that the Publishing workflow in the Content Approval Workflow exercise started automatically and therefore did not require an initiation form. Because the QA Test Assessment workflow required the user to enter the supervisor's e-mail address, SPD automatically generated an ASP.NET web form called QA Test Assessment.aspx.

❏ **Task edit form:** This form allows workflow users to interact with tasks in the Tasks list on a WSS site. From the SPD Custom Task Wizard, custom form fields are created and added to a custom task form. When you finish designing the workflow, SPD automatically generates the ASP.NET forms for your custom tasks (QA Test Result.aspx). When the workflow executes and the tasks are created, the user navigates to the Tasks list and modifies the list item as appropriate. The workflow responds to those changes as defined by the rules specified in the workflow and can utilize the data in the item as a source of information.

Once SPD automatically generates the ASP.NET forms, SPD can also be utilized to customize these forms. Workflow forms are ASP.NET pages with a Data Form Web Part and a master page applied to it, and they can be customized like any other aspx file. The web forms are stored in the Workflows document library on the WSS site along with the workflow source files.

SPD does not create association forms because the workflow is associated with the WSS list as part of the authoring process. Modification forms cannot be created because SPD-designed workflows cannot be modified during execution.

Creating Custom Workflows with Visual Studio Designer

Custom workflows can also be created using Visual Studio 2005 (VS2005). This capability is not native to the Visual Studio 2005 development environment and requires an add-in called Visual Studio Designer for Windows Workflow Foundation (DWF). DWF is available as part of the download titled Visual Studio 2005 Extensions for .NET Framework 3.0 (Windows Workflow Foundation). The Extensions download also contains the WF runtime engine and the WF SDK.

Once installed, DWF provides graphical designer capability, workflow project templates, and toolbox items for building custom workflows. The process for creating a custom workflow is much more involved than what you had to do in SharePoint Designer and requires experience with the development environment and some knowledge of WF. This process generally includes the following:

1. Author the workflow and write any custom .NET code required to complete the workflow.

2. Create any forms that are required for the workflow.

3. Create the two XML files that represent the feature definition and workflow definitions. These forms contain information about the workflow assembly, the binding to any forms utilized, and deployment characteristics.

4. Compile the workflow assembly and create a strong-named assembly.

5. Create a package to deploy the completed workflow to the WSS server using the SharePoint Features functionality.

6. Test and debug the workflow using the debugging capability of VS2005 and DWF.

7. Redeploy the completed workflow after any errors are fixed.

Next, you will use DWF to create a custom workflow. The focus of the exercise is to build a real-world workflow and to compare the process to that used when building custom workflows using SPD. You will look at the WF details as you build the workflow.

> In addition to installing the Visual Studio Extensions for .NET Framework 3.0, you should also install the WSS and MOSS Software Development Kits. The MOSS SDK will install SharePoint workflow templates as well as some code snippets for creating Features to deploy custom workflows to your SharePoint sites. The SDKs are not required but will definitely improve the custom workflow authoring experience.

Content Approval Workflow Scenario

Following are the steps involved in this example: A content author submits a document for approval to a SharePoint document library. The author launches a workflow that creates a task for the content approver and posts the task to the Tasks list within the SharePoint site. The approver updates the specific task item in the Tasks list and approves or rejects the document after the content has been reviewed.

Creating the Content Approval Workflow:
Exercise and Tutorial

This exercise creates a custom workflow using DWF that models the scenario just described. The workflow utilizes three custom InfoPath 2007 forms:

❑   The **association form** will allow the administrator to configure the individual who approves the content.

❑   The **initiation form** will allow comments to be entered by the author prior to starting the workflow.

❑   The **task edit form** will provide the approver with the capability to approve or reject the content.

The emphasis is on using the graphical designer, writing custom .NET code, and introducing the WF object model. Experience using Visual Studio 2005 and C# is assumed.

The code and forms created in this exercise are available in the downloadable code for this book available at www.wrox.com. Because a detailed discussion of InfoPath has not been included in this book, the reader is not expected to create these forms as part of the exercise. Readers can utilize the completed forms from the downloadable code to test the workflow. For those interested, detailed instructions for creating the InfoPath forms are included in the download. The following exercise focuses on creating the actual workflow:

1.   Open Visual Studio 2005 and create a new SharePoint Sequential Workflow project from the SharePoint Server project types. Call the workflow project **ApprovalFromScratch** and save it to a folder location called `c:\DevProjects`.

This project will add the proper workflow activities to the toolbox and configure the necessary workflow assembly and namespace references. The `System.Workflow` namespace and all the classes it contains represent the WF runtime. For creating custom workflows, a key set of classes reside in the `System.Workflow.Activities` and `System.Workflow.Rules` namespaces. The `Microsoft.SharePoint.Workflow` namespace is another key set of classes that provides functionality for managing WSS workflows, and collectively represents the WSS workflow object model. Generally these classes are not a direct part of the authoring experience unless, for example, modifying an executing workflow is necessary.

By default, the workflow contains an `OnWorkflowActivated` activity that can be seen in designer view. All WSS workflows must start with this type of activity. `OnWorkflowActivated` provides the initialization of the workflow when it is called for the first time by the WSS host environment. As part of the initialization, workflow properties are set using the design-time values, and SharePoint passes information into the workflow that configures the workflow's properties. SharePoint's information includes such things as the site and list associated with the workflow and any data collected by an initialization form. Initialization form data is stored within the workflow for local use.

2. Open the `Workflow1.cs` file in code view and make the following class-level declarations:

```
public string supervisor = default(string);
public string comments = default(string);
public string taskStatus = default(string);
```

These fields represent workflow custom properties that are passed from one of the workflow's custom forms as an XML string. The workflow then extracts these values and stores them as local properties.

3. Open the `Workflow1.cs` file in design mode and create an `onWorkflowActivated` event handler by right-clicking the `onWorkflowActivated1` activity and selecting Generate Handlers. The following code needs to be added to the generated handler:

```
workflowId = workflowProperties.WorkflowId;
```

This method is called when the workflow is activated. The incoming initiation data is read from the `workflowProperties` object variable. The `workflowProperties` object is initialized when an instance is created, and includes properties common to all workflows. Custom properties can also be passed in from a custom form, as you'll see in this exercise.

Add the following code to retrieve the workflow's properties from the initiation form:

```
XmlDocument doc = new XmlDocument();
doc.LoadXml(workflowProperties.InitiationData);
```

Add the following code to extract each individual property:

```
XmlNamespaceManager nsmgr = new XmlNamespaceManager(doc.NameTable);
nsmgr.AddNamespace("my", "http://schemas.microsoft.com/office/infopath/2003/
myXSD/2007-01-03T18:31:17");
supervisor = doc.SelectSingleNode("my:myFields/my:Supervisor", nsmgr).InnerText;
comments = doc.SelectSingleNode("my:myFields/my:Comments", nsmgr).InnerText;
```

Now it's time to add activities that define the operation and flow of the workflow. These activities include creating a new task, assigning it to the specified user, waiting until the task has been completed, and then terminating the workflow. These activities will also use custom InfoPath forms that gather information from users and submit the data to the executing workflow.

4. Information about the task properties needs to be stored in the workflow. Add the following declarations at the class level:

```
public Guid taskID = default(System.Guid);
public Microsoft.SharePoint.Workflow.SPWorkflowTaskProperties taskProperties = new
Microsoft.SharePoint.Workflow.SPWorkflowTaskProperties();
public Microsoft.SharePoint.Workflow.SPWorkflowTaskProperties beforeProperties =
new Microsoft.SharePoint.Workflow.SPWorkflowTaskProperties();
public Microsoft.SharePoint.Workflow.SPWorkflowTaskProperties afterProperties = new
Microsoft.SharePoint.Workflow.SPWorkflowTaskProperties();
```

5. Open `Workflow1.cs` in design mode. From the Toolbox, drag a `CreateTask` activity onto the workflow and drop it immediately under the `onWorkflowActivated1` activity. Within the Properties windows, set the `CorrelationToken` property to `taskToken` and the OwnerActivityName to `Workflow1`. Set the TaskId to `taskId` and TaskProperties to `taskProperties`.

The correlation token, taskToken, establishes a relationship between the executing workflow and the corresponding task so that MOSS can properly route data to and from the workflow. `taskID` is a GUID that identifies the task within the workflow instance, and `taskProperties` contains the properties to initialize the task. Now it's time to assign values to these properties.

6. Right-click CreateTask1 in design mode and select Generate Handlers. Add the following code to populate the task's properties:

```
taskID = Guid.NewGuid();
taskProperties.AssignedTo = supervisor;
taskProperties.Description = "Approve the document";
taskProperties.Title = "Document Approval From Scratch";
taskProperties.ExtendedProperties["Supervisor"] = supervisor;
taskProperties.ExtendedProperties["Comments"] = comments;
```

The task is created and assigned to the supervisor. At this point, there may be a time lag before the supervisor approves or rejects the content, so you must utilize a `Wait` activity.

7. From the Toolbox, drag a `While` activity onto the workflow and drop it immediately under the `createTask1` activity. From the Properties window, set Condition to `Code Condition` and the subproperty Condition to `notFinished`.

The `While` activity loops until a certain condition evaluates to true. This workflow will use it to loop around the task change event until the supervisor explicitly completes the task. Setting the Condition property to `Code Condition` specifies that the `While` activity call a function to determine its complete state. The `Condition` subproperty specifies that the `notFinished` method will define completion. Next you will add an `OnTaskChanged` activity to the `while1` activity to determine when the `while1` activity is assessed. For this scenario, you want the `while1` activity to loop every time the task is edited. Therefore, you need an activity that handles the task change event.

8. Drag an `OnTaskChanged` activity from the toolbox and place it on top of the `while1` activity. Set the CorrelationToken to `taskToken` and the OwnerActivityName to `Workflow1`. Set the TaskId to `taskId`, BeforeProperties to `beforeProperties`, and AfterProperties to `afterProperties`.

The workflow is paused while the `OnTaskChanged` activity waits for the task change event, and then it is reactivated when the task is changed. When the task is changed, you need to determine whether the supervisor approved or rejected the document. If so. then the task is complete. You need to add a variable that holds a Boolean value that determines whether the task is completed.

9. Declare the following class-level variable that will hold the Boolean value determining the completion of the workflow:

```
private bool isFinished = false;
```

10. Create a handler for the `onTaskChanged1` activity. Inside this method, add the following code:

```
if (this.afterProperties.ExtendedProperties["TaskStatus"].ToString() == "Approved"
| this.afterProperties.ExtendedProperties["TaskStatus"].ToString() == "Rejected")
{
    isFinished = true;
}
```

This method is invoked when the workflow receives the task change event. Inside this method you need to retrieve the contents of the `TaskStatus` variable, which holds the value of the supervisor's assessment (Approved or Rejected). The `afterProperties` variable represents the task properties after the task change event has occurred. You will set a variable `isFinished` that specifies whether the task is complete based on the value of `TaskStatus`.

11. The `notFinished` method should have been automatically created in step 7, but if not, go ahead and add it to the `workflow1.cs` file:

```
private void notFinished(object sender, ConditionalEventArgs e)
{
    e.Result = !isFinished;
}
```

This method is invoked by the `while1` activity each time the task is changed to determine whether its condition is met. As long as the `Result` property of the `ConditionalEventArgs` object evaluates to true, the `while1` activity continues to wait.

Now, each time the user edits the task, the `onTaskChanged1` activity handles the task changed event. It invokes the `onTaskChanged` method, which examines the task properties and sets the `isFinished` variable to indicate whether the supervisor completed the task. The `while1` activity then invokes the `notFinished` method, which sets the result of the event to the opposite of the `isFinished` variable. If `isFinished` is equal to `false`, then the event result is set to true, and the `while1` activity keeps waiting for task changes; if `isFinished` is equal to `true`, then the event result is set to false, and the `while1` activity completes and the workflow continues to the next activity. The last step in the workflow is to complete the task.

12. From the Toolbox, drag a `CompleteTask` activity onto the workflow and drop it immediately under the `while1` activity. Set the CorrelationToken to `taskToken` and OwnerActivityName to `Workflow1`. Set TaskId to `taskId` and TaskOutcome to `TaskStatus`.

The `CorrelationToken` and `TaskId` properties are set to the variables used in the `createTask1` activity. This maintains the binding between this activity and the task created by the `createTask1` activity.

The workflow designer view of the completed workflow is shown in Figure 8-13.

13. Build the ApprovalFromScratch project and fix any errors.

This completes the DWF portion of the workflow. The corresponding InfoPath forms that are utilized as part of the workflow would be created next. This workflow would need three forms: association, initiation, and task edit. The individual steps for creating the InfoPath forms and the workflow are available in the downloadable code at www.wrox.com. Once the workflow and forms are complete, the workflow needs to be deployed, tested, and debugged.

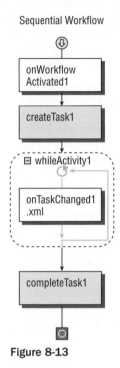

Figure 8-13

The actual deployment steps are not discussed in detail but they are covered in the downloadable material. The primary purpose of this exercise was to illustrate the capability of the Visual Studio Designer for Windows Workflow Foundation, give you a feel for the custom workflow creation process, and compare and contrast the DWF process with that used with SPD. The table toward the end of the chapter summarizes the differences between these two authoring tools and is included as a detailed reference.

Custom Workflow Debugging

Custom workflows built using DWF can take advantage of most of the debugging capabilities of VS2005. Recall that when you built the custom workflow using SPD, you did not have debugging capability. Because the WF engine is hosted by another process, you must attach to that process to accomplish debugging. For WSS workflows, the host process will be the IIS worker process called `w3wp.exe`. Once attached to the worker process, the normal debugging process of setting breakpoints, stepping into and over code operations, and viewing call stack windows can be utilized. Unfortunately, VS2005's complete feature set is not available because debugging of just-in-time (JIT) exceptions is not supported. A complete description of the capability and a step-by-step process for debugging is provided in the WF SDK that is installed with DWF.

Creating the Content Approval Workflow: Review

Building the custom workflow using DWF illustrated the capability of the graphical designer, the general process for utilizing activities, and examples of the custom .NET code necessary to author a real-life workflow. As part of the authoring process, we also covered some of the basics of WF and the WF object model. This exercise demonstrated a lot of the key differences between authoring a workflow with SPD and DWF. These differences are summarized in the following table:

	VS2005	SPD
Authoring	Full development environment with a graphical designer that produces a template which can be *associated* with multiple sites, lists, and content types	Wizard-driven interface that utilizes conditions and actions to produce a template that contains a set of declarative rules and is bound to a specific list
Custom .NET code	Yes	No
Types	Sequential, State Machine	Sequential only
Completed Workflow	Workflow markup file and code-behind files are *compiled* into workflow assembly.	Workflow markup, workflow rules, and supporting files are stored *uncompiled* in a hidden document library on the site and compiled on demand.
Debugging	Yes. Visual Studio 2005 debugging except for JIT exceptions.	No step-by-step debugging available
Deployment	Packaged as a SharePoint feature and deployed to the WSS server by an administrator	Deployed automatically when workflow is completed
Association	Template must be associated with each and every list before it will be available	Association occurs at design time only
Workflow Forms	Can use any forms technology, such as InfoPath 2007 or ASP.NET 2.0 forms	Automatically generates ASP.NET 2.0 forms, which can then be customized
Create Custom Activities and Conditions	Yes	No. Must use a predefined set of activities and conditions.
Workflow Modification	Executing workflows can be modified.	No modification is possible.

In addition to creating custom workflows, DWF can also be used to create custom activities and custom conditions. A condition is a .NET assembly containing a static method that evaluates certain aspects of executing code and returns a Boolean value. Once created, custom activities and conditions are then deployed to the WSS server and utilized by SPD in the custom workflow authoring process. Custom activity creation was beyond the scope of the chapter, but an example is included as part of the downloadable code available at www.wrox.com.

Summary

WF in combination with WSS provides a platform for the information worker to create and utilize workflows in order to improve the effectiveness of document or content-centric business processes. This chapter has provided an introduction to WF, the workflow capability and implementation characteristics of WSS and MOSS, and the hands-on exercises for creating custom workflows using SPD and DWF.

WF provides a runtime engine and a set of services for executing workflows that are hosted for SharePoint by WSS. WSS provides the SharePoint architecture for hosting workflow applications but does not include any workflows itself. MOSS extends the SharePoint workflow capability by installing templates that provide ready-to-use workflows, and MOSS provides workflow integration with the Office 2007 clients.

SPD and DWF are two tools for authoring custom workflows. Each of these tools has its own strengths, and they are targeted at different audiences. SPD is a wizard-driven environment for workflow authoring that doesn't require .NET developer skills, while DWF is a more complete, versatile development environment but requires a .NET developer. DWF can also be used to create custom activities and conditions that can subsequently be used in SPD. A detailed comparison of each of these tools was also provided.

The workflow author should have the knowledge and skills necessary to build custom workflows using SPD with little difficulty based on what was presented in this chapter. Authoring custom workflows with DWF may require additional skill and experience depending on the reader's .NET development background and understanding of the WF and WSS object models. However, regardless of the reader's development skill, one should have a good understanding of how workflow is created and utilized within the SharePoint environment.

Collaboration effectiveness and efficiency can be improved by automating business processes using SharePoint workflow, so ultimately workflow can be another highly useful tool in SharePoint's collaboration arsenal.

9

Web Content Management in Office SharePoint Server

Web Content Management (WCM) in Microsoft Office SharePoint Server 2007 (MOSS) is a set of technologies built on top of the Windows SharePoint Services v3.0 platform. Previously, Microsoft Content Management Server 2002 was the product that delivered WCM capability for the Microsoft platform. This was a separate product from SharePoint Portal Server 2003. Each of these products had features that were desired by most users. WCM combines the functionality of Content Management Server 2002 and SharePoint Portal Server 2003, along with a host of new features. Typically, WCM systems allow non-technical users to author and manage content on the web with minimal involvement from IT.

This chapter explores WCM, and the main objectives for this chapter are:

❑ Understand the architecture, concepts, and capability of MOSS WCM

❑ Create WCM Web intranet and Internet portals using different security configurations

❑ Understand and utilize the WCM authoring and publishing process

❑ Provide an understanding and demonstration of site variations and content deployment

Content Management

Content management describes the process of controlled publishing of information to content repositories; this information is then made accessible via a web browser for universal access. This controlled publishing environment includes capability for content authoring, web site and page branding, content approval, and subsequent management of the published information. There must be capability to reuse content and customize the content that is rendered to individual users. A common look-and-feel is typically achieved through some type of page templating architecture, cascading style sheets, and visual navigation. The content repositories include management capability such as version control, check-in and check-out, workflow, and security. SharePoint's content management has historically included lists and document libraries. MOSS introduces new content management capability in the form of *Web Content Management* (WCM).

Web Content Management

WCM involves content authoring, web page and site branding, and controlled publishing. As mentioned previously, this capability was the core feature set of a separate product called Microsoft Content Management Server 2002, which is being discontinued. WCM is only one of the features of *Enterprise Content Management (ECM)* provided by MOSS.

ECM, sometimes also referred to as Electronic Content Management, includes records management, WCM, and document management. Enterprise content is managed using workflows, retention and expiration policies, and auditing. WCM is focused on content authoring and publishing, relying on workflows to manage the business processes associated with authoring and approval, as we discussed in Chapter 8. ECM and WCM can both be delivered as an internal-only function via an intranet web portal or externally through an Internet or extranet portal. MOSS contains a comprehensive set of features for managing documents and web content.

Key Web Content Management Features

WCM provides the following features:

- ❏   Simple page authoring capability
- ❏   Separation of page presentation from content for rapid creation and easy modification
- ❏   Flexible and consistent look and feel capability based on content types, master pages, and page layouts
- ❏   Controlled publishing using Windows Workflow Foundation
- ❏   Site management tools that provide a drag-and-drop interface
- ❏   An extensible platform based on ASP.NET v2

Historically, one of SharePoint's key strengths is *site provisioning*. In this process, a *site template* is used to create and configure a web site based on a set of predetermined features. This process is complete in roughly seconds versus tasking a development effort for every request for a separate web site or web page. The use of templates is being reapplied to WCM through the use of master pages and page layouts. These two templates work together to define the look and feel of the WCM web site. This simplifies the authoring process and allows content creators to focus on content versus presentation. Content authors use page layouts to create new web content pages that are then subject to an approval process before the page is published.

WCM is built on top of WSS v3 and, therefore, can utilize the workflow capability of Windows Workflow Foundation to manage the business processes associated with content creation, approval, and publishing. MOSS includes a WCM approval workflow that can be used as is, or custom workflows can be created, as shown in Chapter 8.

WCM content can be created using the web browser, Word 2007, or InfoPath. These three authoring environments are available out of the box along with a document converter framework. A document converter is a software component that converts content from one format into another format. Specifically, the converter framework is designed to read content from an external format and convert it into a MOSS web page.

WCM also includes the capability to publish content in one language while duplicating the content so it can be translated into other languages. This capability is called *site variations*. Site variations are applicable to rendering content for different devices, as well as for multiple languages. When a new content page is created and published, this page is automatically copied and inserted into the structure of the other sites as well.

MOSS WCM provides a rich feature set for authoring and publishing Web content. In combination with all the other features, this delivers a comprehensive platform for building web portals. Since MOSS is built on top of WSS v3, and WSS v3 is built on top of ASP.NET v2, MOSS can be readily customized and extended using the same code one would create for a standalone ASP.NET Web application.

SharePoint has historically been used as an intranet collaboration platform. The new authentication capability in WSS enables MOSS to be used more readily as an Internet or extranet portal. This creates a new challenge for SharePoint, since Internet-facing web sites are used differently than their intranet counterparts. Internet sites tend to provide static content and a read-only environment. WCM is the technology that fills this challenge. WCM extends the SharePoint collaborative portal environment to include the capability for creating and publishing web content. These publishing features are activated using new publishing site templates mentioned previously and through the use of *Features*.

Publishing Templates and Publishing Sites

As indicated, one of SharePoint's strengths is the ability to create (provision) web sites with little or no coding, and site templates are the tool for delivering this capability. MOSS includes several new templates, one of which is the new publishing template. MOSS and WSS both provide templates, with the former focusing on enterprise portals and the latter with collaborative team sites. The following WCM discussion assumes that MOSS is installed, since WCM requires MOSS.

The first step toward utilizing WCM is to enable publishing capability. Publishing capability is introduced by provisioning a site from a publishing template or by directly activating the publishing feature. The publishing site templates automatically activate the publishing feature. The following list reviews these options in a little more detail.

❑ **Create a new site collection using a publishing template.** There are two different publishing templates available, the Collaboration Portal and the Publishing Portal. During the site collection provisioning process, the site collection creator chooses one of these two templates on the Create Site Collection web page. A complete description of each of these templates can be viewed by navigating to the Create Site Collection web page from Central Administration ➪ Application Management, but a brief description is shown in the table that follows for convenience.

	Description
Collaboration Portal	A template for creating a corporate intranet
Publishing Portal	A template for creating an Internet site or a corporate portal

❑ If upon navigation to the Create Site Collection web page you don't see the Publishing tab in the Template Section, then it's likely you have a WSS-only installation, and you will need to install MOSS to utilize WCM.

❑ **Create a new site using a publishing template.** A publishing site collection contains three different options for creating a site with publishing capability. These templates are shown along with their descriptions under Template Selection on the New SharePoint Site web page during the creation of a new site. Brief descriptions of each are summarized in the following table.

	Description
Publishing Site	Creates a site for publishing web pages
Publishing Site with Workflow	Creates a site for publishing web pages on a schedule by using approved workflows
News Site	A site for publishing news articles and links to news articles

❑ When a site collection is provisioned with a publishing template, one of the functions of the template is to activate a feature called the Office SharePoint Server Publishing Infrastructure. The status of this feature can be viewed from the Site Collection Features web page. Creating a child site with one of the preceding publishing templates activates the Office SharePoint Publishing feature. If a site has been provisioned without using a publishing template, the publishing feature can be enabled directly and thus provide publishing capability.

❑ **Activate the publishing feature directly.** The publishing feature is not active by default for non-publishing sites. Feature activation is accomplished from the Site Collection Features page. For example, if you have provisioned a site collection using one of the enterprise templates that doesn't include publishing capability, this capability can be enabled after the fact.

Once publishing capability has been enabled for the SharePoint site, you are ready to begin utilizing these features. The next part of this chapter illustrates this process from the beginning by creating a new web application that will contain a publishing site collection.

Creating a WCM Web Portal: Exercise and Tutorial

In this exercise, we will demonstrate how to create a new publishing web site that is accessible from the Internet and from the intranet. WCM is a set of features and therefore is available for intranet, Internet, or extranet portals. The major difference between these scenarios would likely be the authentication mechanism and, therefore, the configuration of the web site.

The web site will allow internal users to author, approve, and publish web content. These users will utilize their Active Directory credentials for authentication. The web site will also be configured to allow anonymous users to browse selective content from the Internet, but some content will not be viewable to the anonymous user. This scenario will utilize two different zones to achieve the dual authentication requirements for the web application (Windows authentication and anonymous users).

Part 1 — Creating a New Web Application and a New Publishing Site Collection

1. Open the SharePoint 3.0 Central Administration web site and navigate to the Application Management page.

2. Click on Create Or Extend A Web Application from within the SharePoint Web Application Management section.

3. On the Create Or Extend A Web Application page, click Create A New Web Application.

4. From the Create A New Web Application page, add the following properties:

   a. Enter **Litware Internet Publishing** for the description.

   b. Make sure NTLM is the Authentication provider.

   c. Make sure Anonymous Access is set to No.

   d. Create a new application pool called **Litware Pool**.

   e. Configure the application pool user account to use **Litwareinc\administrator** and enter an appropriate password.

   f. Select IIS to automatically restart.

   Note that many properties were left with their default values, including the port number. It is also possible to use port 80 in combination with host headers.

5. Click the OK button to create the web application.

6. On the Application Created page, click the link Create Site Collection.

7. On the Create Site Collection page, enter the following values:

   a. **Litware Publishing** in the title field.

   b. Under template selection, choose the Publishing Portal template after clicking on the Publishing tab.

   c. Enter **Litwareinc\administrator** for the Primary Site Collection Administrator.

8. Click OK to create the new site collection with publishing features activated.

9. From the Top-Level Site Successfully Created page, open the new site by clicking on the URL for the new site. Add the home page of the site to your favorites with the name Litware Publishing Intranet. The home page for the Litware Publishing portal web site is shown in Figure 9-1.

At this point we have created an intranet web site that is configured for Windows authentication. Internal users will author and publish content from this web site using the Active Directory accounts. This intranet access represents the *default zone* of the web application. The published content will be viewable from the Internet by anonymous users via the *Internet zone*. Since each zone is associated with a new IIS site, we must extend our web application into the Internet zone.

If you want content within the web application to be included in search results, ensure that at least one zone is configured to use NTLM authentication. NTLM authentication is required by the indexing technology to index the Web content.

Figure 9-1

Part 2 — Extending the Web Application into the Internet Zone and Configuring Anonymous Access

1. From SharePoint 3.0 Central Administration home page, navigate to the Application Management page.

2. Click on Create Or Extend A Web Application from within the SharePoint Web Application Management section.

3. On the Create Or Extend A Web Application page, click Extend An Existing Web Application.

4. On the Extend Web Application To Another IIS Web Site page, click the drop-down to select a web application and click Change Web Application.

5. In the Select Web Application page, choose the Litware Internet Publishing application previously created.

6. Enter the following values on the Extend Web Application to Another IIS Web Site page:

   a. Make sure Create A New IIS Web Site is selected.

   b. Enter **Litware Public Internet** for the Description, with the port number included in parentheses.

   c. Select Internet from the Zone drop-down list under the Load Balanced URL section.

   Anonymous access will be configured in step 8. Make sure you make a note of the port and URL to ensure you don't have to look for it later.

7. Click OK to extend the web application.

8. To enable anonymous access, go to Application Management (this screen should be open since completing step 7) and under Application Security click the Authentication Provider link.

9. On the Authentication Provider page, click the link for the Internet zone. If the Internet zone link is not present, you may need to choose a different web application from the drop-down list box on the right-hand side of the page.

10. On the Edit Authentication page, check the box to Enable Anonymous Access and click the Save button.

This enables anonymous access for the Internet zone only. Next, we'll need to provide our administrator account or equivalent account with full control to make administrative changes to the Internet zone. This is done using Web Application Policies. Web Application Policies are a new WSS capability. These policies override any other permission setting within the web site itself and are configured from Central Administration.

11. From Application Management, click Policy For Web Application under the Application Security section.

12. Ensure the correct Web Application (Litware Internet Publishing) is selected on the Policy For Web Application page and click Add Users from the toolbar.

13. On the Add Users page, make sure All Zones is selected in the Zones drop-down box and click the Next button.

If we had created a specific account that was to have full control of only the Internet zone, we would have selected the Internet zone instead of All Zones.

14. Enter **LitwareInc\administrator** in the Users box and check the Full Control permission check box. This provides the administrator account full control access to both the default and Internet zones.

15. Click the Finish button.

16. Open the Internet site using the Internet-specific URL that you made a note of previously in step 6. Add this site to your favorites with the name **Litware Publishing Internet**.

The final step is configuring the Internet site to allow anonymous access to the top-level Internet site in the site collection. On the home page of the out-of-the-box Internet site, there is a shortcut to the administration page where you have the option to give anonymous users authorization to access content on the whole Internet site.

17. From the home page of the Internet site, click on the link Enable Anonymous Access. If the Enable anonymous access link is not present, skip to step 19.

18. From the Change Anonymous Access Settings: Litware Publishing page, select Entire Web Site for Anonymous Users Can Access and click OK. Skip steps 19–22.

19. From the Internet web site, choose Site Settings and then Modify All Site Settings.

20. On the Site Settings page, select Advanced Permissions.

21. On the Permissions: Litware Publishing page, select Settings, and then select Anonymous Access.

22. On the Change Anonymous Access Settings: Litware Publishing page, select Entire Web Site and click OK.

23. Open a new browser session or log out by using the Welcome control at the top of the site.

24. Navigate to the Internet Litware Publishing web site. The site should open and be viewable as an anonymous user. You can confirm that you aren't authenticated because there's a Sign In link in the upper right-hand corner of the site where the Welcome control usually resides. This can be seen in Figure 9-2, which shows the Litware Publishing Internet site as viewed by an anonymous user.

Figure 9-2

Part 3 — Restricting the Viewability of Specific Web Site Content from the Internet Zone

Keep in mind that anonymous access can be configured per site in the site collection. For example, you can hide one or more sites from anonymous users and make those sites available only to authenticated users.

1. Log on to the Internet site by clicking the Sign In link.

2. Select Site Actions and then click Manage Content And Structure.

3. On the Site Content And Structure page, select Press Releases from the left-hand navigation tree. From the drop-down context menu adjacent to Press Releases, select Advanced Permissions.

4. Select Actions and then click Edit Permissions. Click OK when you are prompted to accept your changes. This breaks inheritance from the parent site.

5. Click Anonymous Access from the Settings drop-down list.

6. From the Change Anonymous Access Settings page, select Nothing and click OK.

7. Open the Internet site from a new browser session and notice that the Press Releases section is absent from the horizontal navigation (see Figure 9-3). Click on the Sign In link in the upper right-hand corner and notice that the Press Releases section now appears since you have been authenticated.

Figure 9-3

That completes the creation and configuration of the Internet Web portal. This configuration could be extended to include other methods of authentication if required. For example, you may want to configure forms authentication for the Internet users so they can access more secure content from the Internet zone. In the previous version of SharePoint, this was not possible because SharePoint relied completely on IIS to authenticate users, and this required visitors to have a valid Windows account. WSS v3.0 provides new capability and user accounts can be stored in an alternate data store because of the support for the ASP.NET v2.0 membership and role provider model.

WCM Site Content and Structure

Part 3 of the previous exercise illustrated the use of the new Site Manager tool. As shown in Figure 9-4, Site Manager provides a two-pane view of the site collection and provides the capability to manage an individual site collection's taxonomy and content. From this view, the hierarchy of sites and pages in a site collection can be copied, moved, deleted, etc. The left pane shows a typical tree-control view of the sites, libraries, and pages in the collection. Selecting an item in the tree expands its constituent site elements in the right pane. The scope of this tool includes all sites in the site collection. It cannot be used to display multiple site collections at the same time and cannot be used to move content between site collections.

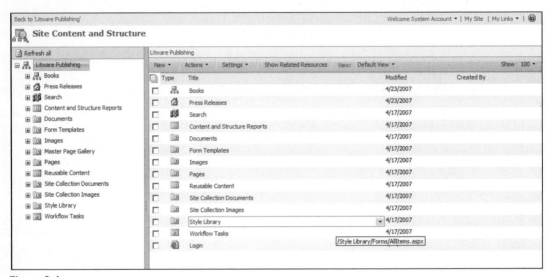

Figure 9-4

This tool is accessible in two ways:

❑ From the Site Actions menu, select Manage Content And Structure.

❑ From the Site Settings Web page, select Content And Structure under the Site Administration column.

Site Manager provides for an easy drag-and-drop interface for managing navigation, sites, site hierarchies, and content.

Site Manager shows the actual structure of sites and their contents. A few points are worth noting here regarding the structure of publishing sites and how they can be managed using this tool:

❑ **Each site is represented as a node in the tree, providing a single view of all sites in the site collection.** For this Litware Publishing site, you'll notice the Books and Press Releases sites in the upper-left navigation pane of the tree. Expand the view for each of these two sites by clicking on the plus sign next to their name. Each site contains a set of document libraries named Documents, Images, Pages, and Workflow Tasks. These libraries are created as part of activating the publishing site capability.

❑ **Sites and their content can be managed from this tool.** Place your cursor over the Books site and click on the drop-down arrow. A context menu is displayed that shows several options for managing this site. By clicking on the name of the site, its content is displayed in the navigation pane on the right. Click on one of the document libraries. The document library options of New, Actions, and Settings are displayed, allowing you to manage these features without navigating to the library directly.

❑ **Each site has its own pages library.** The pages library contains the web content pages that are creating during the authoring process. Each web page is represented as a document in this library. Therefore, all the WSS features available to document libraries, such as versioning, are available for web content as well.

Web Content Management Architecture and Page Structure

Delivering a robust authoring and publishing infrastructure comprises several different requirements, including the following:

1. The ability to provide a common look and feel across the web site

2. The ability to create new content easily and rapidly while controlling the authoring process

3. The ability to update existing content quickly

WCM fulfills all these requirements via the architectural and page structure components discussed in this section.

Templates

To satisfy these requirements, a technical strategy that has proven effective in the past is to separate the look and feel (presentation) from the content and provide a template mechanism to control the structure, navigation, and ease of reuse. This approach allows the content to be updated independently by content owners without involving page designers or developers. Likewise, page designers and developers can modify the presentation and implement a new look and feel without affecting content. The WCM authoring and publishing process is achieved using a set of templates. The two different types of templates include *master pages* and *page layouts*.

Master Pages

Master pages determine the look and feel and navigation features of your web site and web pages. These would include logos, headers, footers, and any other branding common to all your web pages. Master pages allow existing and new content pages to inherit and reuse a common design, which simplifies development and maintenance. They contain controls that are shared across multiple page layouts; changes to the master page will propagate to all pages in the site and therefore provide a re-branding mechanism. Custom master pages can also be created and applied to SharePoint sites. Master pages also include reference to Cascading Style Sheets (CSS). Each site must reference a single master page. To achieve a common branding across the site collection, all sites could use the same master page. However, multiple master pages could be utilized for different sites in the collection. Every site displays a link to its own master page gallery from the Site Settings page. (Master pages are discussed thoroughly in Chapter 5.) The WCM authoring process also uses page layouts to provide more granular control.

Page Layouts

Page layouts are separate from the underlying content and represent a template for the content. They control the arrangement of the authored content, the look and feel, and the type of content being authored. This helps to enforce consistency across content authors. Multiple page layouts can be created to accommodate several different types of content pages (Press Releases, Job Announcements, Products, etc.). A page layout is an ASP.NET page with special controls called *field controls*. Page layouts may also contain Web Part Zones and Web Parts. Master pages and page layouts collectively govern the overall look and feel of your SharePoint site, as illustrated in Figure 9-5.

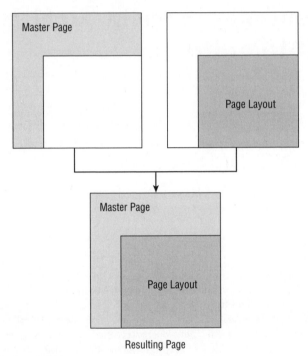

Figure 9-5

All page layouts are bound to a single WSS content type, and they reference a master page. When a content page is requested, the master page and page layout are merged with the content to produce the rendered output displayed in the browser. Master pages and page layouts are stored in the Master Page and Page Layout Gallery at the root of the site collection.

Master Page and Page Layout Gallery

The top-level site in the site collection contains a special document library called the Master Page and Page Layout Gallery, which stores all page layouts and master pages. Two types of master pages are stored in the Master Page Gallery, each identified by a unique icon:

❑ **System master pages** define the appearance of form, view, and Web Part pages. For example, compare the page for a document library (such as the Master Page Gallery) with the home page of your Internet site.

❑ **Site master pages** define the look and feel of the pages published within the site. These pages, like the home page, are the ones your visitors see and, therefore, are more important to customize than the system master pages.

Because it's a standard document library, it supports versioning, workflow, etc. Although MOSS creates a Master Page Gallery for every site, new content pages can use only the page layouts stored in the Master Page Gallery of the top-level site in the site collection.

Other Key WCM Components

Templates such as page layouts provide a mechanism for creating and reusing capability as described in the previous sections. Page layouts utilize field controls to define the location and type of information that authors will create and content types to describe the overall structure of the content, including the field controls. These additional WCM architecture components are briefly discussed in this section.

Field Controls

Field controls are the content containers on a page; they render the content inside the page layout. Their dual function includes displaying content to site users and providing editing capability during authoring. For example, consider authoring a job announcement — the job title is rendered as a heading while being viewed by site users and also rendered as an editable text box for a content author. Each field control displays data from one or more columns defined by the layout's content type.

Content Types

A single content management site may hold a range of different file types, such as documents, images, and spreadsheets, each characteristic of a specific type of content. Each different type of content can now be classified according to *content type*. Content types are new to SharePoint and are part of the WSS v3 feature set.

A content type defines the metadata and a group of reusable settings that describes the specific type of content. A content type is an abstract entity. It is somewhat analogous to a class in object-oriented programming. To use a class you typically create an instance, which is represented by an object. The object represents a real-life entity whose characteristics are described within the class structure. Likewise, an instance of a content type is a list item or a document. An example of the Page content type is shown in Figure 9-6. The Page content type is used in the creation and publishing of content pages. Content types are hierarchical. As the figure shows, the Page content type inherits from System Page.

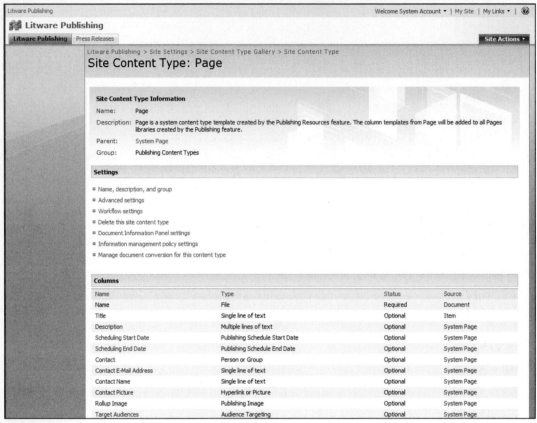

Figure 9-6

For example, and as can be seen from the figure, a content type can specify the following settings:

- ❑ The site columns (metadata) and corresponding data types for this content
- ❑ The document template on which to base new items of this type
- ❑ The custom New, Edit, and Display forms to use with this content type
- ❑ Workflows that you can associate with the content type
- ❑ Information management policies that apply to this content
- ❑ The types of document conversation available for content of this type

In previous versions of SharePoint, metadata was defined at the specific list or document library level. If multiple types of content existed in a single list, all items or documents contained all of the same metadata.

Content types introduce a new degree of flexibility because they allow you to define the structure and characteristics of list items outside the context of a specific list instance. Once defined, a content type can be used across multiple lists or document libraries. In addition, lists and document libraries can be enabled to support multiple content types, each with their own characteristics. This is a huge advantage that was not previously available.

Page Content

Page content is stored in a WSS list (remember that document libraries are lists). The actual "page" is represented by a list item or document stored in the Pages document library. The list item information is merged with the page layout and master page at rendering. This list has columns (metadata) that are described in the content type of the page, as previously discussed. Representing pages as list items enables all of the WSS features associated with lists to be made available for web content pages.

Authoring, Approving, and Publishing New Pages

WCM content can be created using the Web browser, Microsoft Word 2007, and InfoPath. Irrespective of how the content is created, workflows can be utilized to manage the submission, approval, and publication of content.

Authoring

A user with the Contributor permission level can create and edit pages, as summarized in Figure 9-7. MOSS provides a series of page layout templates for new page creation. These are shown in Figure 9-8. When authors create a new page or edit an existing one, the changes must be approved by an authorized user before the page can be published. The author can save their work as a draft and any member of the contributors group, including the author, can check it out to continue the authoring process. When the authoring process is complete, the page can be submitted for approval. The page is not published until it is approved by a member of the Approver group. Until approved, the page is considered a minor draft and is assigned the version number 0.1.

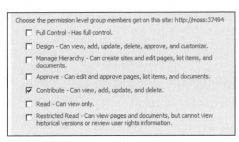

Figure 9-7

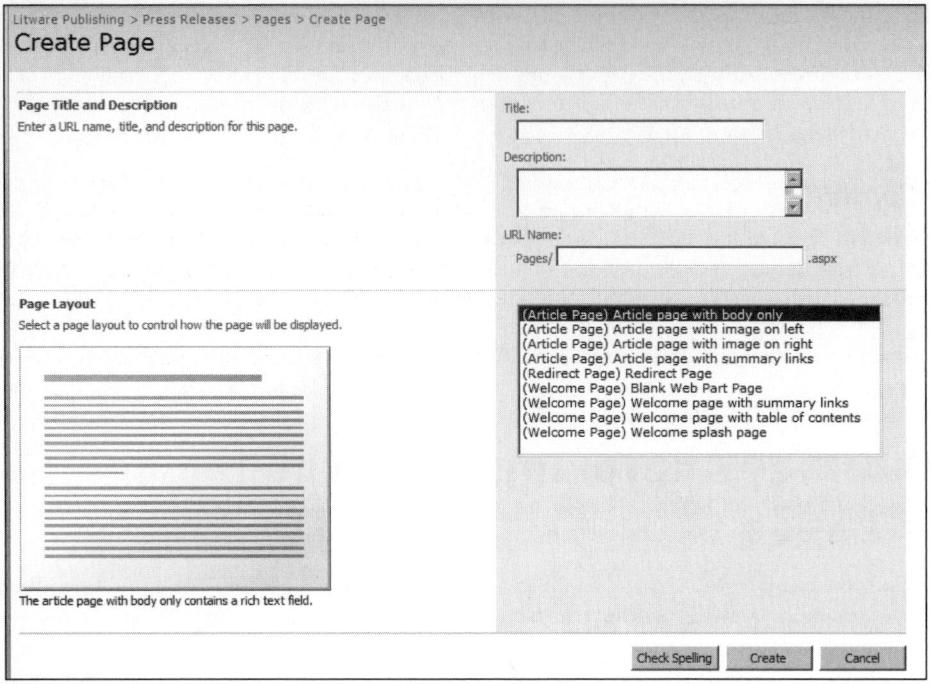

Figure 9-8

Content Approval

Approval by a member of the Approver group who has the Approve permission level or Owner group who has the full control permission level publishes the submitted page and updates the page version number to a major number, such as 1.0. Publishing may not occur immediately because the page may be under the control of a time or date restriction. The publishing process, especially in a multitier environment, will almost certainly have scheduled content updates.

Authoring, Approving, and Publishing: Exercise and Tutorial

The exercise will illustrate the authoring, approval, and publishing process for new content. Specifically, this will include the following:

1. Creating a new content type that contains new site columns

2. Creating a new page layout template

3. Authoring a new page using the custom page layout and illustrating the out-of-the-box content approval process for publishing new page content

We could have demonstrated the page authoring process by using a predefined page layout, but this would not have shown all of the necessary steps to create new content. So, we will choose to start from scratch with everything except the master page.

Part 1 will describe the process for creating a new content type. This content will be used for the new page layout created in Part 2. Part 3 will demonstrate the authoring, approval, and publishing process. So let's get started.

Part 1 — Creating a Custom Content Type

Content types use columns to define the field types that will be used in the page layout. These columns are referred to as *site columns*. Site columns offer reusability and are new to WSS v3. You can define a column at the level of the site collection and then associate this column with any of the lists, libraries, or content types on the site. Changes to the site column are, by default, pushed forward to any list, library, or content type that uses the specific site column. This can be disabled per site column.

When creating a new content type, you must decide on the desired scope of your content type. The scope can be limited to one of the subsites in the site collection. Each site has its own content type gallery. By creating your content type within a subsite, you limit its scope to that level and below. A much broader scope is the whole site collection. A content type created at the level of the site collection will be available to all sites. This can be accomplished by navigating to the Site Content Type Gallery from the Site Settings administration page from the top-level site in the site collection. The gallery has already been populated with content types due to the provisioning process, as shown in Figure 9-9.

New content types can be created by clicking Create in the toolbar. This will display the New Site Content Type page, as shown in Figure 9-10.

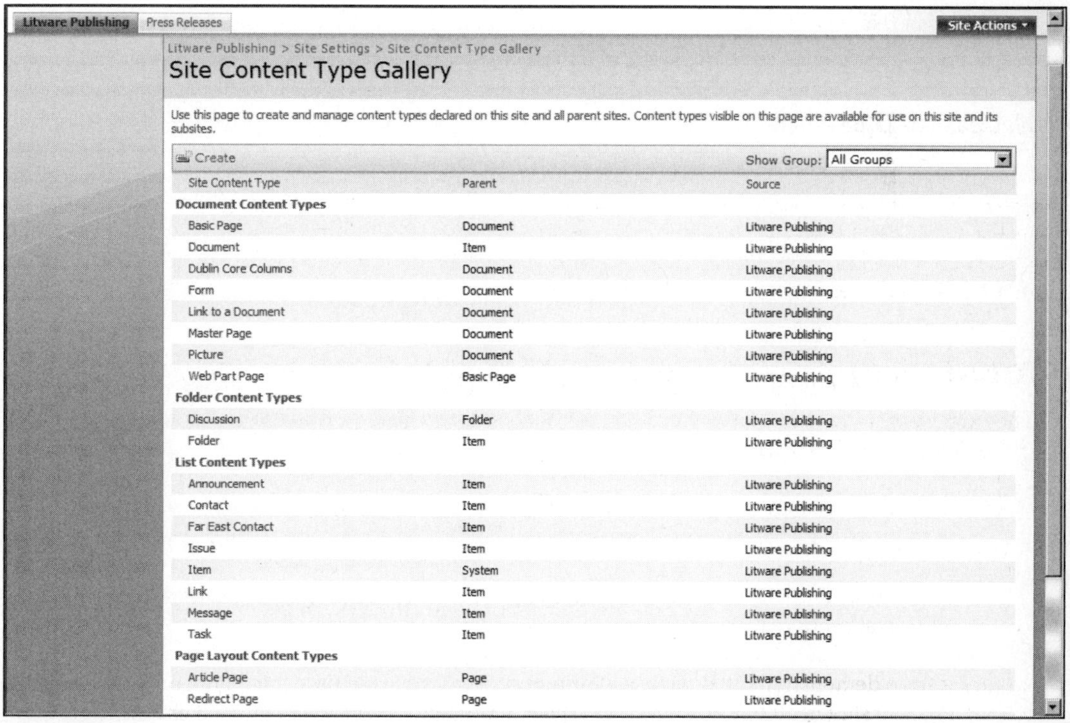

Figure 9-9

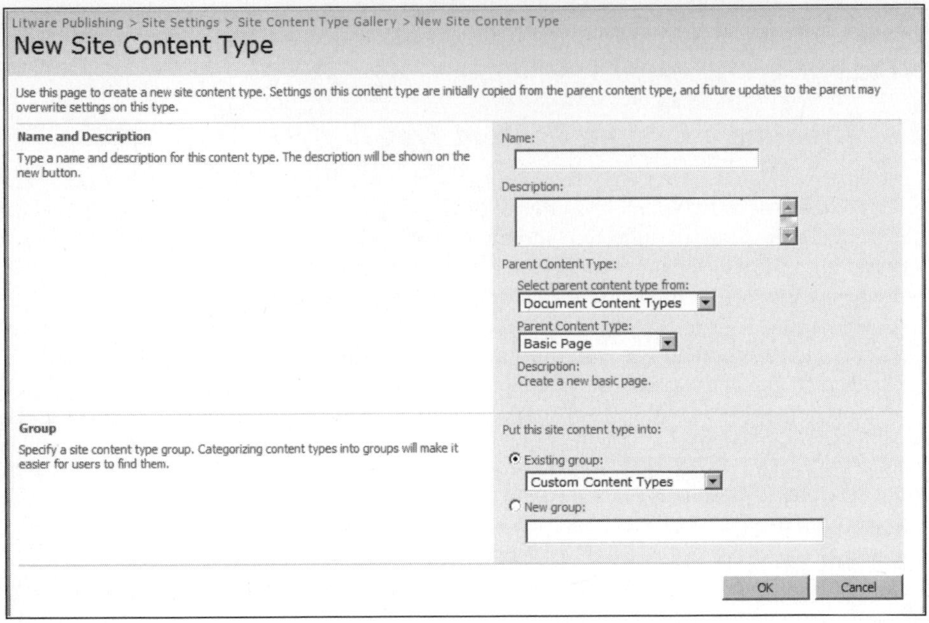

Figure 9-10

The first section in the New Site Content Type page specifies the name, description, and parent of the new content type. Content types are hierarchical, with the top of the hierarchy held by the System content type. We will create a new content type that will have scope across the site collection and be based on several new site columns.

1. Open the browser to the Litware Publishing intranet site.

2. Navigate to the Site Content Type Gallery from the Site Settings page and click Create on the toolbar to create a new content type.

3. Enter **Book Summary** for the Name and **"Content type used by the Book Summary Page Layout"** for the Description.

   If you select Page Layout Content Types from the drop-down list labeled "Select parent content type" from: on the New Site Content Type page, the Parent Content Type drop-down list contains all of the content types that are the basis for all of the page layouts included with MOSS.

4. Select Publishing Content Types from the drop-down list and choose Page. This content type will provide the necessary features for enabling WCM. Our Book Summary content type will inherit from Page.

5. Create a new group called **Litware Publishing** and click OK. In general, this step is optional, but it provides an organizational benefit since there are many out-of-the-box content types.

   From the Site Content Type: Book Summary page, you can view the default columns associated with this content type, as shown in Figure 9-11. Notice the names of the columns, their types and the sources. The Sources column demonstrates the effect of inheritance for the content type. In the next page you define the details for your content type. The columns you define here become field controls in the page layout.

Columns			
Name	Type	Status	Source
Name	File	Required	Document
Title	Single line of text	Optional	Item
Description	Multiple lines of text	Optional	System Page
Scheduling Start Date	Publishing Schedule Start Date	Optional	System Page
Scheduling End Date	Publishing Schedule End Date	Optional	System Page
Contact	Person or Group	Optional	System Page
Contact E-Mail Address	Single line of text	Optional	System Page
Contact Name	Single line of text	Optional	System Page
Contact Picture	Hyperlink or Picture	Optional	System Page
Rollup Image	Publishing Image	Optional	System Page
Target Audiences	Audience Targeting	Optional	System Page

- Add from existing site columns
- Add from new site column
- Column order

Figure 9-11

6. Click the Add From New Site column link at the bottom of the page. Enter **Book Title** for the column name and click OK.

7. Click the Add From New Site column link. Enter **Book Description** for the Column name and choose Full HTML content with formatting and constraints for publishing as the column type. Click OK.

8. Click the Add From New Site column link. Enter **Book Category** as the Title and select the type Choice. Display the choices using check boxes. Add the following choices: ASP.NET, SharePoint, and SQL Server. Click OK.

9. Click the Add From Existing Site columns. Choose Comments and click the Add > button. Click OK.

That completes the creation of the content type that will be used by the custom page layout.

Part 2 — Creating a Page Layout

As discussed previously, page layouts are one of two templates used to author new content pages. A page layout contains field controls that originate from the column definitions and data types contained in the content type. Every page layout must be bound to one and only one content type at any given time. Multiple page layouts can use the same content type.

New page layouts can be created from the Master Page Gallery, which is accessible via the Site Settings page of your top-level site. From the Site Settings page, click the Master Pages and Page Layouts link in the Galleries group. Once you complete the process via the web browser, you will need to open SharePoint Designer and complete the process. Instead of using the browser for part of the process and then using SharePoint Designer to finish the page layout, you can use SharePoint Designer for the whole process. This is what we will do here in Part 2.

1. Open the Litware Publishing intranet site in SharePoint Designer (SPD). Make sure to open the site at the root of the site collection. If SPD opens with a site already loaded, close the site and then open the intranet site.

2. Expand the _catalogs node in the Folder List and right-click the master page (Master Page Gallery) node. Point to New and click SharePoint Content. In the dialog box, click SharePoint Publishing and select Page Layout.

3. Under Options on the right-hand side of the dialog box, use the drop-down lists to select your content type group (Litware Publishing) and then the content type (Book Summary) that you created in the previous steps. Enter **Books** for the URL Name and **Books Page Layout** for the Title. Click OK.

    SPD should be displaying the `Books.aspx` file, which is your new page layout. Switch to Design mode so that you can see the effects of the master page on the look and feel of the new page layout (Figure 9-12). Switch to Code view and review the page code. Notice the `PlaceHolderMain asp:Content` tag and then switch back to Design view. The next step in creating the page layout is to add the field controls that will control how and where the content is actually positioned on the page.

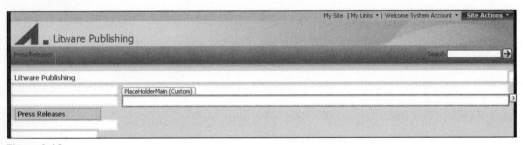

Figure 9-12

4. Place the cursor directly beneath PlaceHolderMain (Custom). Click on Table in the menu bar and choose Insert Table. Insert a table containing one column and eight rows.

5. In the first row of the table, enter **Book Title**.

6. In the second row of the table, drag-and-drop the Book Title field control from the Toolbox. A snapshot of the Toolbox, showing available SPD controls, is shown in Figure 9-13. If the Toolbox is not visible, it can be displayed from the Task Panes menu item.

7. Enter **Book Description** in the third row and drag-and-drop the Book Description field control from the Toolbox and place it in the fourth row.

Figure 9-13

8.   Enter **Book Category** in the fifth row and drag-and-drop the Book Category field control from the Toolbox and place it in the sixth row.

9.   Add the text **Comments** and the corresponding Comments field control to the next two rows in a similar way the previous steps were performed.

Feel free to add any styling you would like to the text or field controls. This can be done through the Apply Styles window or the Properties window.

10.   Save the changes you have made to `Books.aspx`. From the Folder List view, check-in the `Books.aspx` file by right-clicking on the file and choosing Publish a major version from the Check In dialog box.

11.   Click the Yes button when prompted about content approval. Right-click on the Books.aspx file from the gallery web page and approve the content.

This completes the creation of the custom page layout, and it is available for use by content authors.

Part 3 — Authoring a Web Page Using a Custom Page Layout and Approving the Content

We have completed the custom content type and created a new page layout based on the custom content type. This provides the necessary template for authoring new web page content. Prior to authoring new content, we will create a new site that will host our content. Creating a new site within the Litware Publishing portal would not be required but helps provide for better organization of content in our scenario. For example, our fictitious company likely publishes many types of content such as books, magazines, etc. Once our new site is created, we will walk through the page authoring and approval process.

1.   Navigate to the Litware Publishing intranet portal.

2.   Create a new site called **Books** from the Site Actions menu. Once created, the new site `default.aspx` page should be displayed and look like the page displayed in Figure 9-14.

As can be seen from the figure, the new site has been created and `default.aspx` will need to be submitted for approval once it has been modified to meet the design needs. We are going to leave this page blank, so we can proceed with the approval process.

3.   Click the Submit For Approval button on the Page Editing Toolbar.

Figure 9-14

4. Familiarize yourself with the Start "Parallel Approval": default page. Click the Start button.

5. Navigate to the All Site Content page from the Site Actions menu by choosing View All Site Content. Click the Pages document library to reveal the library with the `default.aspx` page item displayed.

6. Drag the mouse over the `default.aspx` item, select the drop-down arrow and choose Approve/Reject from the context menu. Choose Approve and click OK.

This completes the approval of the new site.

7. Navigate to the Litware Publishing intranet portal home page. Click Site Actions and then click View All Site Content. Under the Lists section, click Reusable Content.

The Reusable Content list contains content that can be inserted into web pages during the authoring process. This list does not have any content by default. The following steps create content that will be used later in the exercise when a new web page is being created.

8. Click the drop-down arrow next to the New button and notice that you have the option of creating Reusable HTML or Reusable Text, as well as a new folder. Click Reusable Text.

9. On the Reusable Content: New Item web page, enter **Copyright Notice** for the title. Notice the Content Category drop-down box that does not contain any choices other than None. Review the Automatic Update section but leave this enabled.

Enabling Automatic Update makes the text of HTML fragment read-only. If a text or HTML fragment has been created and inserted into a web page by a page author, the content of the fragment will be updated automatically. This helps maintain a common set of content that can be updated in a single location and have the changes applied universally across published content.

The Content Category property is represented by a column of data type Choice. If the authoring environment is going to rely on reusable text or HTML fragments then you might want to consider populating this column with other entries. This will improve organization and help with reusability. To add new entries, click List Settings from the Settings menu and open up the Content Category properties information by clicking on Content Category under the Columns section. Add each new entry to the Additional Column Settings section.

10. Enter **Copyright 2006 - 2007, Litware Publishing** in the Reusable Text section. Click OK.

This completes the steps for authoring a reusable content fragment. Next, a new web page will be created that uses this fragment.

11. Navigate to the Books site default page. Click Site Actions and click Create Page. Provide **Wikinomics** as the name and an optional description, and then select the Books Page Layout. Click Create. The new page is displayed in Figure 9-15 with the field controls that were placed on the page layout using SPD in the previous part of the exercise.

12. For Book Title, enter **Wikinomics: How Mass Collaboration Changes Everything**.

13. Click inside the Book Description field control and notice the HTML editor control. This control can be moved to other locations on the page by placing the cursor over the left-most edge until the cross-hairs become visible. Once visible, click the mouse button and move the control to a more preferred location. Enter **This book goes into detail about the impact of collaboration using wikis**. At the end of the description, hit the Enter key twice to create some additional space below the description you just entered.

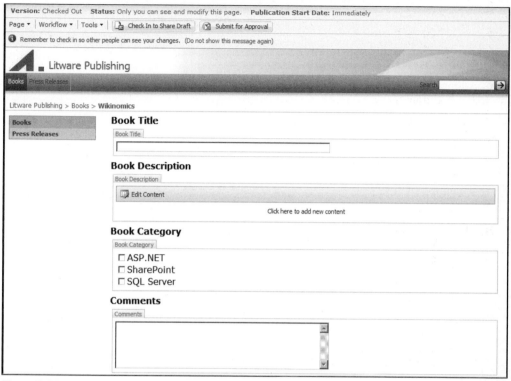

Figure 9-15

14. Place the cursor over each of the icons in the editor control to familiarize yourself with the functionality. Click on the icon to Insert Reusable Content that is located between the Insert Image and Insert Table icons. This should display the Select Reusable Content dialog box, which shows the Copyright Notice fragment created previously. Click on the Copyright Notice item to select it and then click the OK button at the bottom of the page. This should return you to the web page for additional authoring.

15. Under Book Category, check the SharePoint choice. Add a comment such as **Pretty good book** to the Comments field control. The completed page can be viewed without publishing by selecting the Preview In New Windows option from the Tools button in the Page Editing Toolbar. Because versioning is enabled by default when the Pages library is provisioned, this is a much better option to preview the completed page since it will generate fewer versions than using the Check In To Share Draft option or publishing the page.

This completes the authoring process for this web page. Submit this page for approval by clicking the Submit For Approval button in the Page Editing Toolbar at the top of the page. If the toolbar is not visible, it can be made visible from the Site Actions menu. A user who is a member of the Approvers group can approve the page, which makes it accessible by everyone visiting the site. An example of a completed page once it has been approved is shown in Figure 9-16.

As part of the development process or as you investigate the authoring process, it may be beneficial to disable content approval. This is accomplished from within the Versioning Settings Web page for the Pages document library.

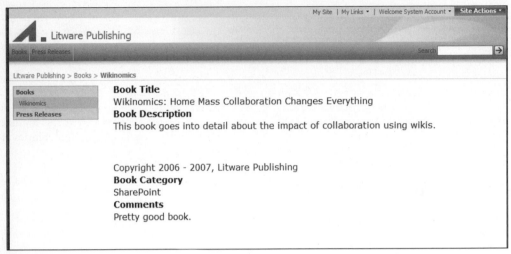

Figure 9-16

Site Variations

SharePoint provides new capability for authoring scenarios that require content to be present in more than one language or the same content rendered to different types of devices. This capability is called *site variations*. Some of the key features include:

❑ MOSS can host different sites in different languages or locales within the same site collection. Documents authored in different languages can be stored in the same document library for a given site.

❑ A user's request will be redirected to the appropriate variation based on the language settings of their browser client.

❑ Content is synchronized between different sites provisioned in different languages. It is important to note that site variations *do not* translate the content into different languages. This must be done separately.

❑ The language translation process can be managed using the new workflow capability.

A few key points regarding multilingual capability:

❑ MOSS supports a single language per server farm during the installation process. All servers in the farm must be installed with the same language.

❑   For multilingual capability, additional language packs must be installed on all servers in the farm.

❑   Site collections can contain child sites in different languages from the parent site and from each other.

In general, variations can be used much more broadly, including scenarios where different sites have different branding requirements. Variations require the assignment of a single source and at least one target site. Modifications to the source site are duplicated to any variations of the site. For example, when a new page is authored and approved in a source site, a corresponding page that has the same content is automatically created in all the target sites that participate in the variation hierarchy. The variations are created only once the source content has been approved. This includes any resources like images; they, too, must be approved before they are copied to the variation targets. Once the new content is created in the target site(s), the owner(s) of the target variations are notified and the approval process for that site begins. By default, only publishing content is copied to the site targets; general lists and libraries are not part of the out-of-the-box variation capability.

Preparing for Implementing Variations

There are several aspects of variations that should be considered prior to actually implementing this capability. Planning to implement variations is critical to a successful deployment. These topics are discussed in the following list.

1. **Establish your variation site hierarchy and define which content within the site hierarchy will be synchronized**. Variations are created and implemented within a single site collection. A decision needs to be made to establish what level in the site collection hierarchy will represent the variation home page. The home page location is critical since once the hierarchy's home site is defined, the user will never actually visit the home location since they will be redirected to the target site that matches their preferred language. For an Internet portal scenario, the home is typically the top-level web site of the site collection, with each child site or variation representing the content in different languages.

   Once the hierarchy is created, the variations architecture replaces the default page of the variation's home site with a special page called VariationRoot.aspx. *Each time a user navigates to the variation's home they are redirected to the variation target site based on the user's browser language settings (HTTP_ACCEPT_LANGUAGE). This process can be further customized by replacing the* VariationRoot.aspx *page with an aspx page containing the additional requirements.*

2. **Define the type and number of languages that will be supported**. Each language is represented as a *variation label*. This label is specific to a site and a specific language. Multiple labels can be created by the site collection administrator.

3. **Define the site that will be the source of the variation**. Only one site can be the variation source. This source represents the origin of the content that will be copied to each of the sites in the variation hierarchy. There is also some flexibility in the structure of the variation sites. Keep in mind that different page layouts and different master pages can be utilized versus those used for the source. Master pages and page layouts can be customized at the target site level. Make sure subsites under the source site have their publishing feature enabled if their content is to be synchronized with the target sites.

4. **Define the process for translating content**. There are two options for content translation. First, they can translate the content directly in the target site by using the Web content editor tool. Second, content owners could export the content to a variation package, which is especially useful for translation by an independent organization or vendor.

Content Publishing with Variations Enabled

Once planning for implementing variations has occurred and variations has been configured for a specific site collection, the variation process is launched after new content has been authored and approved on the source site. The variation process is as follows:

1. A New content page is created or a new version of an existing page is created on the source site and approved for publication. Publishing the page launches the variation process. The variation content will then be copied throughout the hierarchy.

   Content approval is not required to use the variation process. If content approval is disabled on the source site, all new content created on the source site will immediately go live, which will launch the variation process.

2. The variation infrastructure copies the content throughout the variation hierarchy; this is a one-way process from source to targets. This copying process, which is better known as a *variation job*, checks for published pages at the source site every 20 seconds. Every time a variation process occurs, it copies the entire set of changes from the source site to all target sites in the variation hierarchy. Pages copied to each of the variations need to be approved before they go live, unless content approval has been disabled. During the copy process, external content on the page is reconfigured based on the Variations feature settings. The target page's settings are indicative of the specific label, including the master page and page layouts, which can be configured specifically for the variation site. The actual page content exists in the language of the source and will need to be translated.

3. Content within the site variations may need to be translated. This can be done directly by having the content translated directly within the site variation, since this content is not available until approved, or the content can be sent outside the organization and then the translated content imported. This is accomplished via the creation of a content migration package.

4. The translated content is ready for approval and publishing.

Creating Site Variations: Exercise and Tutorial

The planning, implementing, and process details of variations have been discussed. It's time to create a variation deployment to illustrate each of these steps. The following steps walk you through this process.

1. By default, site variations are not enabled. Open the Litware Publishing intranet site.
2. On the Site Actions menu, choose Site Settings and click Modify All Site Settings.
3. On the Site Settings page, under Site Collection Administration, click Variations.

   Familiarize yourself with all of the options on the Variation Settings page. As discussed previously, the first decision you need to make is where you want to start varying the site. This can begin from the top-level site ("/") or from one of the subsites in the site collection. You can use the Browse button to select a subsite.

4. Click the Browse button and select the Press Releases site. Click OK.

The next step is to create the variation labels, one for each of the languages (variations) you want to support.

5. On the Site Settings page, click Variation Labels.

6. On the Variation Labels page, click New Label.

7. On the Variation Labels page, enter the text **English** for the name of the label and provide a description if you wish. Add the following values to the properties on the page:

Display Name = English

Locale = English (United States)

Set this variation to be the source variation.

Select Publishing Site as the template.

Click OK.

8. Next, create all of the other site variation labels. On the Variation Labels page, click New Label.

9. From the **Create Variation Label** page, notice that there is a much more limited set of options for the target sites. Enter **Spanish** as the Label Name, select Spanish (Spain) as the locale, and click OK.

10. Click Create Hierarchies to create the variations infrastructure.

11. Navigate to the Press Releases site and see the finished site, as shown in Figure 9-17.

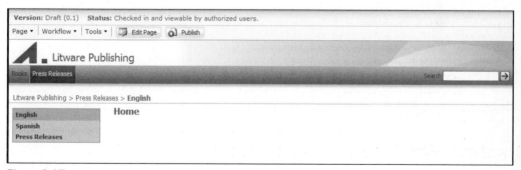

Figure 9-17

The home page of the site shows the two different variations and the Press Releases site from the left-hand side navigation. As can be seen, the home pages of the site variations need to be approved.

12. Approve the English and Spanish variation home pages.

13. Navigate to the English site and create and publish a new page. This is shown in Figure 9-18.

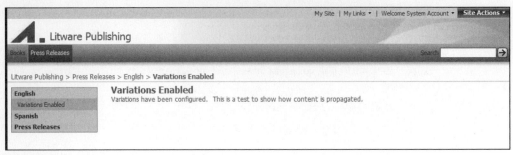

Figure 9-18

14. Navigate to the Spanish variation and notice that the previously created page has been copied but has not been published, as shown in Figure 9-19.

Figure 9-19

Content Deployment

MOSS includes capability to publish content across site collections. These site collections can be in the same SharePoint farm or across farms. The latter would be an example of a staging environment that uses a development and testing farm to test content prior to publishing to a production farm. The process for configuring content deployment within a single farm is discussed in this section. Keep in mind that content deployment is a MOSS feature, and therefore, WSS cannot publish or accept incoming deployments.

1. Enable content deployment on the SharePoint farm. From Central Administration, click on the Operations tab. From the Operations tab, click on the Content Deployment Settings link to display the Content Deployment Settings Web page, which is shown in Figure 9-20. Click the radio button named Accept Incoming Content Deployment jobs. If you were configuring deployment across physical farms then this step would need to be configured on the import farm.

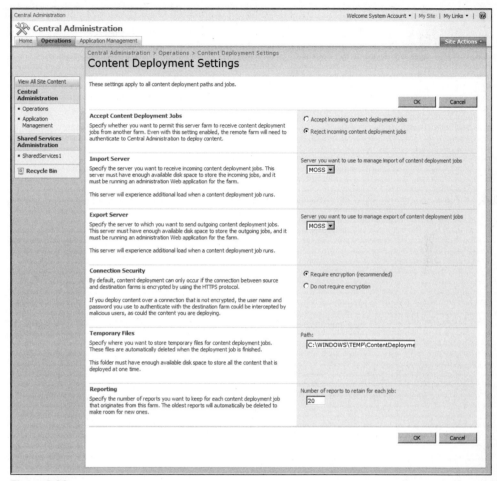

Figure 9-20

2. Configure the import and export servers. In our scenario, these options would be the same server. If you were configuring deployment across physical farms, then the import and export servers would be different choices. As noted under the Import and Export Server sections, keep in mind that the servers must have available disk space for temporarily storing the content for the job and it must be running an administration web application for the farm.

3. Leave the Require Encryption option selected under the Connection Security section. Encryption is recommended since user name and password is transmitted as part of connecting to the destination farm. To test this deployment within our test environment, choose Do Not Require Encryption. Accept the default values for Temporary Files and Reporting. Click OK.

4. The next step is to create a Path from the Content Deployment Paths and Jobs link under the Content Deployment section on the Operations page. Paths represent the relationship between site collections so that one can be identified as the source and the other as the destination. Jobs represent the content that will be sent from the source to the destination. Click the New Path button to reveal the Create Content Deployment Path web page, as shown in Figure 9-21.

Enter a name and a description for the path. Typically you will want to be as descriptive as possible in your naming conventions. Choose the source web application, for example, Litware Publishing, and the source site collection of "/". Enter the URL of your server's central admin-istration site, such as **moss:6147**. Make sure to include the port number in the URL as appropri-ate. Click the Connect button. If you are configured correctly, you will receive a Connection Successful message next to the Connect button, as shown in Figure 9-22.

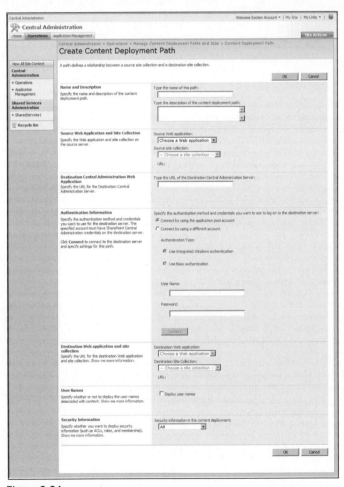

Figure 9-21

Next, select the destination web application and destination site collection that are appropriate for your test environment. You may want to create a new blank site collection just in order to test the deployment capability. Leave the other options with their default values and click OK. This creates a Quick Deploy job that can be used to update this path in the future.

5. On the Manage Content Deployment Paths And Jobs web page, click the New Job button. Enter a name and description as appropriate. Select the path that you previously created in step 4. Accept the default values for all the remaining properties, but review each section to familiarize yourself with the capability. Click OK.

6. From the Manage Content Deployment Paths And Jobs web page, place your cursor over the job created in step 5 and click the drop-down arrow. From this menu, select Run Now to launch the content deployment job. Once the job is complete, the outcome is displayed in the Status column of the job, as shown in Figure 9-23.

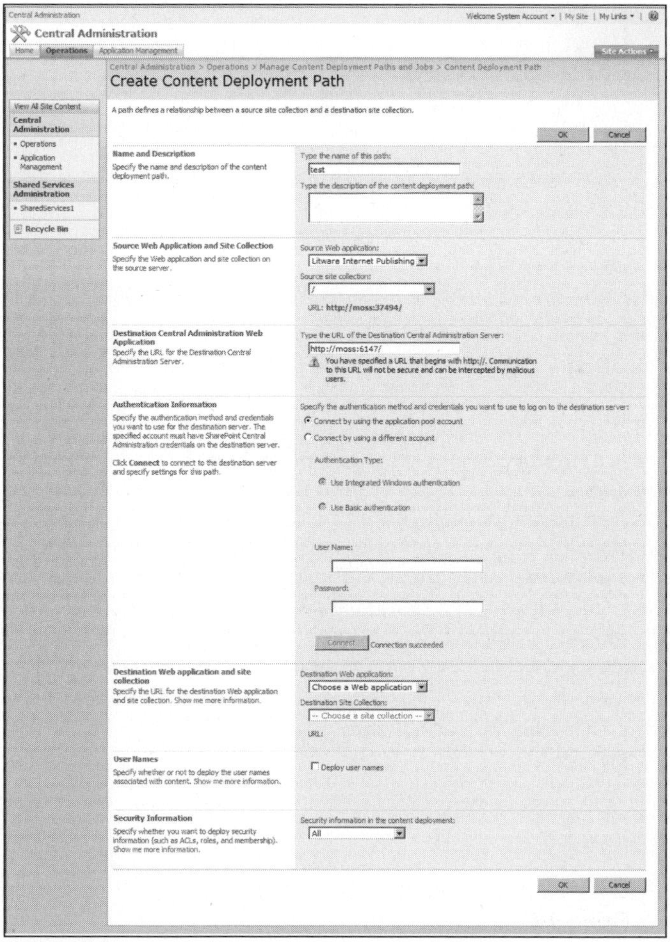

Figure 9-22

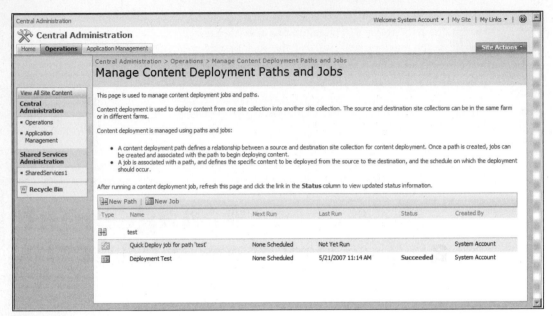

Figure 9-23

That completes the configuration and execution of a content deployment job.

Summary

MOSS WCM provides a feature-rich authoring and publishing environment for Web content. Publishing portals can be created for intranet web sites or Internet-facing web sites. Master pages and page layout templates provide a mechanism for achieving a consistent look and feel across web sites and pages. New web content can be created using a web browser, Word 2007, or InfoPath. WCM integrates the workflow capability from Windows Workflow Foundation for approving content. There is an out-of-the-box approval workflow for new content approval, but custom workflows can also be created if more complex approval scenarios are necessary. Site variations provide the ability to copy published content to other sites to enable multilanguage scenarios and multidevice capability. Content deployment provides capability to publish WCM content across site collections in the same farm or across farms enabling staging environments for testing content prior to releasing into production. MOSS WCM provides a rich feature set for authoring and publishing web content.

10

Content Management and Workflow Scenario

In this chapter, we'll pull together several topics from this book by building a content management solution using MOSS's core web content management framework. We'll also include a front-end process for managing Word documents that feed into the web site. This will show how different aspects of the MOSS platform can be used together to build a solution that extends from the desktop through to the web farm. The following topics will be covered in this scenario:

- ❑ Creating content types and document workflows
- ❑ Defining the structure of documents for enterprise content management
- ❑ Building document metadata input functions into SharePoint documents
- ❑ Creating a feature to deploy the solution to a SharePoint site or farm
- ❑ Developing document conversion processes for web content management
- ❑ Using workflow to extend content management and publishing processes

Our project will streamline the process for creating and publishing press releases at Contoso Enterprises. Press releases are important documents and are used in differently depending on the medium. In our scenario, they'll be published internally to Contoso Enterprises on the PR department's MOSS site, but they'll also be published externally on their web site.

Document Workflow for the Press Release

We'll work on the internal part of the press release process first. This process contains the template for the press release document, the MOSS content type for press releases for standardization, a library to store and manage press releases, and finally, a workflow process for approving press releases.

Content types give us the advantage of consistency, even when documents are stored across different sites and lists. This makes the deployment model more flexible, allowing for decentralization if there's a business need to keep separate repositories of press releases.

Defining the Press Release Content Type

From the Site Settings administration menu on the root site, the "Site content types" option lists the existing content types on the site and has the option to create a new content type. Choosing that option displays the "New Site Content Type" page, which we fill in as shown in Figure 10-1.

Figure 10-1

This creates a new Content Type with only a Title property, so the first step is to expand this Content Type by adding a few properties. Content Types use a shared set of column definitions. We'll start by defining some new column types by clicking the "Add from new site column" link in the Columns section in the Site Content Type Properties page. The following columns need to be created here:

Column Name	Type	Required/Optional
Subtitle	Single line of text	Optional
Publish Date	Date and time (date only)	Optional
PR Contact	Person or group (person only)	Optional

In addition to these, "Date Created" and "Date Modified" columns are added using the Add From Existing Site Columns link on the Site Content Type Properties page. Since these columns were already standardized and defined at the site level, we will reuse them here.

After all of these properties are added, the columns section should look like Figure 10-2.

Now that the Press Release content type has its metadata columns defined, we can start creating the template document that defines the structure and layout of each press release document.

Columns			
Name	Type	Status	Source
Name	File	Required	Document
Title	Single line of text	Optional	Item
Subtitle	Single line of text	Optional	
Publish Date	Date and Time	Optional	
PR Contact	Person or Group	Optional	
Date Created	Date and Time	Optional	
Date Modified	Date and Time	Optional	

Figure 10-2

Building the Initial Press Release Document Template

The press release document itself is built in Microsoft Word, using content controls to define the editable parts of the document. (We cover Content Controls for Word in Chapter 4.) This will make it easier to extract and modify content in that document at a later point. It also enforces a style and structure in the document to ensure consistency. Finally, Content Controls let you lock the document down so users cannot make edits outside of the designated fields.

The first step in creating this template is to start with a blank document in Microsoft Word; add just a heading for now (Figure 10-3). We'll add the remaining elements, but first we have to associate the document template with the content type and inherit the content type's properties into the document template.

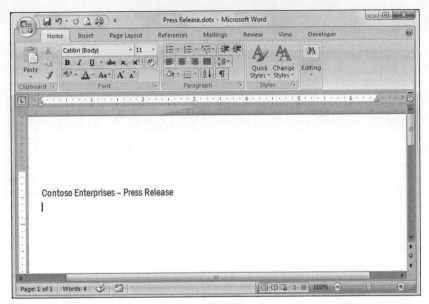

Figure 10-3

Uploading the Document Template to the Content Type Definition

Before we can pull the content type's metadata into the document template, this template needs to be uploaded to the Content Type definition. This is done through the content type advanced properties page with the "Upload a new document template" section (Figure 10-4) and selecting the `dotx` file we just created in the previous step.

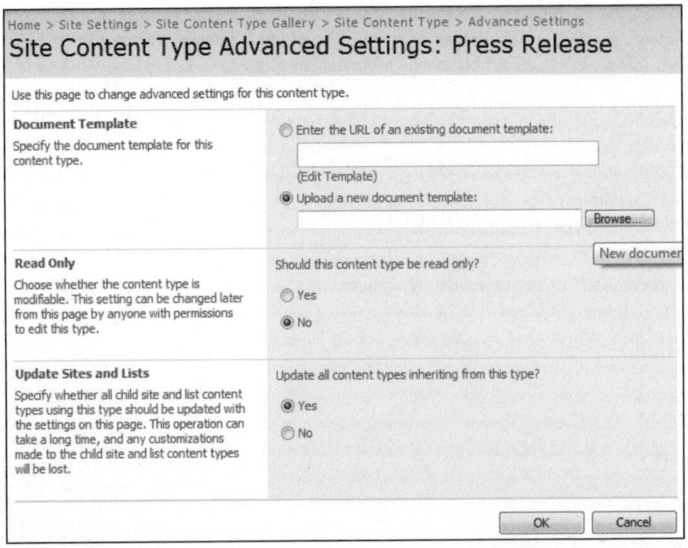

Figure 10-4

Once we click OK, the document template is now stored in the site and is associated with the Press Release content type. Unfortunately, the document template has not yet inherited the content type's properties. We have to build out a document library using this content type before we can finish customizing the template document.

Creating a Document Library Based on the Press Release Content Type

Now we'll create a document library that uses the Press Release content type we just defined.

1. First, we start by creating a normal document library. This can be anywhere on the site that contains the Press Release content type.

2. Once that document library is created, go to the Advanced Settings section on the document library settings and enable management of content types in that library (Figure 10-5).

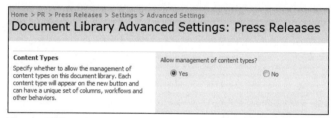

Figure 10-5

3. Once this is enabled, the document library settings page shows a list of content types supported by this document library. It defaults to only the Document content type, so we'll add the Press Release content type as well by clicking Add from existing site content types. From there, we can select the content type we created earlier (Figure 10-6).

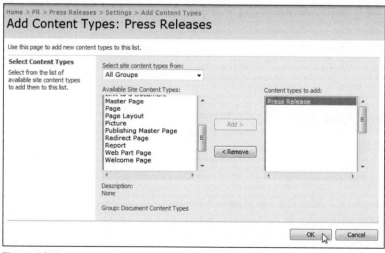

Figure 10-6

Now that the Press Release content type is supported, the document library is ready to let us complete the template document.

Embedding Metadata Properties in the Template Document

Although we uploaded the initial document template to the content type, that document template does not yet contain the metadata properties associated with the Press Release content type. The way we'll add the content type columns into the document is to create a new document from the document library using the Press Release content type. This will add the metadata properties for the content type to template document we created early in this chapter. This document can then be used as the new document template.

The detailed steps are as follows:

1. Navigate to the new Press Releases document library and from the New document library menu, select Press Release, which, after a security prompt, will open a new document with the header we created earlier in the chapter. The difference is that now this document has the content type's metadata properties associated with it. These properties should be visible in a server properties pane in Word (Figure 10-7).

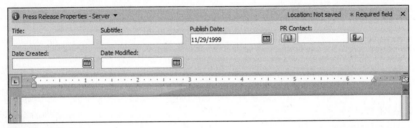

Figure 10-7

2. Save this document as a document template (`dotx`) format on a local drive.

3. Navigate back to the settings for the Press Release content type and go to the Advanced Settings page.

4. Select Upload A New Document Template and use the browse button to select the file we just saved locally and upload that as the new document template by clicking the OK button. By doing this, we replace the document template with this new version that has metadata properties.

Completing the Template Document

Now that the document template associated with the Press Release content type has the correct metadata properties embedded, we'll go back to the Advanced Settings page for the Press Release content type, and this time we'll click the Edit Template link to fill out the document template.

The structure of the press release template will be built using Content Controls, some of which are associated with metadata properties and some of which are not.

1. The title is the first block of content in the document. This is a metadata property, so this will be added from the Insert tab ⇨ Quick Parts ⇨ Document Property ⇨ Title menu selection (Figure 10-8). The result is a Content Control linked to the Title document metadata property.

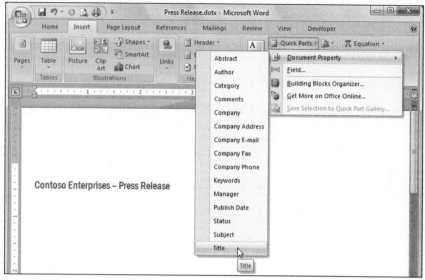

Figure 10-8

2. Set formatting options on the Title Content Control by clicking inside of that control and then, from the Developer ribbon tab, clicking the Properties button in the Controls panel. This opens the Content Control Properties dialog box; check the option for Use A Style To Format Contents and choose Heading 1 as the style for this control (Figure 10-9).

3. Add the Subtitle property Content Control on the next line the same way, repeating steps 1 and 2. Choose the Heading 2 style for this control.

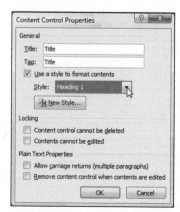

Figure 10-9

4. Add the Publish Date property on the next line. Notice that this time instead of a plain text control, this action creates a Date Picker content control.

5. The next section to add is the Body control that will actually contain the text of the press release. This control is used to structure content, but this will not be linked to document metadata properties like the three previous controls. Since that is the case, this field is added using the Developer ribbon tab, by clicking the Rich Text button in the Controls pane. Then click the Properties button in the same pane to bring up the Content Control Properties and set the control title to Body.

6. The next section of the press release is the company background section. This is handled the same way as the Body section in the previous step — just name the control BackgroundBody instead.

7. The final section is another body block, this one called MoreInfo. Repeat the process in step 5 using this control name.

8. Now the document template is complete. We can get a better look at it by toggling in the Design Mode button in the Control pane in the Developer ribbon tab. The complete template should look like the example in Figure 10-10.

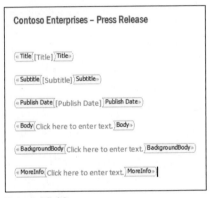

Figure 10-10

9. Save the document template back to the server by clicking the Save button in Word. This will commit this new template to the content type.

Now this document template will maintain the properties displayed in SharePoint for that item. If we go back to the document library and choose the menu option to create a new press release and start filling it in, we'll see the document properties pane appear again. If text is entered in the Title content control in the document, that text value will also appear in the document properties panel under the Title field (Figure 10-11).

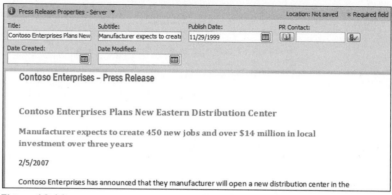

Figure 10-11

Creating a Custom Document Information Panel

Having the standard document information panel is useful, but by default, we are seeing redundant data. The Title and Subtitle are maintained in the document, so a user could be confused by having that information also appear in the document information panel. It would be better to keep the custom metadata fields for Publish Date and PR Contact, but leave the others out. For this reason, next we'll create a streamlined custom property pane for this content type.

1. We'll start by editing the generated document information panel. Opening the Document Information Panel Settings page gives a link to Create A New Custom Template, which we'll follow (see Figure 10-12).

2. This opens InfoPath and a short wizard that will create a template form using the properties for the current content type. The default InfoPath form is a series of horizontal region container controls, each with a label and a data control (or series of data controls) (Figure 10-13).

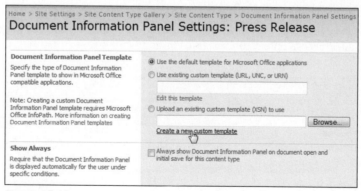

Figure 10-12

Figure 10-13

Removing the Title and Subtitle fields is a simple matter of selecting their container horizontal regions and hitting delete.

3.   To publish the form, we need to use the File/Publish menu, which launches the Publishing Wizard. The publishing wizard needs to save a local copy of this file before publishing it to the server, so we need to assign a local file name. Because this form was created by following a link from the template create link in the content type properties, the wizard provides the option for publishing a Document Information Panel. This is the option we'll use (Figure 10-14).

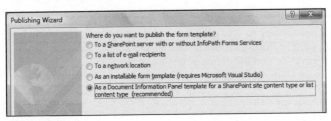

Figure 10-14

4.   The next (and final) step of this wizard is just a confirmation page. Clicking Finish initiates the publishing process and indicates if this publish was successful. Now that this template is associated with the content type, we can edit it in the future from the same Document Information Panel Settings page we used to create the new one.

Now the property page on the Press Release content type is streamlined without the properties displayed within the document using content controls (see Figure 10-15).

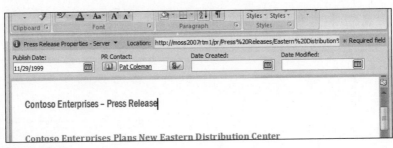

Figure 10-15

296

Now we have a rich, structured form with an integrated metadata panel for collecting information that should not print with the form itself. The next step is to put this structured form in motion using Share-Point's workflow capabilities.

Defining Approval Workflow

In a highly customized environment for managing press releases, the workflow for approving and routing press releases would be built using a custom workflow constructed in Visual Studio. This has several advantages over using built-in workflows, such as:

- ❑ The ability to use content type properties to control the workflow (for example, route the approval request to the PR Contact first, and then possibly a business owner for final sign-off)

- ❑ Integration with the Press Release content type (workflows built in SharePoint Designer, although capable of working with metadata, cannot link to a content type — just to a specific list)

- ❑ More control over how audit logs are maintained

- ❑ Custom forms to collect more information at approval time beyond just comments and approve/reject status

For more detail about creating a custom workflow in Visual Studio, see Chapter 8.

Building out such a custom workflow is beyond the scope of this chapter, but we will add a simple approval workflow from SharePoint's built-in workflow templates. This workflow is configured in a two-page form initiated from the Workflow settings link in the Content Type Settings page.

1. Click on the Add A Workflow link to start, select the Approval workflow and name it **Press Release Approval** (see Figure 10-16).

2. From here, we go to the next page by clicking Next and then configure the workflow to be cancelled if the document is rejected and changed. We also have this workflow drive the approval status by checking Update in the approval status check box. (Figure 10-17).

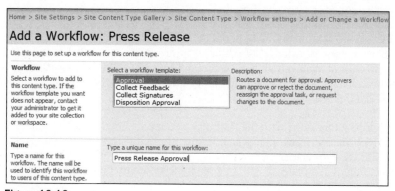

Figure 10-16

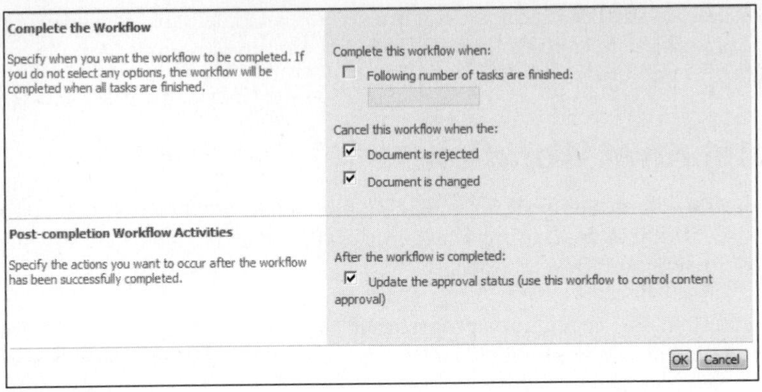

Figure 10-17

Enabling this simple workflow will give our Press Release content type a standard approval mechanism. Now, regardless of where press releases are stored on the site, the workflow for their approval will have more consistency. In a future upgrade, if a custom workflow is added, the existing instances of this content type will also inherit that upgrade. This is one advantage of deploying this capability through a content type instead of just building a one-off custom list.

Packaging the Content Type

Up to this point, we built out a custom content type using the SharePoint user interface, but there are some situations where it is important to have the content type definition stored in a way that allows it to be reused. One example is a multisite deployment where that same content type needs to be deployed across sites. Recreating the definition manually across sites would be error prone and keeping them in sync would be nearly impossible.

Also, enterprises typically have change control processes that demand a way to archive and version solution assets, not just any source code associated with the project. Automated build and deployment processes also benefit from this capability.

SharePoint provides a solution in the form of *Features*. Features can be used to package up many SharePoint definitions including custom lists, custom columns, and content types. Features are stored as a set of XML documents and managed with the command-line STSADM tool. In fact, the SharePoint installation process itself makes heavy use of SharePoint Features because SharePoint's built-in types and structures are installed this way.

Every SharePoint Feature starts with the `feature.xml` document that defines the overall structure of the Feature. The `feature.xml` contains identifying information for that Feature and a list of other XML documents that define the actual contents of the Feature. The Feature package may also contain other assets such as InfoPath templates or ASPX pages associated with that Feature type.

Features are discussed in detail in Chapter 5 of this book as well.

For our press release scenario, we will create a Feature that stores the custom columns associated with our content type and then the content type definition itself.

The folder structure of the Feature is as follows:

PressRelease/	Root folder of the press release Feature
PressRelease/feature.xml	The Feature manifest file
PressRelease/sharedColumns/	A subfolder for the shared column definition
PressRelease/sharedColumns/pressReleaseSharedColumns.xml	The shared column definition file
PressRelease/sharedColumns/contentType/	The subfolder for the content type definition
PressRelease/sharedColumns/contentType/pressReleaseContentTypes.xml	The content type definition file

The `feature.xml` format is designed to pull together the various aspects of a Feature. Here, it pulls together the custom column definition and the content type.

```xml
<?xml version="1.0" encoding="utf-8"?>
<Feature xmlns="http://schemas.microsoft.com/sharepoint/"
  Id="F1BCF8AA-E3F1-4dac-9591-2D68AAE27E47"
  Title="Press Release 2"
  Description="Press Release 2 content type and associated shared columns"
  Scope="Site">

  <ElementManifests>
    <ElementManifest
      Location="sharedColumns\pressReleaseSharedColumns.xml"/>
    <ElementManifest
      Location="contentType\pressReleaseContentTypes.xml"/>
  </ElementManifests>

</Feature>
```

The `<feature>` element contains an ID attribute with a generated GUID to identify this Feature. The title and description attributes are used to enable user-friendly install tools. The scope attribute is important here — content types cannot be defined for an entire farm, so here we're tying it to the `Site` level. (Other options for scope are `Farm`, `WebApplication`, `Site`, and `Web`.) The `<ElementManifest>` elements define the actual contents of the Feature. In this case, they are in order of dependency. In addition to manifests, these elements can also be individual files associated with the solution.

> *Field IDs, content type IDs and other GUID-based identifiers will be different every time these underlying objects are created, so if you are following along with these examples, you will substitute the appropriate values for these identifiers.*

This next listing shows the element definitions for the press release.

```xml
<?xml version="1.0" encoding="utf-8"?>
<Elements xmlns="http://schemas.microsoft.com/sharepoint/">
```

```
   <Field ID="{A4152135-B58A-4977-9834-F37753E3D4D5}"
        Group="Press Release 2"
        Type="User"
        Name="PR Contact"
        StaticName="PR Contact"
        DisplayName="PR Contact" />
  <Field ID="{C42D9B54-E12D-4a1b-BEC2-90CCA2D25C4B}"
        Group="Press Release 2"
        Type="Text"
        Name="Subtitle"
        StaticName="Subtitle"
        DisplayName="Subtitle" />

</Elements>
```

The PR Contact, defined as type User, also has a unique ID column. The <Field> element supports dozens of possible attributes to control all aspects of behavior including validation, maximum size, sorting behavior, and read-only. The IDs are important because that is how the content type links back to these shared columns.

This next listing (pressReleaseContentType.xml) contains the actual content type definition.

```
<?xml version="1.0" encoding="utf-8"?>
<Elements xmlns="http://schemas.microsoft.com/sharepoint/">
  <ContentType ID="0x010100D355729F0DB24b6e8D08230035BDD2E5"
      Name="Press Release 2"
      Group="Document Content Types"
      Description="Press Release 2"
      Version="0">
    <FieldRefs>
      <FieldRef ID="{A4152135-B58A-4977-9834-F37753E3D4D5}" Name="PR Contact" />
      <FieldRef ID="{fa564e0f-0c70-4ab9-b863-0177e6ddd247}" Name="Title" />
      <FieldRef ID="{C42D9B54-E12D-4a1b-BEC2-90CCA2D25C4B}" Name="Subtitle" />
      <FieldRef ID="{8c06beca-0777-48f7-91c7-6da68bc07b69}" Name="Created"/>
      <FieldRef ID="{28cf69c5-fa48-462a-b5cd-27b6f9d2bd5f}" Name="Modified"/>
    </FieldRefs>
  </ContentType>
</Elements>
```

Again, a unique ID is used as an identifier, but this one is different than the GUIDs used in previous elements. The ContentType ID is actually a hierarchical identifier that shows the inheritance structure of the content type. The inheritance is defined in two byte steps. There are two forms of Content Type ID: an inheritable format that allows new content types to be built on top of it, and a sealed content type that does not allow subtypes. Our example here is a sealed content type. The ID breaks out as follows:

❑   0x01: Base content type (Item)

❑   01: Document

❑   00: Delimiter — indicates that this content type will not inherit beyond this point; remaining digits are a unique identifier

❑   D355729F0DB24b6e8D08230035BDD2E5: Unique GUID identifier for this content type

The `<FieldRefs>` section contains links to the columns for this content type. Some of these are new custom columns defined earlier in this Feature, and some are built-in SharePoint types. The GUIDs for the built-in types can be found in the Feature directory for the built-in types stored in `<drive>\program files\common files\microsoft shared\web server extensions\12\TEMPLATE\FEATURES\ fields\fieldswss.xml`.

These field GUIDs are fixed and will be stable across SharePoint installations, unlike the GUIDs for user-created types.

Features are deployed in the following steps:

1. Copy the Feature files into a directory off of SharePoint's feature directory (`<drive>\program files\common files\microsoft shared\web server extensions\12\TEMPLATE\ FEATURES`). In our case, we'll create a `PressRelease2` directory there and copy our Feature directory structure inside of it so that `feature.xml` is one level below SharePoint's FEATURES directory.

2. Use the `stsadm` command line to install the Feature:

```
>stsadm -o installfeature -filename PressRelease2\feature.xml

Operation completed successfully.
```

Installing makes the Feature available for sites, but for site-level Features, it does not actually deploy the Feature. We'll use a second `stsadm` command to do that.

3. Use `stsadm` to deploy the feature to the target site:

```
>stsadm -o activatefeature -name PressRelease2 -url http://moss2007rtm1/

Operation completed successfully.
```

The `activatefeature` command accepts a parameter for `-url` that identifies the deployment target location.

Now this content type is deployed to the site and is available for other sites by repeating the `activatefeature` command. Another useful benefit is that the content type can be deactivated for a site and uninstalled from the server by using the `deactivatefeature` and `uninstallfeature` commands, respectively.

Other aspects of a custom type can be packaged in a Feature, including the document template, a custom document information panel, and custom workflow definitions. When all this is combined, you have a powerful deployment mechanism. Several third parties make tools that extend this deployment tool, including GUI installers and tools that streamline deployments across farms.

Publishing Web Content from Documents

Now that we've examined the process for producing the press release content and leveraging workflow for collaborating on and producing the document, we'll walk through a publishing solution for web content and converted documents. To support the web content management scenario for publishing the Contoso press release document, we'll examine the process for setting up a web publishing site, configuring document conversions for the SharePoint infrastructure, converting document-based content to web content, and then managing that content through a publishing process. To get started, we'll take a look at setting up the web publishing site.

Configuring a Web Publishing Site

To support the press release publishing site, we'll start by creating a web publishing site and then walking through the site configuration. This will provide the foundation to publish new web pages, store documents for conversion, and support the workflows required to support the content management functions. To get started, create a new site and select the Publishing Site With Workflow template, as shown in Figure 10-18.

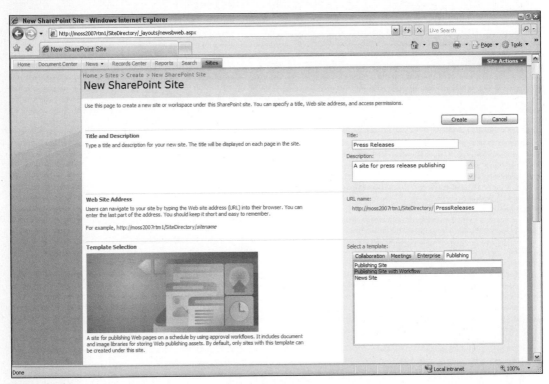

Figure 10-18

Once the site is created, you'll have a blank starting page and the automatic integration with the default Approval workflow for web content management along with the controls to perform page edits. The starting page is shown in Figure 10-19. To set up the press releases site, we'll add page content as shown in Figure 10-20 (following the Edit Page link in the toolbar).

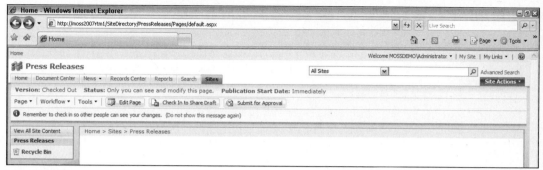

Figure 10-19

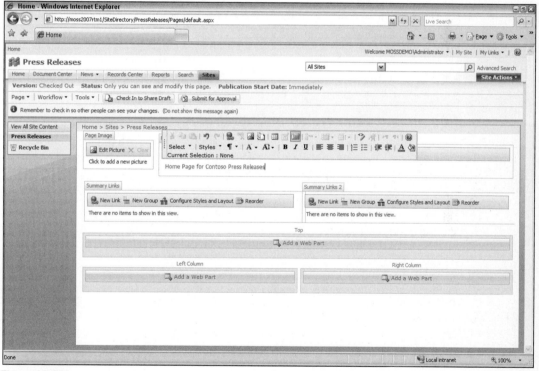

Figure 10-20

Once the content is updated, confirm that the site is set up correctly by submitting the page updates for approval. If you're designated as a content approver, you'll be able to approve the page for immediate publishing. Once the publishing site is working, we'll integrate a publishing site for press release documents in .docx format for conversion to .html files.

Document Converters for Web Publishing

Part of the press release publishing scenario will involve converting documents to web-based content to make it easier for users to publish existing press releases to the press release web site, simplifying the process of building content for web-based publishing. To get started, we'll briefly review document converters in Microsoft Office SharePoint Server.

To support the conversion of documents, SharePoint supports an object model to implement document-conversion utilities into the publishing process. This supports processes such as a document being published in a Word format and automatically converted to a PowerPoint document in a SharePoint site, or, in the case of this scenario, publishing a Word document as a web page. The document converter program is invoked using a command-line syntax that provides arguments specifying the input location, any special settings that are defined in an XML settings file, an output location, and a log to write the result of the conversion. From a deployment perspective, there are a number of options for deploying this kind of document conversion service to process content synchronously, asynchronously, and by a single instance of the conversion service or through a load-balanced configuration. The document conversion is most typically deployed as a batch process that's configured to support standard web content management publishing procedures on a daily basis (or as needed).

To proceed with document conversions for the press releases scenario, we'll need to look at the administrative functions required for document conversions and then test the publishing process.

Configuring Document Conversion in SharePoint

In order to support document conversions in SharePoint, specific services need to be running. To build out this scenario, make sure that both the `DocConversionLoadBalancerService` and `DocConversionLauncherService` services are running in your SharePoint farm. You can check the status of this on the home page of the SharePoint Central Administration web console.

To configure the document conversion services for a SharePoint site, use the Central Administration console's Application Management link to find the External Service Connections configuration settings. The Configure Document Conversions page for configuration settings is shown in Figure 10-21.

The document conversion settings are made at the site level. This allows you to configure the specific out-of-the-box document converters, such as `.docx` to `.html`, and to change individual attributes such as document timeouts, maximum retries, and size limits on documents to convert automatically. To build out the press release conversion scenario, set the maximum document size to 50KB. With the document conversion settings in place, we'll look at publishing a press release to the library and then converting that document into the publishing process.

Publishing Documents for Conversion

Each web publishing site includes a resource site for document publishing and resource files required for the content management processes. With our document workflow for the press release production having copied that document into our content management site already, we can then review those documents and use the standard web content management workflows to support their publishing.

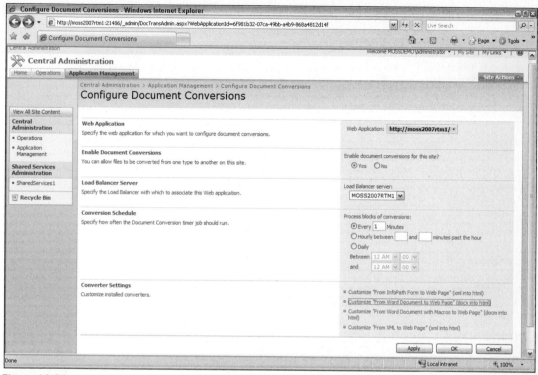

Figure 10-21

Go back to the press release document for Contoso built in the first half of this chapter. With that document resident in the press release site's document library, we can configure the processing of documents for publishing. First, we'll take the document and walk through the manual conversion process. When a document is published to a publishing site with web content management approval workflows, those items, even if published to a document library under the site, must be approved. In the case of the Contoso press release that's sent to the site via workflow, that document is now added to the workflow approval tasks. We'll look at the status of one of these workflow instances from the Workflow menu in the content management toolbar, as shown in Figure 10-22.

At this point, the document can have changes requested or be approved as is. This enables the content authors and editors to work within the Office applications, such as Word, to perform the content management processes, including writing and revisions, for the press release. All changes will be saved back to the document library, and the workflows for content approval that govern the library for new items or changes will apply. Once approved, the content can then be slated for manual or automatic document conversion and web publishing. Once again, the workflows that govern content publishing for the site will apply, and all new published items must be approved. To help you understand this better, we'll walk through the manual and automated options for deployment of the Word document as a web page.

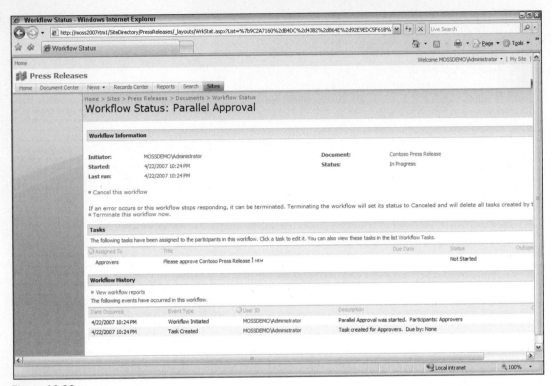

Figure 10-22

Manually Converting Document Content

When developing a web content management application in SharePoint, the idea of a "manual" document conversion may not make sense at first. If you're building the application to do content management, shouldn't conversion be automatic? Well, not necessarily. First, all of the web publishing features in SharePoint for publishing sites that support workflow are going to require some form of approval. This means that even for content that is automatically converted upon being posted, some review will need to take place. In addition, the authors and editors may prefer a smart client such as Word as the collaboration tool for producing content. By reviewing the documents manually in the queue and then performing publishing functions, this in essence becomes part of the human workflow of the content management process.

To start the manual press release publishing process, the only thing you need to do is to view the document library where the Contoso press release is published as a `.docx` file. From there, the menu option as shown in Figure 10-23 enables you to select the document conversion process for publishing.

From there, you can provide a page title and description and can even customize the URL name. Also, you can select the option to go directly to the page on publishing or to queue the page up for conversion and publishing in the background while returning to the original document library. This allows you to work through a list of documents and to have a seamless workflow for review and publishing of document-based content that's being converted to the web format. This publishing form is shown in Figure 10-24.

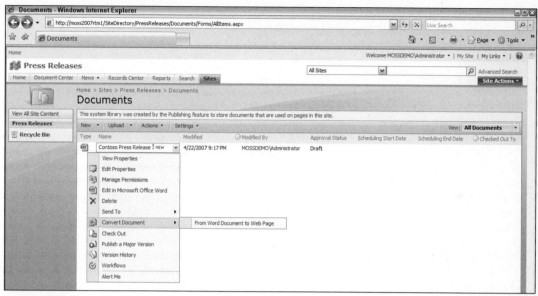

Figure 10-23

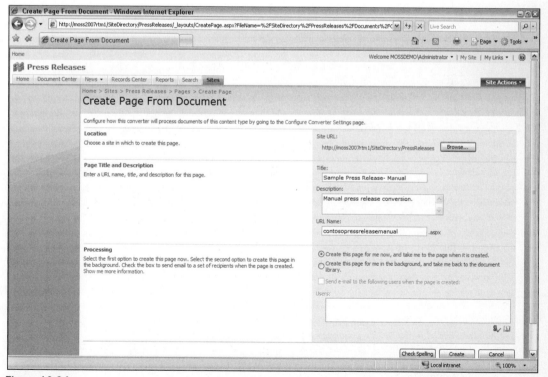

Figure 10-24

While manual document conversion based on the standard document library features is a reasonable solution for some requirements, many content management scenarios will require enhanced workflow for both document processing and content-related functions in addition to the approval workflows that are supported out of the box for web publishing. Next, we'll extend the press release publishing scenario to support document workflow.

Integrating Workflow for Document Conversions

One of the benefits of leveraging SharePoint's workflow to support an integrated document and web content management scenario is the ability to attach workflow to any component of the process or element of data. For producing the Contoso press releases, we'll add two elements of workflow to the process. Instead of using existing reports to drive the workflow, we'll append data to a custom list when new items are added to the document library for press releases. In addition, we'll send a copy of the document to another library for an intranet application that employees of Contoso can use to find press releases and sales literature to provide for customers. We'll create this document library up front so it's ready before we start creating the workflow.

We'll use SharePoint Designer to create the workflow. We'll open up the site for press release web content, in this case `http://moss2007rtm1/SiteDirectory/PressReleases/`. From there, we'll walk through the creation of a new workflow using the following steps:

1. Initiate a new workflow called Press Release Conversions. Attach the workflow to the Documents list in the Press Releases site (created by default). Set the workflow to allow for a manual start and an automatic start for new items, as shown in Figure 10-25.

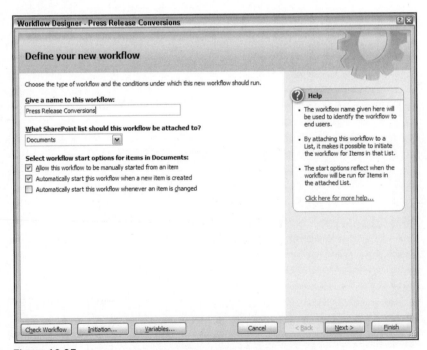

Figure 10-25

2. Create a workflow step called Verify Document Type, as shown in Figure 10-26. The workflow will validate that the file type is a Word document and, if so, will post it to a list that will be used for internal publishing on the Contoso intranet. If a type other than .docx is posted, the content administrator will receive an e-mail and the item will be deleted.

For the posting step, the Name field will be prepopulated on the field list and required. Open Properties on the destination list and set the name field to use the Name field from the current item (Figure 10-27).

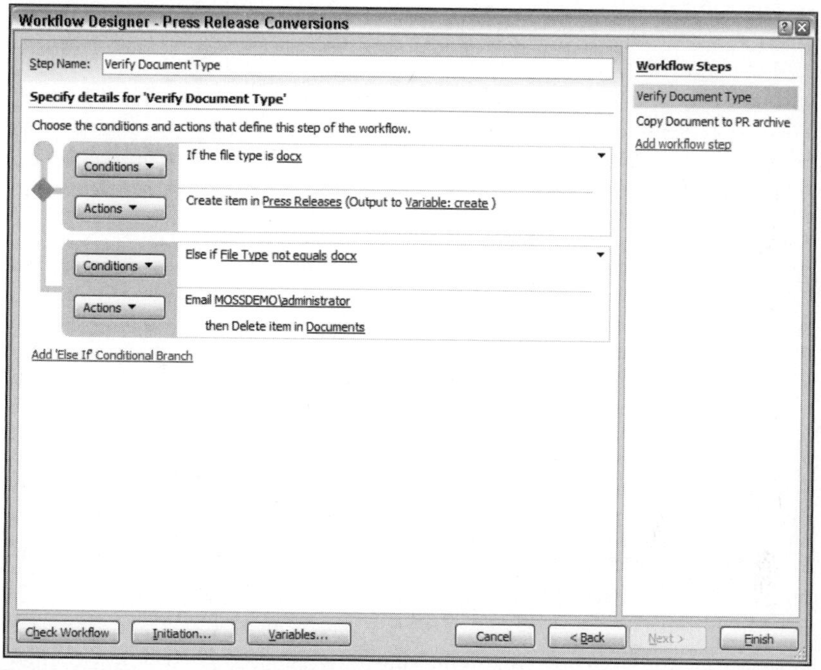

Figure 10-26

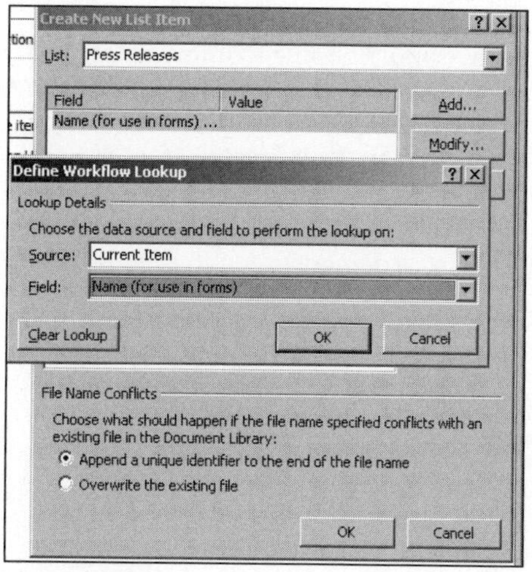

Figure 10-27

3. Create another workflow step named Copy Document To PR Archive, as shown in Figure 10-28. This workflow will copy the current item from the standard Documents library of the web publishing site to a custom library called Press Release Archive. From there, additional workflows can be put in place, and other document management processes can be implemented.

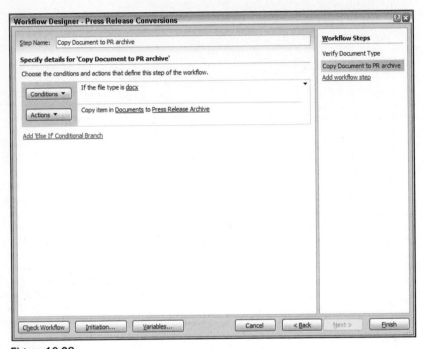

Figure 10-28

4. Finish the workflow to upload it to the site. You can test the workflow from the site itself for an existing document by selecting the Workflows option and then going through the process, or you can upload new documents.

When you run the workflow, you should be able to immediately verify the results by viewing all site content and seeing a summary of each list, as shown in Figure 10-29. As you can see, the Documents library contains the Contoso press release file. In addition, new items have been added to the Press Release Archive document library and to the Press Releases list as shown in Figure 10-30. This enables you as a developer of a SharePoint content management application to build additional lists and data sources to branch processes, all of which can have their content management functions, workflows, or custom list processing applications, as demonstrated in Chapter 3.

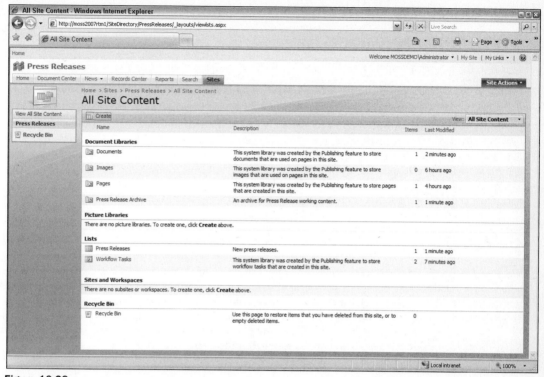

Figure 10-29

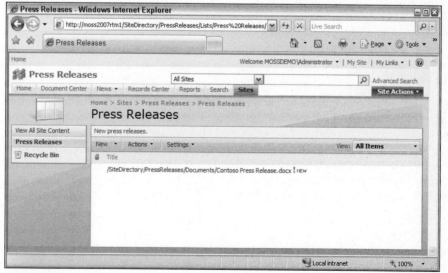

Figure 10-30

Extending a Document Publishing Scenario

There are a number of possibilities that you can consider for extending this kind of document publishing scenario. As you start to build out these scenarios for your web content applications, you'll want to look deeper at the workflow capabilities through Visual Studio, the document converter object model, and the ability to convert documents other than Office and XML formats, and the integration of business data into your web content solutions.

In addition to the manual web content conversions for documents and the SharePoint Designer workflow actions to support the content management process, you can also write code to automate publishing functions. As shown in Chapter 8, you can build custom workflows in Visual Studio that support activities beyond the scope of those supported in SharePoint Designer. Using the SharePoint object model, you can use the `PublishingPage` class to automatically convert document content to web content. The `PublishingPageCollection` object exposes an `Add` method enabling you as a developer to specify the conversion of a document based on a page layout. This functionality can be implemented as part of a custom workflow to facilitate document conversions to web pages. Other custom extensions can be implemented as part of a workflow to extend this kind of scenario.

Another potential scenario is the integration of the Business Data Catalog (BDC) and supporting APIs to integrate business data into your web content scenario. For example, you can access data from a business system through the BDC object model to expose sales results or other business statistics closer to real time. In the content creation process, you can link existing data elements in a document to external BDC services using techniques similar to those shown in Chapter 4. This kind of integration would help to ensure that accurate business data was represented through the content management process while minimizing the fact checking and validation processes for content authors and approvers.

When looking at the tools and technologies supported in SharePoint 2007 to drive content authoring, workflow, content management, and publishing functions, the developer's toolbox includes many options. If you prefer to create a highly customized process based on custom code, those options are extensive. If you'd prefer to build a solution based solely on out-of-the-box functionality that can be configured through SharePoint and SharePoint Designer, those options are there as well. The choices are up to you.

Summary

This chapter focused on scenarios for document collaboration and publishing based on a real-world example — developing and publishing press releases for Contoso Enterprises.

Microsoft's content services features are based on the integration of ASP.NET into SharePoint and the support for Web Content Management and portal and enterprise content management features. Microsoft Office SharePoint Server 2007 provides a set of document services to help users produce and manage content as part of their workflow to do things like generate press release documents or to perform other business functions without requiring a separate set of tools to author content, review it, develop solutions based on that content, and then perform publishing functions.

SharePoint's web content management features, integrated with Office 2007, enable a rich platform for developing document-centric solutions for content management, workflow, and business process automation. Coupled with the capabilities of the .NET platform and tools like SharePoint Designer and Visual Studio, the Microsoft Office SharePoint Server platform, complemented by Office 2007, provides developers with lots of capabilities for automating business functions and changing how end-users interact with their data, documents, and applications.

Index

Index

methods

ClearContent, 112

ClearHeaders, 112

Query, 41

QueryEx, 41

Update, 43

Microsoft Windows SharePoint Services, 1

Microsoft.SharePoint namespace, 42–43

Microsoft.SharePoint.Administration namespace, 42–43

Microsoft.SharePoint.MobileControls namespace, 47

mobile application services, 46–47, 49

MOSS (Microsoft Office SharePoint Server), 1

N

namespaces

Microsoft.SharePoint, 42–43

Microsoft.SharePoint.Administration, 42–43

Microsoft.SharePoint.MobileControls, 47

O

object model, 50–51

Office 2007

integration, 49

document formats, 17–18

enterprise content management, 20–21

InfoPath, 19

OpenXML

file formats, 92–118

OPC (Open Packaging Convention), 93

Office Fluent Ribbon user interface, 150–151

offline cube support, 204–205

OLAP (Online Analytical Processing) cube, 191

OpenXML

Excel

SpreadsheetML, 113–118

/xl/sharedStrings.xml, 117–118

/xl/workbook.xml.rels, 114–115

/xl/worksheets/sheet1.xml, 115–117

OPC (Open Packaging Convention), 93

outputting in ASP.NET, 111–113

PowerPoint, 118

Word

content types, 96–97

document metadata, 99

formatted text, 101–102

hyperlinks, 102

relationship files, 97–99

report generation, 104–113

tables, 103–104

/word/document.xml, 100–101

Word format, 94–104

P

PivotChart reports, 203

portal services, 7–8

PowerPoint, OpenXML, 118

PresentationML, 118

press release

approval workflow definition, 297–298